FUZZY CONTROL SYSTEMS DESIGN AND ANALYSIS

FUZZY CONTROL SYSTEMS DESIGN AND ANALYSIS

A Linear Matrix Inequality Approach

KAZUO TANAKA and HUA O. WANG

A Wiley-Interscience Publication
JOHN WILEY & SONS, INC.
New York / Chichester / Weinheim / Brisbane / Singapore / Toronto

This book is printed on acid-free paper. ∞

Copyright ©2001 John Wiley & Sons, Inc. All rights reserved.

Published simultaneously in Canada.

No part of this publication may be reproduced, stored in a retrieval system or transmitted in any form or by any means, electronic, mechanical, photocopying, recording, scanning or otherwise, except as permitted under Section 107 or 108 of the 1976 United States Copyright Act, without either the prior written permission of the Publisher, or authorization through payment of the appropriate per-copy fee to the Copyright Clearance Center, 222 Rosewood Drive, Danvers, MA 01923, (978) 750-8400, fax (978) 750-4744. Requests to the Publisher for permission should be addressed to the Permissions Department, John Wiley & Sons, Inc., 605 Third Avenue, New York, NY 10158-0012, (212) 850-6011, fax (212) 850-6008, E-Mail: PERMREQ@WILEY.COM.

For ordering and customer service, call 1-800-CALL-WILEY

Library of Congress Cataloging in Publication Data:

Tanaka, Kazuo, 1962-
 Fuzzy control systems design and analysis : a linear matrix inequality approach / Kazuo Tanaka and Hua O. Wang.
     p. cm.
 Includes bibliographical references and index.
 ISBN 0-471-32324-1 (cloth : alk. paper)
 1. Linear control systems. 2. Fuzzy systems. I. Wang, Hua O.

TJ220.T36 2001
629.8'32--dc21

200 024238

Printed in the United States of America

10 9 8 7 6 5 4 3 2 1

CONTENTS

PREFACE   xi

ACRONYMS   xiii

1 INTRODUCTION   1

   1.1 A Control Engineering Approach to Fuzzy Control / 1
   1.2 Outline of This Book / 2

2 TAKAGI-SUGENO FUZZY MODEL AND PARALLEL DISTRIBUTED COMPENSATION   5

   2.1 Takagi-Sugeno Fuzzy Model / 6
   2.2 Construction of Fuzzy Model / 9
       2.2.1 Sector Nonlinearity / 10
       2.2.2 Local Approximation in Fuzzy Partition Spaces / 23
   2.3 Parallel Distributed Compensation / 25
   2.4 A Motivating Example / 26
   2.5 Origin of the LMI-Based Design Approach / 29
       2.5.1 Stable Controller Design via Iterative Procedure / 30
       2.5.2 Stable Controller Design via Linear Matrix Inequalities / 34

2.6 Application: Inverted Pendulum on a Cart / 38
   2.6.1 Two-Rule Modeling and Control / 38
   2.6.2 Four-Rule Modeling and Control / 42
Bibliography / 47

3 LMI CONTROL PERFORMANCE CONDITIONS AND DESIGNS   49

3.1 Stability Conditions / 49
3.2 Relaxed Stability Conditions / 52
3.3 Stable Controller Design / 58
3.4 Decay Rate / 62
3.5 Constraints on Control Input and Output / 66
   3.5.1 Constraint on the Control Input / 66
   3.5.2 Constraint on the Output / 68
3.6 Initial State Independent Condition / 68
3.7 Disturbance Rejection / 69
3.8 Design Example: A Simple Mechanical System / 76
   3.8.1 Design Case 1: Decay Rate / 78
   3.8.2 Design Case 2: Decay Rate + Constraint on the Control Input / 79
   3.8.3 Design Case 3: Stability + Constraint on the Control Input / 80
   3.8.4 Design Case 4: Stability + Constraint on the Control Input + Constraint on the Output / 81
References / 81

4 FUZZY OBSERVER DESIGN   83

4.1 Fuzzy Observer / 83
4.2 Design of Augmented Systems / 84
   4.2.1 Case A / 85
   4.2.2 Case B / 90
4.3 Design Example / 93
References / 96

5 ROBUST FUZZY CONTROL   97

5.1 Fuzzy Model with Uncertainty / 98
5.2 Robust Stability Condition / 98
5.3 Robust Stabilization / 105
References / 108

6 OPTIMAL FUZZY CONTROL 109

6.1 Quadratic Performance Function and Stabilization Control / 110
6.2 Optimal Fuzzy Controller Design / 114
Appendix to Chapter 6 / 118
References / 119

7 ROBUST-OPTIMAL FUZZY CONTROL 121

7.1 Robust-Optimal Fuzzy Control Problem / 121
7.2 Design Example: TORA / 125
References / 130

8 TRAJECTORY CONTROL OF A VEHICLE WITH MULTIPLE TRAILERS 133

8.1 Fuzzy Modeling of a Vehicle with Triple-Trailers / 134
    8.1.1 Avoidance of Jack-Knife Utilizing Constraint on Output / 142
8.2 Simulation Results / 144
8.3 Experimental Study / 147
8.4 Control of Ten-Trailer Case / 150
References / 151

9 FUZZY MODELING AND CONTROL OF CHAOTIC SYSTEMS 153

9.1 Fuzzy Modeling of Chaotic Systems / 154
9.2 Stabilization / 159
    9.2.1 Stabilization via Parallel Distributed Compensation / 159
    9.2.2 Cancellation Technique / 165
9.3 Synchronization / 170
    9.3.1 Case 1 / 170
    9.3.2 Case 2 / 179
9.4 Chaotic Model Following Control / 182
References / 192

10 FUZZY DESCRIPTOR SYSTEMS AND CONTROL 195

10.1 Fuzzy Descriptor System / 196
10.2 Stability Conditions / 197
10.3 Relaxed Stability Conditions / 206
10.4 Why Fuzzy Descriptor Systems? / 211
References / 215

11 NONLINEAR MODEL FOLLOWING CONTROL — 217

11.1 Introduction / 217
11.2 Design Concept / 218
    11.2.1 Reference Fuzzy Descriptor System / 218
    11.2.2 Twin-Parallel Distributed Compensations / 219
    11.2.3 The Common B Matrix Case / 223
11.3 Design Examples / 224
References / 228

12 NEW STABILITY CONDITIONS AND DYNAMIC FEEDBACK DESIGNS — 229

12.1 Quadratic Stabilizability Using State Feedback PDC / 230
12.2 Dynamic Feedback Controllers / 232
    12.2.1 Cubic Parametrization / 236
    12.2.2 Quadratic Parameterization / 243
    12.2.3 Linear Parameterization / 247
12.3 Example / 253
Bibliography / 256

13 MULTIOBJECTIVE CONTROL VIA DYNAMIC PARALLEL DISTRIBUTED COMPENSATION — 259

13.1 Performance-Oriented Controller Synthesis / 260
    13.1.1 Starting from Design Specifications / 260
    13.1.2 Performance-Oriented Controller Synthesis / 264
13.2 Example / 271
Bibliography / 274

14 T-S FUZZY MODEL AS UNIVERSAL APPROXIMATOR — 277

14.1 Approximation of Nonlinear Functions Using Linear T-S Systems / 278
    14.1.1 Linear T-S Fuzzy Systems / 278
    14.1.2 Construction Procedure of T-S Fuzzy Systems / 279
    14.1.3 Analysis of Approximation / 281
    14.1.4 Example / 286

14.2 Applications to Modeling and Control of Nonlinear Systems / 287
    14.2.1 Approximation of Nonlinear Dynamic Systems Using Linear Takagi-Sugeno Fuzzy Models / 287
    14.2.2 Approximation of Nonlinear State Feedback Controller Using PDC Controller / 288
Bibliography / 289

15 FUZZY CONTROL OF NONLINEAR TIME-DELAY SYSTEMS 291

15.1 T-S Fuzzy Model with Delays and Stability Conditions / 292
    15.1.1 T-S Fuzzy Model with Delays / 292
    15.1.2 Stability Analysis via Lyapunov Approach / 294
    15.1.3 Parallel Distributed Compensation Control / 295
15.2 Stability of the Closed-Loop Systems / 296
15.3 State Feedback Stabilization Design via LMIs / 297
15.4 H_∞ Control / 299
15.6 Design Example / 300
References / 302

INDEX 303

PREFACE

The authors cannot acknowledge all the friends and colleagues with whom they have discussed the subject area of this research monograph or from whom they have received invaluable encouragement. Nevertheless, it is our great pleasure to express our thanks to those who have been directly involved in various aspects of the research leading to this book. First, the authors wish to express their hearty gratitude to their advisors Michio Sugeno, Tokyo Institute of Technology, and Eyad Abed, University of Maryland, College Park, for directing the research interest of the authors to the general area of systems and controls. The authors are especially appreciative of the discussions they had with Michio Sugeno at different stages of their research on the subject area of this book. His remarks, suggestions, and encouragement have always been very valuable.

We would like to thank William T. Thompkins, Jr. and Michael F. Griffin, who planted the seed of this book. Thanks are also due to Chris McClurg, Tom McHugh, and Randy Roberts for their support of the research and for the pleasant and fruitful collaboration on some joint research endeavors.

Special thanks go to the students in our laboratories, in particular, Takayuki Ikeda, Jing Li, Tadanari Taniguchi, and Yongru Gu. Our extended appreciation goes to David Niemann for his contribution to some of the results contained in this book and to Kazuo Yamafuji, Ron Chen, and Linda Bushnell for their suggestions, constructive comments, and support. It is a pleasure to thank all our colleagues at both the University of Electro-Communications (UEC) and Duke University for providing a pleasant and stimulating environment that allowed us to write this book. The second author is also thankful to the colleagues of Center for Nonlinear and

Complex Systems at Huazhong University of Science and Technology, Wuhan, China, for their support. We also wish to express our appreciation to the editors and staff of John Wiley and Sons, Inc. for their energy and professionalism.

Finally, the authors are especially grateful to their families for their love, encouragement, and complete support throughout this project. Kazuo Tanaka dedicates this book to his wife, Tomoko, and son, Yuya. Hua O. Wang would like to dedicate this book to his wife, Wai, and daughter, Catherine.

The writing of this book was supported in part by the Japanese Ministry of Education; the Japan Society for the Promotion of Science; the U.S. Army Research Office under Grants DAAH04-93-D-0002 and DAAG55-98-D-0002; the Lord Foundation of North Carolina; the Otis Elevator Company; the Cheung Kong Chair Professorship Program of the Ministry of Education of China and the Li Ka-shing Foundation, Hong Kong; and the Center for Nonlinear and Complex Systems at Huazhong University of Science and Technology, Wuhan, China. The support of these organizations is gratefully acknowledged.

KAZUO TANAKA
HUA O. WANG

Tokyo, Japan
Durham, North Carolina
May 2001

ACRONYMS

ARE	Algebraic Riccati equation
CFS	Continuous fuzzy system
CMFC	Chaotic model following control
CT	Cancellation technique
DFS	Discrete fuzzy system
DPDC	Dynamic parallel distributed compensation
GEVP	Generalized eigenvalue minimization problem
LDI	Linear differential inclusion
LMI	Linear matrix inequality
NLTI	Nonlinear time-invariant operator
PDC	Parallel distributed compensation
PDE	Partial differential equation
TORA	Translational oscillator with rotational actuator
TPDC	Twin parallel distributed compensation
T-S	Takagi-Sugeno
T-SMTD	T-S model with time delays

CHAPTER 1

INTRODUCTION

1.1 A CONTROL ENGINEERING APPROACH TO FUZZY CONTROL

This book gives a comprehensive treatment of model-based fuzzy control systems. The central subject of this book is a systematic framework for the stability and design of nonlinear fuzzy control systems. Building on the so-called Takagi-Sugeno fuzzy model, a number of most important issues in fuzzy control systems are addressed. These include stability analysis, systematic design procedures, incorporation of performance specifications, robustness, optimality, numerical implementations, and last but not the least, applications.

The guiding philosophy of this book is to arrive at a middle ground between conventional fuzzy control practice and established rigor and systematic synthesis of systems and control theory. The authors view this balanced approach as an attempt to blend the best of both worlds. On one hand, fuzzy logic provides a simple and straightforward way to decompose the task of modeling and control design into a group of local tasks, which tend to be easier to handle. In the end, fuzzy logic also provides the mechanism to blend these local tasks together to deliver the overall model and control design. On the other hand, advances in modern control have made available a large number of powerful design tools. This is especially true in the case of linear control designs. These tools for linear systems range from elegant state space optimal control to the more recent robust control paradigms. By employing the Takagi-Sugeno fuzzy model, which utilizes local linear system description for each rule, we devise a control methodology to fully take advantage of the advances of modern control theory.

2 INTRODUCTION

We have witnessed rapidly growing interest in fuzzy control in recent years. This is largely sparked by the numerous successful applications fuzzy control has enjoyed. Despite the visible success, it has been made aware that many basic issues remain to be addressed. Among them, stability analysis, systematic design, and performance analysis, to name a few, are crucial to the validity and applicability of any control design methodology. This book is intended to address these issues in the framework of the Takagi-Sugeno fuzzy model and a controller structure devised in accordance with the fuzzy model.

1.2 OUTLINE OF THIS BOOK

This book is intended to be used either as a textbook or as a reference for control researchers and engineers. For the first objective, the book can be used as a graduate textbook or upper level undergraduate textbook. It is particularly rewarding that using the approaches presented in this book, a student just entering the field of control can solve a large class of problems that would normally require rather advanced training at the graduate level.

This book is organized into 15 chapters. Figure 1.1 shows the relation among chapters in this book. For example, Chapters 1–3 provide the basis for Chapters 4–5. Chapters 1–3, 9, and 10 are necessary prerequisites to

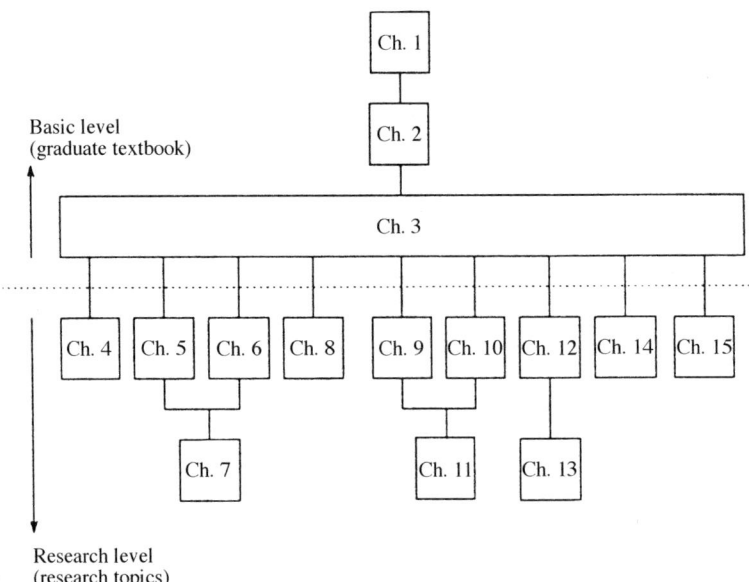

Fig. 1.1 Relation among chapters.

understand Chapter 11. Beyond Chapter 3, all chapters, with the exception of Chapters 7, 11, and 13, are designed to be basically independent of each other, to give the reader flexibility in progressing through the materials of this book. Chapters 1–3 contain the fundamental materials for later chapters. The level of mathematical sophistication and prior knowledge in control have been kept in an elementary context. This part is suitable as a starting point in a graduate-level course. Chapters 4–15 cover advanced analysis and design topics which may require a higher level of mathematical sophistication and advanced knowledge of control engineering. This part provides a wide range of advanced topics for a graduate-level course and more importantly some timely and powerful analysis and design techniques for researchers and engineers in systems and controls.

Each chapter from 1 to 15 ends with a section of references which contain the most relevant literature for the specific topic of each chapter. To probe further into each topic, the readers are encouraged to consult with the listed references.

In this book, $S > 0$ means that S is a positive definite matrix, $S > T$ means that $S - T > 0$ and $W = 0$ means that W is a zero matrix, that is, its elements are all zero.

To lighten the notation, this book employs several particular notions which are listed as follow:

$$i < j \text{ s.t. } h_i \cap h_j \neq \phi,$$

$$i \leq j \text{ s.t. } h_i \cap h_j \neq \phi.$$

For instance, the condition (2.31) in Chapter 2 has the notation,

$$i < j \leq r \text{ s.t. } h_i \cap h_j \neq \phi.$$

This means that the condition should be hold for all $i < j$ excepting $h_i \cap h_j = \phi$ [i.e., $h_i(\mathbf{z}(t)) \times h_j(\mathbf{z}(t)) = 0$ for all $\mathbf{z}(t)$], where $h_i(\mathbf{z}(t))$ denotes the weight of the ith rule calculated from membership functions in the premise parts and r denotes the number of if-then rules. Note that $h_i \cap h_j = \phi$ if and only if the ith rule and jth rule have no overlap.

CHAPTER 2

TAKAGI-SUGENO FUZZY MODEL AND PARALLEL DISTRIBUTED COMPENSATION

Recent years have witnessed rapidly growing popularity of fuzzy control systems in engineering applications. The numerous successful applications of fuzzy control have sparked a flurry of activities in the analysis and design of fuzzy control systems. In this book, we introduce a wide range of analysis and design tools for fuzzy control systems to assist control researchers and engineers to solve engineering problems. The toolkit developed in this book is based on the framework of the Takagi-Sugeno fuzzy model and the so-called parallel distributed compensation, a controller structure devised in accordance with the fuzzy model. This chapter introduces the basic concepts, analysis, and design procedures of this approach.

This chapter starts with the introduction of the Takagi-Sugeno fuzzy model (T-S fuzzy model) followed by construction procedures of such models. Then a model-based fuzzy controller design utilizing the concept of "parallel distributed compensation" is described. The main idea of the controller design is to derive each control rule so as to compensate each rule of a fuzzy system. The design procedure is conceptually simple and natural. Moreover, it is shown in this chapter that the stability analysis and control design problems can be reduced to linear matrix inequality (LMI) problems. The design methodology is illustrated by application to the problem of balancing and swing-up of an inverted pendulum on a cart.

The focus of this chapter is on the basic concept of techniques of stability analysis via LMIs [14, 15, 24]. The more advanced material on analysis and design involving LMIs will be given in Chapter 3.

2.1 TAKAGI-SUGENO FUZZY MODEL

The design procedure describing in this book begins with representing a given nonlinear plant by the so-called Takagi-Sugeno fuzzy model. The fuzzy model proposed by Takagi and Sugeno [7] is described by fuzzy IF-THEN rules which represent local linear input-output relations of a nonlinear system. The main feature of a Takagi-Sugeno fuzzy model is to express the local dynamics of each fuzzy implication (rule) by a linear system model. The overall fuzzy model of the system is achieved by fuzzy "blending" of the linear system models. In this book, the readers will find that many nonlinear dynamic systems can be represented by Takagi-Sugeno fuzzy models. In fact, it is proved that Takagi-Sugeno fuzzy models are universal approximators. The details will be discussed in Chapter 14.

The ith rules of the T-S fuzzy models are of the following forms, where CFS and DFS denote the continuous fuzzy system and the discrete fuzzy system, respectively.

Continuous Fuzzy System: CFS

Model Rule i:

IF $z_1(t)$ is M_{i1} and $\cdots$ and $z_p(t)$ is M_{ip},

$$\text{THEN } \begin{cases} \dot{x}(t) = A_i x(t) + B_i u(t), \\ y(t) = C_i x(t), \end{cases} \quad i = 1, 2, \ldots, r. \quad (2.1)$$

Discrete Fuzzy System: DFS

Model Rule i:

IF $z_1(t)$ is M_{i1} and $\cdots$ and $z_p(t)$ is M_{ip},

$$\text{THEN } \begin{cases} x(t+1) = A_i x(t) + B_i u(t), \\ y(t) = C_i x(t), \end{cases} \quad i = 1, 2, \ldots, r. \quad (2.2)$$

Here, M_{ij} is the fuzzy set and r is the number of model rules; $x(t) \in R^n$ is the state vector, $u(t) \in R^m$ is the input vector, $y(t) \in R^q$ is the output vector, $A_i \in R^{n \times n}$, $B_i \in R^{n \times m}$, and $C_i \in R^{q \times n}$; $z_1(t), \ldots, z_p(t)$ are known premise variables that may be functions of the state variables, external disturbances, and/or time. We will use $z(t)$ to denote the vector containing all the individual elements $z_1(t), \ldots, z_p(t)$. It is assumed in this book that the premise variables are not functions of the input variables $u(t)$. This assumption is needed to avoid a complicated defuzzification process of fuzzy controllers [12]. Note that stability conditions derived in this book can be applied even to the case that the premise variables are functions of the input variables $u(t)$. Each linear consequent equation represented by $A_i x(t) + B_i u(t)$ is called a "subsystem."

Given a pair of $(x(t), u(t))$, the final outputs of the fuzzy systems are inferred as follows:

CFS

$$\dot{x}(t) = \frac{\sum_{i=1}^{r} w_i(z(t))\{A_i x(t) + B_i u(t)\}}{\sum_{i=1}^{r} w_i(z(t))}$$

$$= \sum_{i=1}^{r} h_i(z(t))\{A_i x(t) + B_i u(t)\}, \qquad (2.3)$$

$$y(t) = \frac{\sum_{i=1}^{r} w_i(z(t)) C_i x(t)}{\sum_{i=1}^{r} w_i(z(t))}$$

$$= \sum_{i=1}^{r} h_i(z(t)) C_i x(t). \qquad (2.4)$$

DFS

$$x(t+1) = \frac{\sum_{i=1}^{r} w_i(z(t))\{A_i x(t) + B_i u(t)\}}{\sum_{i=1}^{r} w_i(z(t))}$$

$$= \sum_{i=1}^{r} h_i(z(t))\{A_i x(t) + B_i u(t)\}, \qquad (2.5)$$

$$y(t) = \frac{\sum_{i=1}^{r} w_i(z(t)) C_i x(t)}{\sum_{i=1}^{r} w_i(z(t))}$$

$$= \sum_{i=1}^{r} h_i(z(t)) C_i x(t), \qquad (2.6)$$

where

$$z(t) = [z_1(t)\, z_2(t)\, \cdots\, z_p(t)],$$

$$w_i(z(t)) = \prod_{j=1}^{p} M_{ij}(z_j(t)),$$

$$h_i(z(t)) = \frac{w_i(z(t))}{\sum_{i=1}^{r} w_i(z(t))} \quad (2.7)$$

for all t. The term $M_{ij}(z_j(t))$ is the grade of membership of $z_j(t)$ in M_{ij}. Since

$$\begin{cases} \sum_{i=1}^{r} w_i(z(t)) > 0, \\ w_i(z(t)) \geq 0, \quad i = 1, 2, \ldots, r, \end{cases} \quad (2.8)$$

we have

$$\begin{cases} \sum_{i=1}^{r} h_i(z(t)) = 1, \\ h_i(z(t)) \geq 0, \quad i = 1, 2, \ldots, r, \end{cases} \quad (2.9)$$

for all t.

Example 1 Assume in the DFS that

$$p = n,$$
$$z_1(t) = x(t), z_2(t) = x(t-1), \ldots, z_n(t) = x(t-n+1).$$

Then, the model rules can be represented as follows.

Model Rule i:

IF $x(t)$ is M_{i1} and $\cdots$ and $x(t-n+1)$ is M_{in},

THEN $\begin{cases} x(t+1) = A_i x(t) + B_i u(t), \\ y(t) = C_i x(t), \end{cases} \quad i = 1, 2, \ldots, r,$

where $x(t) = [x(t)\, x(t-1)\, \cdots\, x(t-n+1)]^T$.

Remark 1 The Takagi-Sugeno fuzzy model is sometimes referred as the Takagi-Sugeno-Kang fuzzy model (TSK fuzzy model) in the literature. In this book, the authors do not refer to (2.1) and (2.2) as the TSK fuzzy model. The reason is that this type of fuzzy model was originally proposed by Takagi and Sugeno in [7]. Following that, Kang and Sugeno [8, 9] did excellent work on identification of the fuzzy model. From this historical background, we feel that (2.1) and (2.2) should be addressed as the Takagi-Sugeno fuzzy model. On the other hand, the excellent work on identification by Kang and Sugeno is best referred to as the Kang-Sugeno fuzzy modeling method. In this book the authors choose to distinguish between the Takagi-Sugeno fuzzy model and the Kang-Sugeno fuzzy modeling method.

2.2 CONSTRUCTION OF FUZZY MODEL

Figure 2.1 illustrates the model-based fuzzy control design approach discussed in this book. To design a fuzzy controller, we need a Takagi-Sugeno fuzzy model for a nonlinear system. Therefore the construction of a fuzzy model represents an important and basic procedure in this approach. In this section we discuss the issue of how to construct such a fuzzy model.

In general there are two approaches for constructing fuzzy models:

1. Identification (fuzzy modeling) using input-output data and
2. Derivation from given nonlinear system equations.

There has been an extensive literature on fuzzy modeling using input-output data following Takagi's, Sugeno's, and Kang's excellent work [8, 9]. The procedure mainly consists of two parts: structure identification and parameter identification. The identification approach to fuzzy modeling is suitable

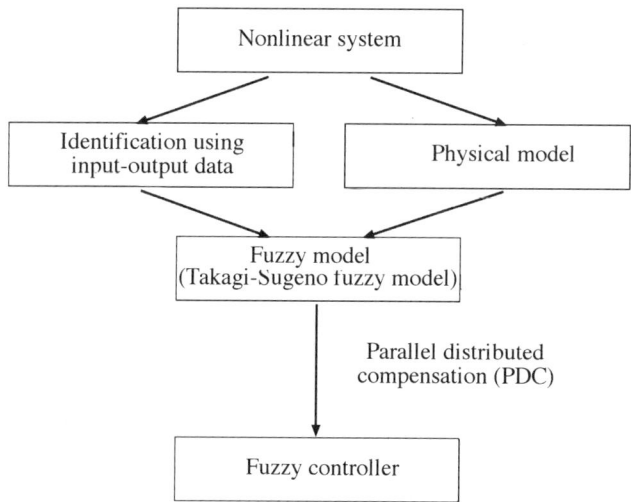

Fig. 2.1 Model-based fuzzy control design.

10   TAKAGI-SUGENO FUZZY MODEL AND PARALLEL DISTRIBUTED COMPENSATION

for plants that are unable or too difficult to be represented by analytical and/or physical models. On the other hand, nonlinear dynamic models for mechanical systems can be readily obtained by, for example, the Lagrange method and the Newton-Euler method. In such cases, the second approach, which derives a fuzzy model from given nonlinear dynamical models, is more appropriate. This section focuses on this second approach. This approach utilizes the idea of "sector nonlinearity," "local approximation," or a combination of them to construct fuzzy models.

2.2.1 Sector Nonlinearity

The idea of using sector nonlinearity in fuzzy model construction first appeared in [10]. Sector nonlinearity is based on the following idea. Consider a simple nonlinear system $\dot{x}(t) = f(x(t))$, where $f(0) = 0$. The aim is to find the global sector such that $\dot{x}(t) = f(x(t)) \in [a_1 \ a_2]x(t)$. Figure 2.2 illustrates the sector nonlinearity approach. This approach guarantees an exact fuzzy model construction. However, it is sometimes difficult to find global sectors for general nonlinear systems. In this case, we can consider local sector nonlinearity. This is reasonable as variables of physical systems are always bounded. Figure 2.3 shows the local sector nonlinearity, where two lines become the local sectors under $-d < x(t) < d$. The fuzzy model exactly represents the nonlinear system in the "local" region, that is, $-d < x(t) < d$. The following two examples illustrate the concrete steps to construct fuzzy models.

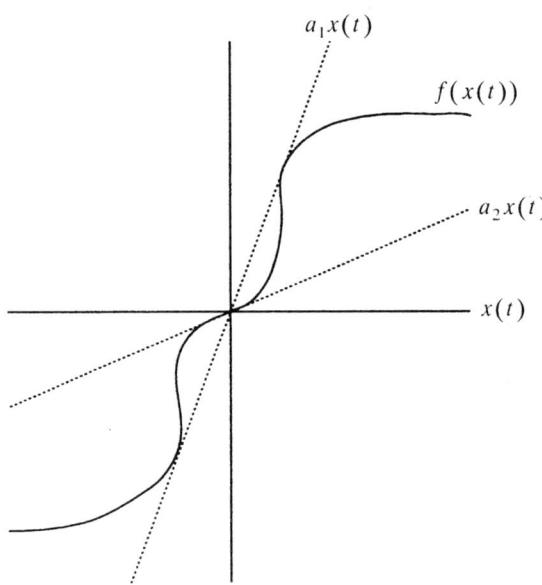

Fig. 2.2   Global sector nonlinearity.

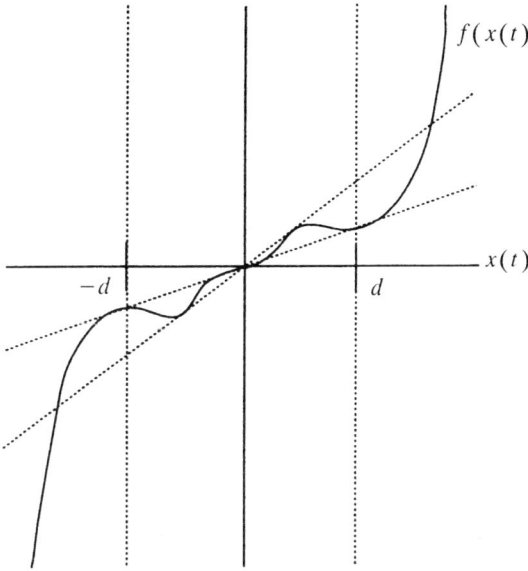

Fig. 2.3 Local sector nonlinearity.

Example 2 Consider the following nonlinear system:

$$\begin{pmatrix} \dot{x}_1(t) \\ \dot{x}_2(t) \end{pmatrix} = \begin{pmatrix} -x_1(t) + x_1(t)x_2^3(t) \\ -x_2(t) + (3 + x_2(t))x_1^3(t) \end{pmatrix}. \quad (2.10)$$

For simplicity, we assume that $x_1(t) \in [-1, 1]$ and $x_2(t) \in [-1, 1]$. Of course, we can assume any range for $x_1(t)$ and $x_2(t)$ to construct a fuzzy model.

Equation (2.10) can be written as

$$\dot{x}(t) = \begin{bmatrix} -1 & x_1(t)x_2^2(t) \\ (3 + x_2(t))x_1^2(t) & -1 \end{bmatrix} x(t),$$

where $x(t) = [x_1(t) \ x_2(t)]^T$ and $x_1(t)x_2^2(t)$ and $(3 + x_2(t))x_1^2(t)$ are nonlinear terms. For the nonlinear terms, define $z_1(t) \equiv x_1(t)x_2^2(t)$ and $z_2(t) \equiv (3 + x_2(t))x_1^2(t)$. Then, we have

$$\dot{x}(t) = \begin{bmatrix} -1 & z_1(t) \\ z_2(t) & -1 \end{bmatrix} x(t).$$

Next, calculate the minimum and maximum values of $z_1(t)$ and $z_2(t)$ under $x_1(t) \in [-1, 1]$ and $x_2(t) \in [-1, 1]$. They are obtained as follows:

$$\max_{x_1(t), x_2(t)} z_1(t) = 1, \quad \min_{x_1(t), x_2(t)} z_1(t) = -1,$$

$$\max_{x_1(t), x_2(t)} z_2(t) = 4, \quad \min_{x_1(t), x_2(t)} z_2(t) = 0.$$

From the maximum and minimum values, $z_1(t)$ and $z_2(t)$ can be represented by

$$z_1(t) = x_1(t)x_2^2(t) = M_1(z_1(t)) \cdot 1 + M_2(z_1(t)) \cdot (-1),$$
$$z_2(t) = (3 + x_2(t))x_1^2(t) = N_1(z_2(t)) \cdot 4 + N_2(z_2(t)) \cdot 0,$$

where

$$M_1(z_1(t)) + M_2(z_1(t)) = 1,$$
$$N_1(z_2(t)) + N_2(z_2(t)) = 1.$$

Therefore the membership functions can be calculated as

$$M_1(z_1(t)) = \frac{z_1(t) + 1}{2}, \quad M_2(z_1(t)) = \frac{1 - z_1(t)}{2},$$

$$N_1(z_2(t)) = \frac{z_2(t)}{4}, \quad N_2(z_2(t)) = \frac{4 - z_2(t)}{4}.$$

We name the membership functions "Positive," "Negative," "Big," and "Small," respectively. Then, the nonlinear system (2.10) is represented by the following fuzzy model.

Model Rule 1:

IF $z_1(t)$ is "Positive" and $z_2(t)$ is "Big,"

THEN $\dot{x}(t) = A_1 x(t)$.

Model Rule 2:

IF $z_1(t)$ is "Positive" and $z_2(t)$ is "Small,"

THEN $\dot{x}(t) = A_2 x(t)$.

Model Rule 3:

IF $z_1(t)$ is "Negative" and $z_2(t)$ is "Big,"

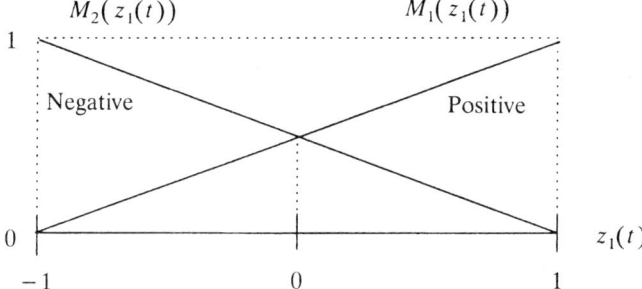

Fig. 2.4 Membership functions $M_1(z_1(t))$ and $M_2(z_1(t))$.

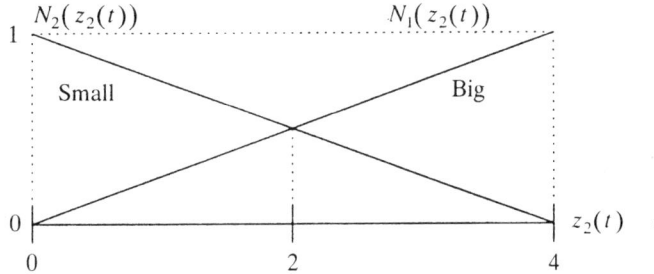

Fig. 2.5 Membership functions $N_1(z_2(t))$ and $N_2(z_2(t))$.

THEN $\dot{x}(t) = A_3 x(t)$.

Model Rule 4:

 IF $z_1(t)$ is "Negative" and $z_2(t)$ is "Small,"

  THEN $\dot{x}(t) = A_4 x(t)$.

Here,

$$A_1 = \begin{bmatrix} -1 & 1 \\ 4 & -1 \end{bmatrix}, \quad A_2 = \begin{bmatrix} -1 & 1 \\ 0 & -1 \end{bmatrix},$$

$$A_3 = \begin{bmatrix} -1 & -1 \\ 4 & -1 \end{bmatrix}, \quad A_4 = \begin{bmatrix} -1 & -1 \\ 0 & -1 \end{bmatrix}.$$

Figures 2.4 and 2.5 show the membership functions. The defuzzification is carried out as

$$\dot{x}(t) = \sum_{i=1}^{4} h_i(z(t)) A_i x(t),$$

where

$$h_1(z(t)) = M_1(z_1(t)) \times N_1(z_2(t)),$$
$$h_2(z(t)) = M_1(z_1(t)) \times N_2(z_2(t)),$$
$$h_3(z(t)) = M_2(z_1(t)) \times N_1(z_2(t)),$$
$$h_4(z(t)) = M_2(z_1(t)) \times N_2(z_2(t)).$$

This fuzzy model exactly represents the nonlinear system in the region $[-1, 1] \times [-1, 1]$ on the x_1-x_2 space.

Example 3 The equations of motion for the inverted pendulum [21] are

$$\dot{x}_1(t) = x_2(t),$$
$$\dot{x}_2(t) = \frac{g \sin(x_1(t)) - amlx_2^2(t) \sin(2x_1(t))/2 - a \cos(x_1(t))u(t)}{4l/3 - aml \cos^2(x_1(t))},$$
(2.11)

where $x_1(t)$ denotes the angle (in radians) of the pendulum from the vertical and $x_2(t)$ is the angular velocity; $g = 9.8 \ m/s^2$ is the gravity constant, m is the mass of the pendulum, M is the mass of the cart, $2l$ is the length of the pendulum, and u is the force applied to the cart (in newtons); $a = 1/(m + M)$.

Equation (2.11) is rewritten as

$$\dot{x}_2(t) = \frac{1}{4l/3 - aml \cos^2(x_1(t))}$$
$$\times \left(g \sin(x_1(t)) - \frac{amlx_2(t) \sin(2x_1(t))}{2} x_2(t) - a \cos(x_1(t))u(t) \right).$$
(2.12)

Define

$$z_1(t) \equiv \frac{1}{4l/3 - aml \cos^2(x_1(t))},$$
$$z_2(t) \equiv \sin(x_1(t)),$$
$$z_3(t) \equiv x_2(t) \sin(2x_1(t)),$$
$$z_4(t) \equiv \cos(x_1(t)),$$

where $x_1(t) \in (-\pi/2, \pi/2)$ and $x_2(t) \in [-\alpha, \alpha]$. Note that the system is uncontrollable when $x_1(t) = \pm \pi/2$. To maintain controllability of the fuzzy model, we assume that $x_1(t) \in [-88°, 88°]$. Equation (2.12) is rewritten as

$$\dot{x}_2(t) = z_1(t)\left\{gz_2(t) - \frac{aml}{2}z_3(t)x_2(t) - az_4(t)u(t)\right\}.$$

As shown in Example 2, we replace $z_1(t) - z_4(t)$ with T-S fuzzy model representation. Since

$$\max_{x_1(t)} z_1(t) = \frac{1}{4l/3 - aml\beta^2} \equiv q_1, \qquad \beta = \cos(88°),$$

$$\min_{x_1(t)} z_1(t) = \frac{1}{4l/3 - aml} \equiv q_2,$$

$z_1(t)$ can be rewritten as

$$z_1(t) = \sum_{i=1}^{2} E_i(z_1(t))q_i, \qquad (2.13)$$

where

$$E_1(z_1(t)) = \frac{z_1(t) - q_2}{q_1 - q_2}, \qquad E_2(z_1(t)) = \frac{q_1 - z_1(t)}{q_1 - q_2}.$$

The membership functions, $E_1(z_1(t))$ and $E_2(z_1(t))$, are obtained from the property of $E_1(z_1(t)) + E_2(z_1(t)) = 1$.

Figure 2.6 shows $z_2(t) = \sin(x_1(t))$ and its local sector, where $x_1(t) \in (-\pi/2, \pi/2)$. From Figure 2.6, we can find the sector $[b_2, b_1]$ that consists of two lines $b_1 x_1$ and $b_2 x_1$, where the slopes are $b_1 = 1$ and $b_2 = 2/\pi$.

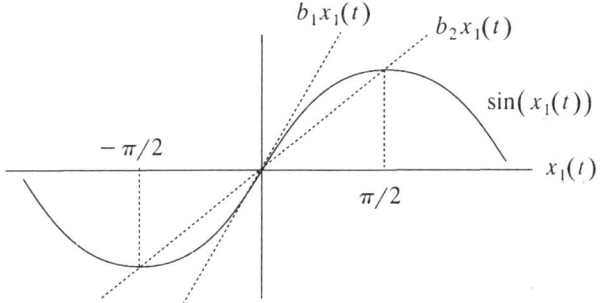

Fig. 2.6 $\sin(x_1(t))$ and its sector.

16 TAKAGI-SUGENO FUZZY MODEL AND PARALLEL DISTRIBUTED COMPENSATION

Therefore, we represent $\sin(x_1(t))$ as follows:

$$z_2(t) = \sin(x_1(t)) = \left(\sum_{i=1}^{2} M_i(z_2(t))b_i\right)x_1(t). \quad (2.14)$$

From the property of membership functions $[M_1(z_2(t)) + M_2(z_2(t)) = 1]$, we can obtain the membership functions

$$M_1(z_2(t)) = \begin{cases} \dfrac{z_2(t) - (2/\pi)\operatorname{Sin}^{-1}(z_2(t))}{(1 - 2/\pi)\operatorname{Sin}^{-1}(z_2(t))}, & z_2(t) \neq 0, \\ 1, & \text{otherwise}, \end{cases}$$

$$M_2(z_2(t)) = \begin{cases} \dfrac{\operatorname{Sin}^{-1}(z_2(t)) - z_2(t)}{(1 - 2/\pi)\operatorname{Sin}^{-1}(z_2(t))}, & z_2(t) \neq 0 \\ 0, & \text{otherwise}. \end{cases}$$

Next, consider $z_3(t) = x_2(t)\sin(2x_1(t))$. Since

$$\max_{x_1(t), x_2(t)} z_3(t) = \alpha \equiv c_1 \quad \text{and} \quad \min_{x_1(t), x_2(t)} z_3(t) = -\alpha \equiv c_2,$$

we can derive in the same way as the $z_1(t)$ case:

$$z_3(t) = x_2(t)\sin(2x_1(t)) = \sum_{i=1}^{2} N_i(z(t))c_i, \quad (2.15)$$

where

$$N_1(z_3(t)) = \dfrac{z_3(t) - c_2}{c_1 - c_2}, \quad N_2(z_3(t)) = \dfrac{c_1 - z_3(t)}{c_1 - c_2}.$$

We take the same procedure for $z_4(t)$ as well. Since

$$\max_{x_1(t)} z_4(t) = 1 \equiv d_1 \quad \text{and} \quad \min_{x_1(t)} z_4(t) = \beta \equiv d_2,$$

we obtain

$$z_4(t) = \cos(x_1(t)) = \sum_{i=1}^{2} S_i(z(t))d_i, \quad (2.16)$$

where

$$S_1(z_4(t)) = \dfrac{z_4(t) - d_2}{d_1 - d_2}, \quad S_2(z_4(t)) = \dfrac{d_1 - z_4(t)}{d_1 - d_2}.$$

CONSTRUCTION OF FUZZY MODEL

From (2.13)–(2.16), we construct the following Takagi-Sugeno fuzzy model for the inverted pendulum:

$$\begin{bmatrix} \dot{x}_1(t) \\ \dot{x}_2(t) \end{bmatrix} = \sum_{i=1}^{2} \sum_{j=1}^{2} \sum_{k=1}^{2} \sum_{l=1}^{2} E_i(z_1(t)) M_j(z_2(t)) N_k(z_3(t)) S_l(z_4(t))$$

$$\times \left(\begin{bmatrix} 0 & 1 \\ g \cdot q_i b_j & -\dfrac{aml}{2} q_i c_k \end{bmatrix} \begin{bmatrix} x_1(t) \\ x_2(t) \end{bmatrix} + \begin{bmatrix} 0 \\ -a \cdot q_i d_l \end{bmatrix} u(t) \right)$$

$$= \sum_{i=1}^{2} \sum_{j=1}^{2} \sum_{k=1}^{2} \sum_{l=1}^{2} E_i(z_1(t)) M_j(z_2(t)) N_k(z_3(t)) S_l(z_4(t))$$

$$\times \{ A_{ijkl} x(t) + B_{ijkl} u(t) \}. \tag{2.17}$$

The summations in (2.17) can be aggregated as one summation:

$$\dot{x}(t) = \sum_{\rho=1}^{16} h_\rho(z(t)) \{ A_\rho^* x(t) + B_\rho^* u(t) \}, \tag{2.18}$$

where

$$\rho = l + 2(k-1) + 4(j-1) + 8(i-1),$$
$$h_\rho(z(t)) = E_i(z_1(t)) M_j(z_2(t)) N_k(z_3(t)) S_l(z_4(t)),$$
$$A_\rho^* = A_{ijkl}, \qquad B_\rho^* = B_{ijkl}.$$

Equation (2.18) means that the fuzzy model has the following 16 rules:

Model Rule 1:

  IF $z_1(t)$ is "Positive" and $z_2(t)$ is "Zero"
  and $z_3(t)$ is "Positive" and $z_4(t)$ is "Big,"

  THEN $\dot{x}(t) = A_1^* x(t) + B_1^* u(t)$.

Model Rule 2:

  IF $z_1(t)$ is "Positive" and $z_2(t)$ is "Zero"
  and $z_3(t)$ is "Positive" and $z_4(t)$ is "Small,"

  THEN $\dot{x}(t) = A_2^* x(t) + B_2^* u(t)$.

Model Rule 3:

  IF $z_1(t)$ is "Positive" and $z_2(t)$ is "Zero"
  and $z_3(t)$ is "Negative" and $z_4(t)$ is "Big,"

  THEN $\dot{x}(t) = A_3^* x(t) + B_3^* u(t)$.

Model Rule 4:

    IF $z_1(t)$ is "Positive" and $z_2(t)$ is "Zero"
        and $z_3(t)$ is "Negative" and $z_4(t)$ is "Small,"

    THEN $\dot{x}(t) = A_4^* x(t) + B_4^* u(t)$.

Model Rule 5:

    IF $z_1(t)$ is "Positive" and $z_2(t)$ is "Not Zero"
        and $z_3(t)$ is "Positive" and $z_4(t)$ is "Big,"

    THEN $\dot{x}(t) = A_5^* x(t) + B_5^* u(t)$.

Model Rule 6:

    IF $z_1(t)$ is "Positive" and $z_2(t)$ is "Not Zero"
        and $z_3(t)$ is "Positive" and $z_4(t)$ is "Small,"

    THEN $\dot{x}(t) = A_6^* x(t) + B_6^* u(t)$.

Model Rule 7:

    IF $z_1(t)$ is "Positive" and $z_2(t)$ is "Not Zero"
        and $z_3(t)$ is "Negative" and $z_4(t)$ is "Big,"

    THEN $\dot{x}(t) = A_7^* x(t) + B_7^* u(t)$.

Model Rule 8:

    IF $z_1(t)$ is "Positive" and $z_2(t)$ is "Not Zero"
        and $z_3(t)$ is "Negative" and $z_4(t)$ is "Small,"

    THEN $\dot{x}(t) = A_8^* x(t) + B_8^* u(t)$.

Model Rule 9:

    IF $z_1(t)$ is "Negative" and $z_2(t)$ is "Zero"
        and $z_3(t)$ is "Positive" and $z_4(t)$ is "Big,"

    THEN $\dot{x}(t) = A_9^* x(t) + B_9^* u(t)$.

CONSTRUCTION OF FUZZY MODEL   19

Model Rule 10:

   IF $z_1(t)$ is "Negative" and $z_2(t)$ is "Zero"
   and $z_3(t)$ is "Positive" and $z_4(t)$ is "Small,"

   THEN $\dot{x}(t) = A_{10}^* x(t) + B_{10}^* u(t)$.

Model Rule 11:

   IF $z_1(t)$ is "Negative" and $z_2(t)$ is "Zero"
   and $z_3(t)$ is "Negative" and $z_4(t)$ is "Big,"

   THEN $\dot{x}(t) = A_{11}^* x(t) + B_{11}^* u(t)$.

Model Rule 12:

   IF $z_1(t)$ is "Negative" and $z_2(t)$ is "Zero"
   and $z_3(t)$ is "Negative" and $z_4(t)$ is "Small,"

   THEN $\dot{x}(t) = A_{12}^* x(t) + B_{12}^* u(t)$.

Model Rule 13:

   IF $z_1(t)$ is "Negative" and $z_2(t)$ is "Not Zero"
   and $z_3(t)$ is "Positive" and $z_4(t)$ is "Big,"

   THEN $\dot{x}(t) = A_{13}^* x(t) + B_{13}^* u(t)$.

Model Rule 14:

   IF $z_1(t)$ is "Negative" and $z_2(t)$ is "Not Zero"
   and $z_3(t)$ is "Positive" and $z_4(t)$ is "Small,"

   THEN $\dot{x}(t) = A_{14}^* x(t) + B_{14}^* u(t)$.

Model Rule 15:

   IF $z_1(t)$ is "Negative" and $z_2(t)$ is "Not Zero"
   and $z_3(t)$ is "Negative" and $z_4(t)$ is "Big,"

   THEN $\dot{x}(t) = A_{15}^* x(t) + B_{15}^* u(t)$.

Model Rule 16:

IF $z_1(t)$ is "Negative" and $z_2(t)$ is "Not Zero"
and $z_3(t)$ is "Negative" and $z_4(t)$ is "Small,"

THEN $\dot{x}(t) = A_{16}^* x(t) + B_{16}^* u(t)$.

Here, $z_1(t)$, $z_2(t)$, $z_3(t)$ and $z_4(t)$ are premise variables and

$$A_1^* = A_{1111} = \begin{bmatrix} 0 & 1 \\ g \cdot q_1 b_1 & -\frac{aml}{2} \cdot q_1 c_1 \end{bmatrix}, \quad B_1^* = B_{1111} = \begin{bmatrix} 0 \\ -a \cdot q_1 d_1 \end{bmatrix},$$

$$A_2^* = A_{1112} = \begin{bmatrix} 0 & 1 \\ g \cdot q_1 b_1 & -\frac{aml}{2} \cdot q_1 c_1 \end{bmatrix}, \quad B_2^* = B_{1112} = \begin{bmatrix} 0 \\ -a \cdot q_1 d_2 \end{bmatrix},$$

$$A_3^* = A_{1121} = \begin{bmatrix} 0 & 1 \\ g \cdot q_1 b_1 & -\frac{aml}{2} \cdot q_1 c_2 \end{bmatrix}, \quad B_3^* = B_{1121} = \begin{bmatrix} 0 \\ -a \cdot q_1 d_1 \end{bmatrix},$$

$$A_4^* = A_{1122} = \begin{bmatrix} 0 & 1 \\ g \cdot q_1 b_1 & -\frac{aml}{2} \cdot q_1 c_2 \end{bmatrix}, \quad B_4^* = B_{1122} = \begin{bmatrix} 0 \\ -a \cdot q_1 d_2 \end{bmatrix},$$

$$A_5^* = A_{1211} = \begin{bmatrix} 0 & 1 \\ g \cdot q_1 b_2 & -\frac{aml}{2} \cdot q_1 c_1 \end{bmatrix}, \quad B_5^* = B_{1211} = \begin{bmatrix} 0 \\ -a \cdot q_1 d_1 \end{bmatrix},$$

$$A_6^* = A_{1212} = \begin{bmatrix} 0 & 1 \\ g \cdot q_1 b_2 & -\frac{aml}{2} \cdot q_1 c_1 \end{bmatrix}, \quad B_6^* = B_{1212} = \begin{bmatrix} 0 \\ -a \cdot q_1 d_2 \end{bmatrix},$$

$$A_7^* = A_{1221} = \begin{bmatrix} 0 & 1 \\ g \cdot q_1 b_2 & -\frac{aml}{2} \cdot q_1 c_2 \end{bmatrix}, \quad B_7^* = B_{1221} = \begin{bmatrix} 0 \\ -a \cdot q_1 d_1 \end{bmatrix},$$

$$A_8^* = A_{1222} = \begin{bmatrix} 0 & 1 \\ g \cdot q_1 b_2 & -\frac{aml}{2} \cdot q_1 c_2 \end{bmatrix}, \quad B_8^* = B_{1222} = \begin{bmatrix} 0 \\ -a \cdot q_1 d_2 \end{bmatrix},$$

$$A_9^* = A_{2111} = \begin{bmatrix} 0 & 1 \\ g \cdot q_2 b_1 & -\frac{aml}{2} \cdot q_2 c_1 \end{bmatrix}, \quad B_9^* = B_{2111} = \begin{bmatrix} 0 \\ -a \cdot q_2 d_1 \end{bmatrix},$$

$$A_{10}^* = A_{2112} = \begin{bmatrix} 0 & 1 \\ g \cdot q_2 b_1 & -\dfrac{aml}{2} \cdot q_2 c_1 \end{bmatrix}, \quad B_{10}^* = B_{2112} = \begin{bmatrix} 0 \\ -a \cdot q_2 d_2 \end{bmatrix},$$

$$A_{11}^* = A_{2121} = \begin{bmatrix} 0 & 1 \\ g \cdot q_2 b_1 & -\dfrac{aml}{2} \cdot q_2 c_2 \end{bmatrix}, \quad B_{11}^* = B_{2121} = \begin{bmatrix} 0 \\ -a \cdot q_2 d_1 \end{bmatrix},$$

$$A_{12}^* = A_{2122} = \begin{bmatrix} 0 & 1 \\ g \cdot q_2 b_1 & -\dfrac{aml}{2} \cdot q_2 c_2 \end{bmatrix}, \quad B_{12}^* = B_{2122} = \begin{bmatrix} 0 \\ -a \cdot q_2 d_2 \end{bmatrix},$$

$$A_{13}^* = A_{2211} = \begin{bmatrix} 0 & 1 \\ g \cdot q_2 b_2 & -\dfrac{aml}{2} \cdot q_2 c_1 \end{bmatrix}, \quad B_{13}^* = B_{2211} = \begin{bmatrix} 0 \\ -a \cdot q_2 d_1 \end{bmatrix},$$

$$A_{14}^* = A_{2212} = \begin{bmatrix} 0 & 1 \\ g \cdot q_2 b_2 & -\dfrac{aml}{2} \cdot q_2 c_1 \end{bmatrix}, \quad B_{14}^* = B_{2212} = \begin{bmatrix} 0 \\ -a \cdot q_2 d_2 \end{bmatrix},$$

$$A_{15}^* = A_{2221} = \begin{bmatrix} 0 & 1 \\ g \cdot q_2 b_2 & -\dfrac{aml}{2} \cdot q_2 c_2 \end{bmatrix}, \quad B_{15}^* = B_{2221} = \begin{bmatrix} 0 \\ -a \cdot q_2 d_1 \end{bmatrix},$$

$$A_{16}^* = A_{2222} = \begin{bmatrix} 0 & 1 \\ g \cdot q_2 b_2 & -\dfrac{aml}{2} \cdot q_2 c_2 \end{bmatrix}, \quad B_{16}^* = B_{2222} = \begin{bmatrix} 0 \\ -a \cdot q_2 d_2 \end{bmatrix}.$$

Figures 2.7–2.10 show the membership functions, that is,

$$E_1(z_1(t)) = \frac{z_1(t) - q_2}{q_1 - q_2}, \quad E_2(z_1(t)) = \frac{q_1 - z_1(t)}{q_1 - q_2},$$

$$M_1(z_2(t)) = \frac{\sin(x_1(t)) - (2/\pi)z_2(t)}{(1 - 2/\pi)z_2(t)}, \quad M_2(z_2(t)) = \frac{x_1(t) - z_2(t)}{(1 - 2/\pi)z_2(t)},$$

$$N_1(z_3(t)) = \frac{z_3(t) - c_2}{c_1 - c_2}, \quad N_2(z_3(t)) = \frac{c_1 - z_3(t)}{c_1 - c_2},$$

$$S_1(z_4(t)) = \frac{z_4(t) - d_2}{d_1 - d_2}, \quad S_2(z_4(t)) = \frac{d_1 - z_4(t)}{d_1 - d_2}.$$

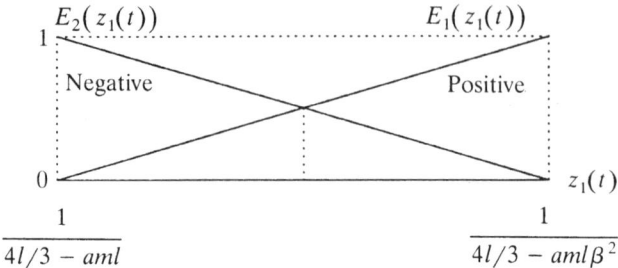

Fig. 2.7 Membership functions $E_1(z_1(t))$ and $E_2(z_1(t))$.

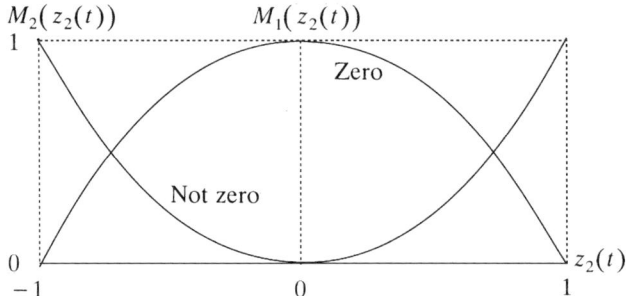

Fig. 2.8 Membership functions $M_1(z_2(t))$ and $M_2(z_2(t))$.

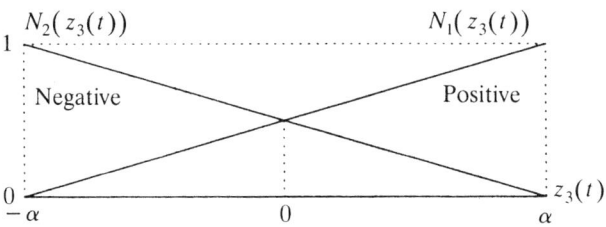

Fig. 2.9 Membership functions $N_1(z_3(t))$ and $N_2(z_3(t))$.

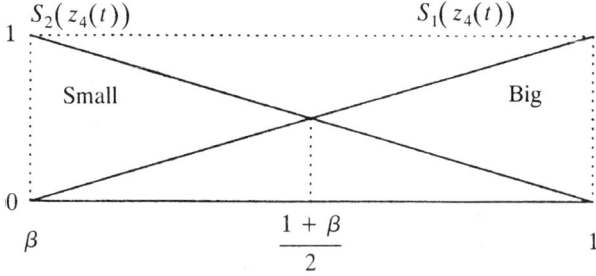

Fig. 2.10 Membership functions $S_1(z_4(t))$ and $S_2(z_4(t))$.

Remark 2 Prior to applying the sector nonlinearity approach, it is often a good practice to simplify the original nonlinear model as much as possible. This step is important for practical applications because it always leads to the reduction of the number of model rules, which reduces the effort for analysis and design of control systems. This aspect will be illustrated in design examples throughout this book. For instance, in the vehicle control described in Chapter 8, a two-rule fuzzy model is obtained. If we attempt to derive a fuzzy model without simplifying the original nonlinear model, 2^6 rules would be needed to exactly represent the nonlinear model. We will see in Chapter 8 that the fuzzy controller design based on the two-rule fuzzy model performs well even for the original nonlinear system.

2.2.2 Local Approximation in Fuzzy Partition Spaces

Another approach to obtain T-S fuzzy models is the so-called local approximation in fuzzy partition spaces. The spirit of the approach is to approximate nonlinear terms by judiciously chosen linear terms. This procedure leads to reduction of the number of model rules. For instance, the fuzzy model for the inverted pendulum in Example 3 has 16 rules. In comparison, in Example 4 a 2-rule fuzzy model will be constructed using the local approximation idea. The number of model rules is directly related to complexity of analysis and design LMI conditions. This is because the number of rules for the overall control system is basically the combination of the model rules and control rules.

Remark 3 As pointed out above, the local approximation technique leads to the reduction of the number of rules for fuzzy models. However, designing control laws based on the approximated fuzzy model may not guarantee the stability of the original nonlinear systems under such control laws. One of the approaches to alleviate the problem is to introduce robust controller design, described in Chapter 5.

Example 4 Recall the inverted pendulum in Example 3. In that example, the constructed fuzzy model has 16 rules. In the following we attempt to construct a two-rule fuzzy model by local approximation in fuzzy partition spaces. Of course, the derived model is only an approximation to the original system. However, it will be shown later in this chapter that a fuzzy controller design based on the two-rule fuzzy model performs well when applied to the original nonlinear pendulum system.

When $x_1(t)$ is near zero, the nonlinear equations can be simplified as

$$\dot{x}_1(t) = x_2(t), \qquad (2.19)$$

$$\dot{x}_2(t) = \frac{gx_1(t) - au(t)}{4l/3 - aml}. \qquad (2.20)$$

When $x_1(t)$ is near $\pm\pi/2$, the nonlinear equations can be simplified as

$$\dot{x}_1(t) = x_2(t), \qquad (2.21)$$

$$\dot{x}_2(t) = \frac{2gx_1(t)/\pi - a\beta u(t)}{4l/3 - aml\beta^2}, \qquad (2.22)$$

where $\beta = \cos(88°)$.

Note that (2.19)–(2.22) are now linear systems. We arrive at the following fuzzy model based on the linear subsystems:

Model Rule 1

IF $x_1(t)$ is about 0,

THEN $\dot{x}(t) = A_1 x(t) + B_1 u(t)$.

Model Rule 2:

IF $x_1(t)$ is about $\pm\pi/2(|x_1| < \pi/2)$,

THEN $\dot{x}(t) = A_2 x(t) + B_2 u(t)$.

Here,

$$A_1 = \begin{bmatrix} 0 & 1 \\ \dfrac{g}{4l/3 - aml} & 0 \end{bmatrix}, \quad B_1 = \begin{bmatrix} 0 \\ -\dfrac{a}{4l/3 - aml} \end{bmatrix},$$

$$A_2 = \begin{bmatrix} 0 & 1 \\ \dfrac{2g}{\pi(4l/3 - aml\beta^2)} & 0 \end{bmatrix}, \quad B_2 = \begin{bmatrix} 0 \\ -\dfrac{a\beta}{4l/3 - aml\beta^2} \end{bmatrix},$$

and $\beta = \cos(88°)$. Membership functions for Rules 1 and 2 can be simply defined as shown in Figure 2.11.

Remark 4 In Example 4, the membership functions are simply defined using triangular types. Note that the fuzzy model is an approximated model. Therefore we may simply define triangular-type membership functions. On the other hand, in the fuzzy model in Example 3, the membership functions are obtained so as to exactly represent the nonlinear dynamics.

The following remark addresses the important issue of approximating nonlinear systems via T-S models.

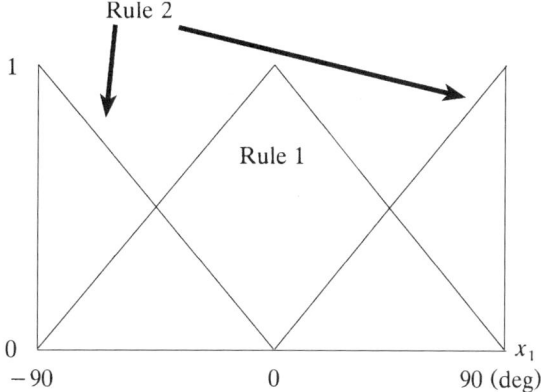

Fig. 2.11 Membership functions of two-rule model.

Remark 5 Section 2.2 presents the approaches to obtain a fuzzy model for a nonlinear system. An important and natural question arises in the construction using local approximation in fuzzy partition spaces or simplification before using sector nonlinearity. One may ask, "Is it possible to approximate any smooth nonlinear systems with Takagi-Sugeno fuzzy models (2.1) having no consequent constant terms?" The answer is fortunately Yes if we consider the problem in C^0 or C^1 context. That is, the original vector field plus its first-order derivative can be accurately approximated. Details will be presented in Chapter 14.

2.3 PARALLEL DISTRIBUTED COMPENSATION

The history of the so-called parallel distributed compensation (PDC) began with a model-based design procedure proposed by Kang and Sugeno (e.g., [16]). However, the stability of the control systems was not addressed in the design procedure. The design procedure was improved and the stability of the control systems was analyzed in [2]. The design procedure is named "parallel distributed compensation" in [14].

The PDC [2, 14, 15] offers a procedure to design a fuzzy controller from a given T-S fuzzy model. To realize the PDC, a controlled object (nonlinear system) is first represented by a T-S fuzzy model. We emphasize that many real systems, for example, mechanical systems and chaotic systems, can be and have been represented by T-S fuzzy models.

In the PDC design, each control rule is designed from the corresponding rule of a T-S fuzzy model. The designed fuzzy controller shares the same fuzzy sets with the fuzzy model in the premise parts. For the fuzzy models

(2.1) and (2.2), we construct the following fuzzy controller via the PDC:

Control Rule i:

$$\text{IF } z_1(t) \text{ is } M_{i1} \text{ and } \cdots \text{ and } z_p(t) \text{ is } M_{ip},$$

$$\text{THEN } u(t) = -F_i x(t), \quad i = 1, 2, \ldots, r.$$

The fuzzy control rules have a linear controller (state feedback laws in this case) in the consequent parts. We can use other controllers, for example, output feedback controllers and dynamic output feedback controllers, instead of the state feedback controllers. For details, consult Chapters 12 and 13, which are devoted to the problem of dynamic output feedback.

The overall fuzzy controller is represented by

$$u(t) = -\frac{\sum_{i=1}^{r} w_i(z(t)) F_i x(t)}{\sum_{i=1}^{r} w_i(z(t))} = -\sum_{i=1}^{r} h_i(z(t)) F_i x(t). \quad (2.23)$$

The fuzzy controller design is to determine the local feedback gains F_i in the consequent parts. With PDC we have a simple and natural procedure to handle nonlinear control systems. Other nonlinear control techniques require special and rather involved knowledge.

Remark 6 Although the fuzzy controller (2.23) is constructed using the local design structure, the feedback gains F_i should be determined using global design conditions. The global design conditions are needed to guarantee the global stability and control performance. An interesting example will be presented in the next section.

Example 5 If the controlled object is represented as the model rules shown in Example 1, the following control rules can be constructed via the PDC:

Control Rule i:

$$\text{IF } x(t) \text{ is } M_{i1} \text{ and } \cdots \text{ and } x(t - n + 1) \text{ is } M_{in},$$

$$\text{THEN } u(t) = -F_i x(t), \quad i = 1, 2, \ldots, r.$$

2.4 A MOTIVATING EXAMPLE

In this chapter, for brevity only results for discrete-time systems are presented. The results, however, also hold for continuous-time systems subject to some minor modifications.

The open-loop system of (2.5) is

$$x(t+1) = \sum_{i=1}^{r} h_i(z(t)) A_i x(t). \tag{2.24}$$

A sufficient stability condition, derived by Tanaka and Sugeno [1, 2], for ensuring stability of (2.24) follows.

THEOREM 1 [1, 2] *The equilibrium of a fuzzy system (2.24) is globally asymptotically stable if there exists a common positive definite matrix P such that*

$$A_i^T P A_i - P < 0, \quad i = 1, 2, \ldots, r, \tag{2.25}$$

that is, a common P has to exist for all subsystems.

This theorem reduces to the Lyapunov stability theorem for (discrete-time) linear systems when $r = 1$.

The stability condition of Theorem 1 is derived using a quadratic function $V(x(t)) = x(t)^T P x(t)$. If there exists a $P > 0$ such that $V(x(t)) = x(t)^T P x(t)$ proves the stability of system (2.24), system (2.24) is also said to be *quadratically stable* and $V(x(t))$ is called a quadratic Lyapunov function. Theorem 1 thus presents a sufficient condition for the quadratic stability of system (2.24).

To check the stability of fuzzy system (2.24), the lack of systematic procedures to find a common positive definite matrix P has long been recognized. Most of the time a trial-and-error type of procedure has been used [2, 23]. In [13] a procedure to construct a common P is given for second-order fuzzy systems, that is, the dimension of state $n = 2$. We first pointed out in [14, 15, 24] that the common P problem can be solved efficiently via convex optimization techniques for LMIs [18]. To do this, a very important observation is that the stability condition of Theorem 1 is expressed in LMIs. To check stability, we need to find a common P or determine that no such P exists. This is an LMI problem. See Section 2.5.2 for details on LMIs and the related LMI approach to stability analysis and design of fuzzy control systems. Numerically the LMI problems can be solved very efficiently by means of some of the most powerful tools available to date in the mathematical programming literature. For instance, the recently developed interior-point methods [19] are extremely efficient in practice.

A question naturally arises of whether system (2.24) is stable if all its subsystems are stable, that is, all A_i's are stable. The answer is no in general, as illustrated by the following example.

Example 6 Consider the following fuzzy system:

Rule 1:

  IF $x_2(t)$ is M_1 (e.g., Small),

   THEN $x(t+1) = A_1 x(t)$.

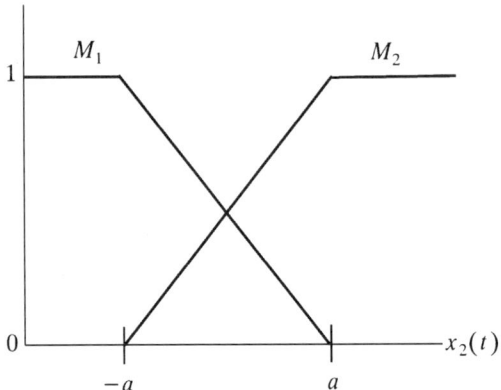

Fig. 2.12 Membership functions of Example 6.

Rule 2:

 IF $x_2(t)$ is M_2 (e.g., Big),

  THEN $x(t + 1) = A_2 x(t)$.

Here, $x(t) = [x_1(t) \ x_2(t)]^T$ and

$$A_1 = \begin{bmatrix} 1 & -0.5 \\ 1 & 0 \end{bmatrix}, \quad A_2 = \begin{bmatrix} -1 & -0.5 \\ 1 & 0 \end{bmatrix}.$$

Figure 2.12 shows the membership functions of M_1 and M_2. Since A_1 and A_2 are stable, the linear subsystems are stable. However, for some initial conditions the fuzzy system can be unstable, as shown in Figure 2.13 for the initial condition $x = [0.90 \ -0.70]^T$. It should be noted that the linearization of the fuzzy system around $\mathbf{0}$ is stable (which implies that the fuzzy system is locally stable). Obviously there does not exist a common $P > 0$ since the fuzzy system is unstable. This can be shown analytically. Moreover this can also be shown numerically by convex optimization algorithms involving LMIs.

Still an interesting question is for what initial conditions the fuzzy system is stable (or unstable). This is determined by studying the basin of attraction of the origin.[1]

Figure 2.14(a) shows the basin of attraction for the case of $a = 1$. The black area indicates regions of instability (horizontal axis is x_1). It is also of interest to consider how the basin of attraction changes as the membership functions vary, for instance, how the basin of attraction would change as a varies for this example. Figures (b), (c), and (d) show the basin of attraction

---

[1]Sugeno mentioned this point in his plenary talk titled "Fuzzy Control: Principles, Practice, and Perspectives" at 1992 IEEE International Conference on Fuzzy Systems, March 9, 1992.

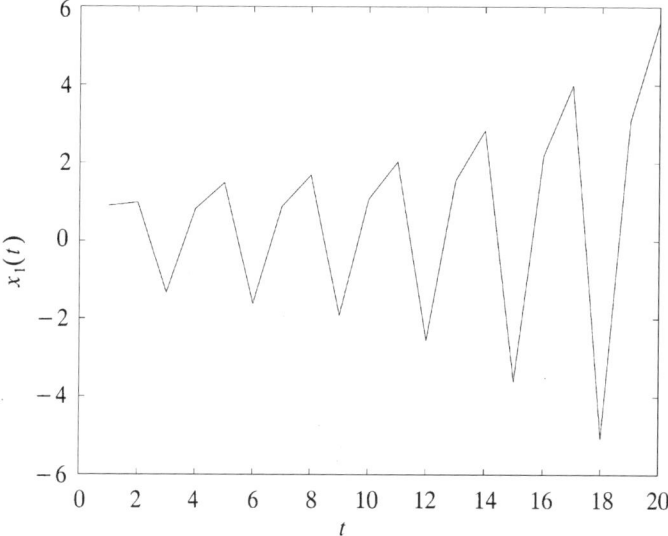

Fig. 2.13 Response of Example 6 ($a = 1$).

for various values of a. It can be seen that as a decreases (increases) from 1, the basin of attraction becomes smaller (larger). Therefore, the basin of attraction for the fuzzy system could be membership function dependent. In the example, when $a = \infty$, the fuzzy system becomes

$$x(t+1) = \frac{A_1 + A_2}{2} x(t),$$

which is linear and globally asymptotically stable.

For this example, an interesting interpretation can be given for the dependence of basin of attraction on membership functions. As a increases (decreases), the inference process tends to be "fuzzier" ("crisper"). Hence a fuzzier decision leads to a larger basin of attraction while a crisper decision leads to a smaller basin of attraction.

As illustrated by the example, we have to take stability into consideration when selecting rules and membership functions. How to systematically select rules and membership functions to satisfy prescribed stability properties is an interesting topic. In the next section, we consider the control design problems via parallel distributed compensations.

2.5 ORIGIN OF THE LMI-BASED DESIGN APPROACH

This section gives the origin of the control design approach, which forms the core subject of this book, that is, the LMI-based design approach. The objective here is to illustrate the basic ideas [24] of stability analysis and

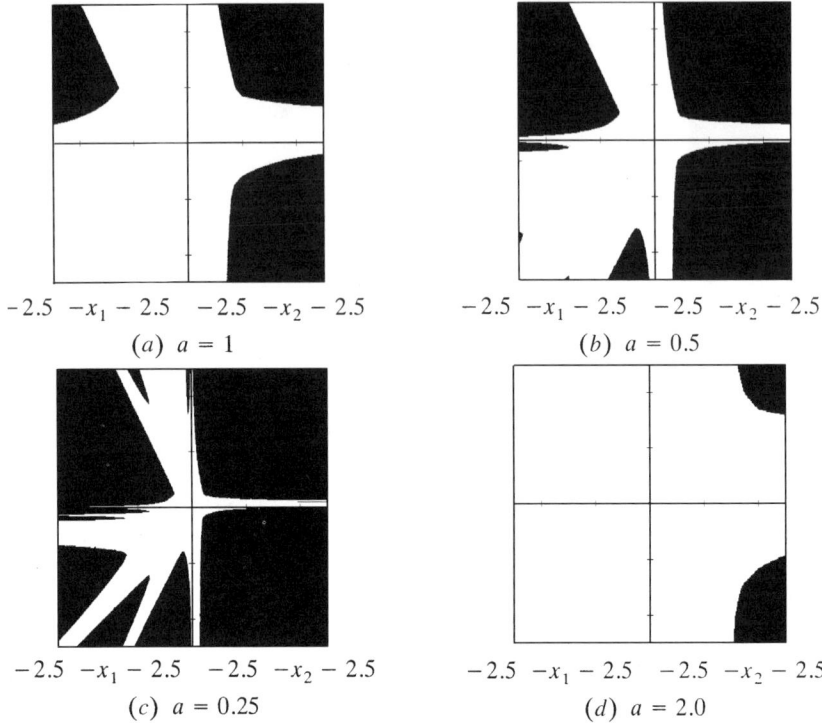

Fig. 2.14 Basin of attraction for Example 6.

stable fuzzy controller design via LMIs. The details will be presented in Chapter 3.

2.5.1 Stable Controller Design via Iterative Procedure

The PDC fuzzy controller is

$$u(t) = -\sum_{i=1}^{r} h_i(z(t)) F_i x(t). \qquad (2.26)$$

Note that the controller (2.26) is *nonlinear* in general.
Substituting (2.26) into (2.5), we obtain

$$x(t+1) = \sum_{i=1}^{r} \sum_{j=1}^{r} h_i(z(t)) h_j(z(t)) \{A_i - B_i F_j\} x(t). \qquad (2.27)$$

Applying Theorem 1, we have the following sufficient condition for (quadratic) stability.

ORIGIN OF THE LMI-BASED DESIGN APPROACH

THEOREM 2 *The equilibrium of a fuzzy control system (2.27) is globally asymptotically stable if there exists a common positive definite matrix P such that*

$$\{A_i - B_i F_j\}^T P \{A_i - B_i F_j\} - P < 0 \qquad (2.28)$$

for $h_i(z(t)) \cdot h_j(z(t)) \neq 0$, $\forall t$, $i, j = 1, 2, \ldots, r$.

Note that system (2.27) can also be written as

$$x(t+1) = \left[\sum_{i=1}^{r} h_i(z(t)) h_i(z(t)) \{A_i - B_i F_i\} x(t) \right.$$

$$\left. + 2 \sum_{i=1}^{r} \sum_{i<j} h_i(z(t)) h_j(z(t)) G_{ij} x(t) \right], \qquad (2.29)$$

where

$$G_{ij} = \frac{\{A_i - B_i F_j\} + \{A_j - B_j F_i\}}{2}, \quad i < j \text{ s.t. } h_i \cap h_j \neq \phi.$$

Therefore we have the following sufficient condition.

THEOREM 3 *The equilibrium of a fuzzy control system (2.27) is globally asymptotically stable if there exists a common positive definite matrix P such that the following two conditions are satisfied*:

$$\{A_i - B_i F_i\}^T P \{A_i - B_i F_i\} - P < 0, \quad i = 1, 2, \ldots, r \qquad (2.30)$$

$$G_{ij}^T P G_{ij} - P < 0, \quad i < j \leq r \text{ s.t. } h_i \cap h_j \neq \phi. \qquad (2.31)$$

For the meaning of the notation $i < j \leq r$ s.t. $h_i \cap h_j \neq \phi$, see Chapter 1.

Remark 7 The conditions of Theorem 3 are more relaxed than those of Theorem 2.

The control design problem is to select F_i ($i = 1, 2, \ldots, r$) such that conditions (2.30) and (2.31) in Theorem 3 are satisfied. Using the notation of quadratic stability, we can also formulate the control design problem as to find F_i's such that the closed-loop system (2.27) is quadratically stable. If there exist such F_i's, the system (2.5) is also said to be *quadratically stabilizable* via PDC design.

In this chapter, we first design a controller for each rule and check whether the stability conditions are satisfied. Recall we can use LMI convex programming techniques to solve this stability analysis problem. If the stability conditions are not satisfied, we have to repeat the procedure. Consult Section 2.5.2 on how LMIs can be used to directly solve the control design problem.

Next consider the common B matrix case, that is, $B_i = B (i = 1, 2, \ldots, r)$. In this case, Theorem 3 reduces to:

THEOREM 4 *When $B_i = B$, $i = 1, \ldots, r$, the equilibrium of the fuzzy control system (2.27) is globally asymptotically stable if there exists a common positive definite matrix P such that*

$$\{A_i - B_i F_i\}^T P \{A_i - B_i F_i\} - P < 0, \quad i = 1, 2, \ldots, r. \quad (2.32)$$

Furthermore, for the common B case, if we can choose F_i such that

$$A_i - BF_i = G, \quad (2.33)$$

where G is a Hurwitz matrix, then the system (2.27) becomes a linear system

$$x(t+1) = Gx(t).$$

This is a global linearization result. We remark that a common G might not always be possible even if (A_i, B_i) are controllable.

Remark 8 As shown in Theorem 4, the stability conditions are simplified in the common B matrix case. The same feature will be observed in all the chapters.

Let us look at some examples.

Example 7 Consider the following fuzzy system:

Model Rule 1:

   IF $x_2(t)$ is M_1,

   THEN $x(t+1) = A_1 x(t) + Bu(t)$.

Model Rule 2:

   IF $x_2(t)$ is M_2,

   THEN $x(t+1) = A_2 x(t) + Bu(t)$.

Here, A_1, A_2 are the same as in Example 6 and

$$B = \begin{bmatrix} 1 \\ 0 \end{bmatrix}.$$

Employ the PDC controller (2.26) and choose the closed-loop eigenvalues to be [0.5 0.35]. We obtain

$$F_1 = [0.15 \ -0.3250],$$
$$F_2 = [-1.85 \ -0.3250],$$

and

$$A_1 - BF_1 = A_2 - BF_2 = G = \begin{bmatrix} 0.85 & -0.1750 \\ 1 & 0 \end{bmatrix}.$$

The closed loop becomes

$$x(t+1) = Gx(t),$$

which is stable since G is stable.

Next we consider the more general case.

Example 8

Consider the following fuzzy system:

Model Rule 1:

  IF $x_2(t)$ is M_1,

  THEN $x(t+1) = A_1 x(t) + B_1 u(t)$.

Model Rule 2:

  IF $x_2(t)$ is M_2,

  THEN $x(t+1) = A_2 x(t) + B_2 u(t)$.

Here, A_1, A_2 are the same as in Example 6 and

$$B_1 = \begin{bmatrix} 1 \\ 1 \end{bmatrix}, \quad B_2 = \begin{bmatrix} -2 \\ 1 \end{bmatrix}.$$

The membership functions of Example 6 ($a = 1$) are used in the simulation. Again choose the closed-loop eigenvalues to be [0.5, 0.35]. We have

$$F_1 = [0.65 \ -0.5],$$
$$F_2 = [0.87 \ -0.11],$$

and
$$A_1 - B_1 F_1 = \begin{bmatrix} 0.35 & 0 \\ 0.35 & 0.5 \end{bmatrix}, \quad A_2 - B_2 F_2 = \begin{bmatrix} 0.74 & -0.72 \\ 0.13 & 0.11 \end{bmatrix},$$

$$G_{12} = \begin{bmatrix} 0.2150 & -0.9450 \\ 0.2400 & 0.3050 \end{bmatrix}.$$

Note that G_{12} is stable.

The PDC controller is given as follows:

Control Rule 1:

  IF $x_2(t)$ is M_1,

  THEN $u(t) = -F_1 x(t)$.

Control Rule 2:

  IF $x_2(t)$ is M_2,

  THEN $u(t) = -F_2 x(t)$.

It can be easily shown that if we choose the positive definite matrix P to be

$$P = \begin{bmatrix} 1.1810 & -0.0614 \\ -0.0614 & 2.3044 \end{bmatrix},$$

the stability conditions (2.30) and (2.31) are satisfied. In other words, the closed-loop fuzzy control system which consists of the fuzzy model and the PDC controller is globally asymptotically stable. The P is obtained by utilizing an LMI optimization algorithm. Figure 2.15 illustrates the behavior of the fuzzy control system for the same initial condition of Figure 2.13.

In the next section, we present an introduction to LMIs as well as the LMI approach to stability analysis and design of fuzzy control systems.

2.5.2 Stable Controller Design via Linear Matrix Inequalities

Recently a class of numerical optimization problems called linear matrix inequality (LMI) problems has received significant attention [18]. These optimization problems can be solved in polynomial time and hence are tractable, at least in a theoretical sense. The recently developed interior-point methods [19] for these problems have been found to be extremely efficient in practice. For systems and control, the importance of LMI optimization stems from the fact that a wide variety of system and control problems can be

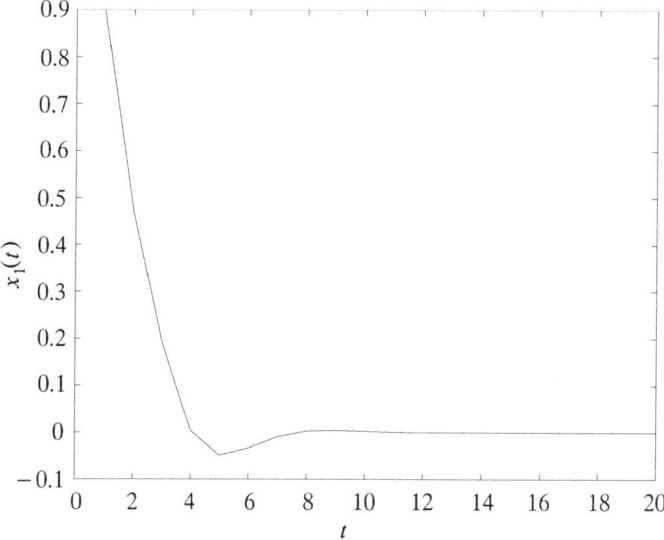

Fig. 2.15 Response of Example 8.

recast as LMI problems [18]. Except for a few special cases these problems do not have analytical solutions. However, the main point is that through the LMI framework they can be efficiently solved *numerically* in all cases. Therefore recasting a control problem as an LMI problem is equivalent to finding a "solution" to the original problem.

DEFINITION 1 [18] *An LMI is a matrix inequality of the form*

$$F(x) = F_0 + \sum_{i=1}^{m} x_i F_i > 0, \tag{2.34}$$

where $x^T = (x_1, x_2, \ldots, x_m)$ is the variable and the symmetric matrices $F_i = F_i^T \in \mathbb{R}^{n \times n}$, $i = 0, \ldots, m$, are given. The inequality symbol > 0 means that $F(x)$ is positive definite.

The LMI (2.34) is a convex constraint on x, that is, the set $\{x | F(x) > 0\}$ is convex. The LMI (2.34) can represent a wide variety of convex constraints on x. In particular, linear inequalities, convex quadratic inequalities, matrix norm inequalities, and constraints that arise in control theory, such as Lyapunov and convex quadratic matrix inequalities, can all be cast in the form of an LMI. Multiple LMIs $F^{(i)} > 0$, $i = 1, \ldots, p$, can be expressed as a single LMI **diag**($F^{(1)}, \ldots, F^{(p)}) > 0$.

Very often in the LMIs the variables are matrices, for example, the Lyapunov inequality

$$A^T P A - P < 0, \tag{2.35}$$

where $A \in \mathbb{R}^{n \times n}$ is given and $P = P^T$ is the variable. In this case the LMI will not be written explicitly in the form $F(x) > 0$. In addition to saving notation, this may lead to more efficient computation [18]. Of course, the inequality (2.35) can be readily put in the form (2.34): take $F_0 = 0$, $F_i = -A^T P_i A + P_i$, where $P_1, \ldots, P_m$ are a basis for symmetric $n \times n$ matrices.

LMI problems [18] Given an LMI $F(x) > 0$, the LMI problem is to find x^{feas} such that $F(x^{\text{feas}}) > 0$ or determine that the LMI is infeasible. This is a convex feasibility problem.

As an example, the simultaneous Lyapunov stability condition in Theorem 1 is exactly an LMI problem: Given $A_i \in \mathbb{R}^{n \times n}$, $i = 1, \ldots, r$, we need to find P satisfying the LMI

$$P > 0, \qquad A_i^T P A_i - P < 0, \qquad i = 1, 2, \ldots, r,$$

or determine that no such P exists.

The LMI problems are tractable from both theoretical and practical viewpoints: They can be solved in polynomial time, and they can be solved in practice very efficiently by means of some of the most powerful tools available to date in the mathematical programming literature (e.g., the recently developed interior-point methods [19]).

The stability conditions encountered in this book are expressed in the form of LMIs. This recasting is significant in the sense that efficient convex optimization algorithms can be used for stability analysis and control design problems. The recasting therefore constitutes solutions to the stability analysis and control design problems in the framework of the Takagi-Sugeno fuzzy model and PDC design.

The design procedure presented in the previous section involves an iterative process. For each rule a controller is designed based on consideration of local performance only. Then an LMI-based stability analysis is carried out to check whether the stability conditions are satisfied. In the case that the stability conditions are not satisfied, the controller for each rule will be redesigned. The iterative design procedure has been very effective in our experience. However, from the standpoint of control design, it is more desirable to be able to directly design a control that ensures the stability of the closed-loop system. This is referred as the control problem in the framework of the Takagi-Sugeno fuzzy model and PDC design. We claim that the control problem can be recast (hence solved) using the LMI approach. Here we only briefly state the ideas of the LMI approach to the control design problem. We show a simple case ($r = 1$), that is, the linear case, below. Fuzzy control case will be presented in Chapter 3.

Consider the case $r = 1$, that is, there is only one IF-THEN rule; (2.5) becomes a linear time-invariant system,

$$x(t+1) = Ax(t) + Bu(t). \qquad (2.36)$$

ORIGIN OF THE LMI-BASED DESIGN APPROACH

For a given control gain F, using standard stability theory for linear time-invariant systems or Theorem 2, the system (2.36) is (quadratically) stable if there exists $P > 0$ such that

$$\{A - BF\}^T P \{A - BF\} - P < 0. \tag{2.37}$$

The control design problem is to find a state feedback gain F such that the closed-loop system is (quadratically) stable. If such a gain F exists, the system is said to be quadratically stabilizable (via linear state feedback). This quadratic stabilizability problem can be recast as an LMI problem.

The condition (2.37) is not jointly convex in F and P. Now multiplying the inequality on the left and right by P^{-1}, and defining a new variable $X = P^{-1}$, we may rewrite (2.37) as

$$X\{A - BF\}^T X^{-1} \{A - BF\} X - X < 0. \tag{2.38}$$

Define $M = FX$ so that for $X > 0$ we have $F = MX^{-1}$. Substituting into (2.38) yields

$$X - \{AX - BM\}^T X^{-1} \{AX - BM\} > 0. \tag{2.39}$$

This nonlinear (convex) inequality can now be converted to LMI form using Schur complements [18]. The resulting LMI is

$$\begin{bmatrix} X & (AX - BM)^T \\ (AX - BM) & X \end{bmatrix} > 0 \tag{2.40}$$

in X and M. Thus the system (2.36) is quadratically stabilizable if there exist $X > 0$ and M such that the LMI (2.40) holds. The state feedback gain is $F = MX^{-1}$.

We can easily extend the LMI-based control design approach to multiple-rule ($r > 1$) cases of the Takagi-Sugeno fuzzy models. For instance, the quadratic stabilizability of the Takagi-Sugeno fuzzy models via a linear state feedback can be cast as the following LMI problem in X and M:

$$X > 0,$$

$$\begin{bmatrix} X & (A_i X - B_i M)^T \\ (A_i X - B_i M) & X \end{bmatrix} > 0, \quad i = 1, 2, \ldots, r,$$

with the state feedback gain $F = MX^{-1}$.

The LMI-based control design approach has also been developed for the control of Takagi-Sugeno fuzzy models via PDC design. For more details, see Chapter 3.

Some important remarks are in order.

Remark 9 The stability conditions presented in this book not only guarantee stability of fuzzy models and fuzzy control systems, they also guarantee stability for related uncertain linear time-varying [linear differential inclusion (LDI)] systems and nonlinear systems satisfying some *global* or *local sector* conditions. Thus a controller that works well with the fuzzy model is likely to work well when applied to the real system. This point is clearly demonstrated by the application in the next section. The theoretical details, however, will be discussed in other chapters.

Remark 10 The stability analysis and control design results presented in this section hold for continuous-time systems as well. Instead of using the Lyapunov inequality for discrete-time systems, we should use the Lyapunov inequality for continuous-time systems,

$$A^T P + PA < 0.$$

In the next section, we apply the PDC approach to a continuous-time system.

2.6 APPLICATION: INVERTED PENDULUM ON A CART

To illustrate the PDC approach, consider the problem of balancing and swing-up of an inverted pendulum on a cart. Recall the equations of motion for the pendulum [21]:

$$\dot{x}_1(t) = x_2(t)$$

$$\dot{x}_2(t) = \frac{g\sin(x_1(t)) - amlx_2^2(t)\sin(2x_1(t))/2 - a\cos(x_1(t))u(t)}{4l/3 - aml\cos^2(x_1(t))}, \quad (2.41)$$

where $x_1(t)$ denotes the angle (in radians) of the pendulum from the vertical and $x_2(t)$ is the angular velocity; $g = 9.8$ m/s^2 is the gravity constant, m is the mass of the pendulum, M is the mass of the cart, $2l$ is the length of the pendulum, and u is the force applied to the cart (in newtons); and $a = 1/(m + M)$. We choose $m = 2.0$ kg, $M = 8.0$ kg, $2l = 1.0$ m in the simulations [20].

2.6.1 Two-Rule Modeling and Control

The control objective of this subsection is to balance the inverted pendulum for the approximate range $x_1 \in (-\pi/2, \pi/2)$. In order to use the PDC approach, we must have a fuzzy model which represents the dynamics of the nonlinear plant. Therefore we first represent the system (2.41) by a Takagi-Sugeno fuzzy model. To minimize the design effort and complexity, we try to

APPLICATION: INVERTED PENDULUM ON A CART

use as few rules as possible. Notice that when $x_1 = \pm\pi/2$, the system is uncontrollable. Hence, as shown in Example 4, we approximate the system by the following two-rule fuzzy model:

Rule 1:

  IF $x_1(t)$ is about 0,

  THEN $\dot{x}(t) = A_1 x(t) + B_1 u(t)$.

Rule 2:

  IF $x_1(t)$ is about $\pm\pi/2$ ($|x_1| < \pi/2$),

  THEN $\dot{x}(t) = A_2 x(t) + B_2 u(t)$.

Here,

$$A_1 = \begin{bmatrix} 0 & 1 \\ \dfrac{g}{4l/3 - aml} & 0 \end{bmatrix}, \quad B_1 = \begin{bmatrix} 0 \\ -\dfrac{a}{4l/3 - aml} \end{bmatrix},$$

$$A_2 = \begin{bmatrix} 0 & 1 \\ \dfrac{2g}{\pi(4l/3 - aml\beta^2)} & 0 \end{bmatrix}, \quad B_2 = \begin{bmatrix} 0 \\ -\dfrac{a\beta}{4l/3 - aml\beta^2} \end{bmatrix},$$

and $\beta = \cos(88°)$.

Membership functions for Rules 1 and 2 are shown in Figure 2.16.

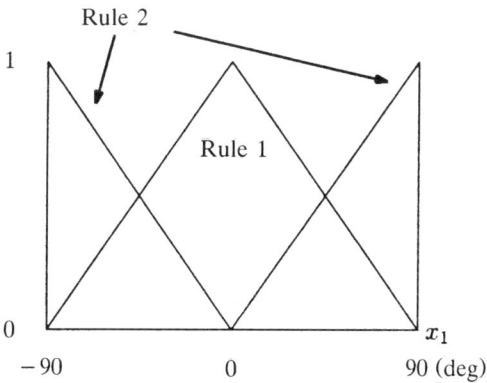

Fig. 2.16 Membership functions of two-rule model.

Choose the closed-loop eigenvalues $[-2, -2]$ for $A_1 - B_1 F_1$ and $A_2 - B_2 F_2$. We have

$$F_1 = [-120.6667 \quad -22.6667],$$
$$F_2 = [-2551.6 \quad -764.0].$$

It follows that

$$A_1 - B_1 F_1 = A_2 - B_2 F_2 = G = \begin{bmatrix} 0 & 1 \\ -4 & -4 \end{bmatrix}$$

and

$$G_{12} = \begin{bmatrix} 0 & 1 \\ -212.1325 & -67.4675 \end{bmatrix}.$$

Note that G_{12} is Hurwitz.

Using an LMI optimization algorithm, we obtain

$$P = \begin{bmatrix} 3.6250 & 0.6250 \\ 0.6250 & 0.2812 \end{bmatrix}. \tag{2.42}$$

It can be easily shown that the following stability conditions are satisfied:

$$\{A_i - B_i F_i\}^T P + P\{A_i - B_i F_i\} < 0, \quad i = 1, 2, \tag{2.43}$$

$$G_{12}^T P + P G_{12} < 0. \tag{2.44}$$

The resulting PDC control law is as follows:

Rule 1:

    IF $x_1(t)$ is about 0,

        THEN $u(t) = -F_1 x(t)$.

Rule 2:

    IF $x_1(t)$ is about $\pm \pi/2$ ($|x_1| < \pi/2$),

        THEN $u(t) = -F_2 x(t)$.

That is,

$$u(t) = -h_1(x_1(t)) F_1 x(t) - h_2(x_1(t)) F_2 x(t). \tag{2.45}$$

The membership values of Rules 1 and 2 are h_1 and h_2, respectively ($h_1 + h_2 = 1$). This (nonlinear) control law guarantees the stability of the fuzzy control system (fuzzy model + PDC control). To assess the effectiveness of the PDC controller, we apply the controller to the original system

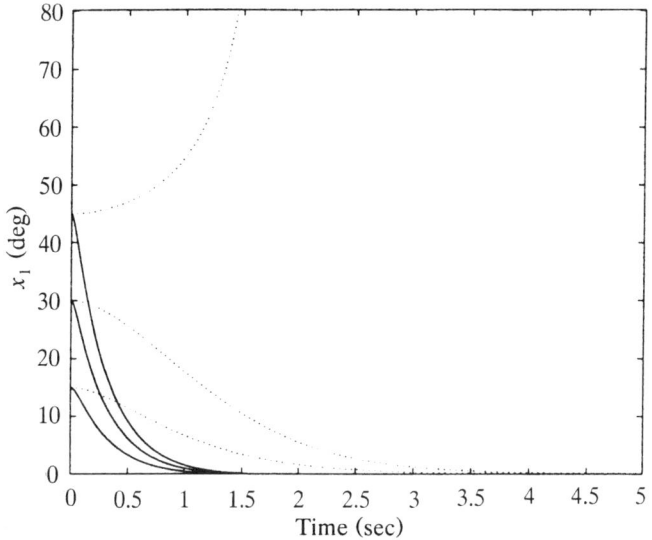

Fig. 2.17 Angle response using linear and two-rule fuzzy control.

(2.41). As pointed out in Remark 3, we may design a robust fuzzy controller (see Chapter 5 for the details) that can compensate the approximation error.

Simulations indicate the control law can balance the pendulum for initial conditions $x_1 \in [-88°, 88°]$ ($x_2 = 0$). In contrast, the linear control alone $u = -F_1 x$ fails to balance the pendulum for initial angles $|x_1| > 45°$. Figure 2.17 shows the response of the pendulum system using linear and fuzzy PDC controls for initial conditions $x_1 = 15°, 30°, 45°$, and $x_2 = 0$. The solid lines indicate responses with the fuzzy controller. The dotted lines show those with the linear controller. Figure 2.18 illustrates the closed-loop behavior of the system with the fuzzy controller for initial conditions $x_1 = 65°, 75°, 85°$, and $x_2 = 0$.

We remark that given the nonlinear plant (2.41) nonlinear control laws can be designed to balance the pendulum for initial angles $x_1 \in (-\pi/2, \pi/2)$. However, such control laws often tend to be quite involved. For example, one such control law is [20] $u = k(x_1, x_2)$, where

$$k(x_1, x_2) = -\frac{g}{a}\tan(x_1) - \frac{4le_1e_2}{3a}\ln[\sec(x_1) + \tan(x_1)]$$
$$+ e_1e_2 ml \sin(x_1)$$
$$+ \frac{(e_1 + e_2)x_2}{a}\left[\frac{4l}{3}\sec(x_1) - aml\cos(x_1)\right] \quad (2.46)$$

and e_1, e_2 are the specified closed-loop eigenvalues.

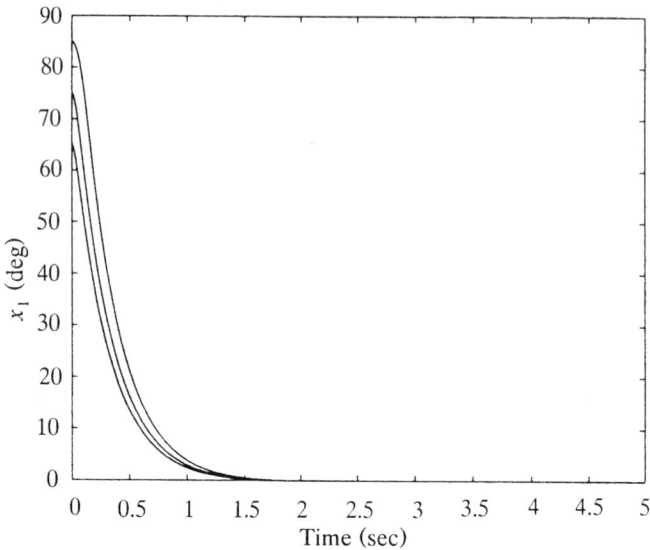

Fig. 2.18 Angle response using two-rule fuzzy control.

In contrast, the PDC design is intuitive and simple. The resulting controller is simple as well.

2.6.2 Four-Rule Modeling and Control

Suppose the pendulum on the cart system is built in such a way that the work space of the pendulum is the full circle $[-\pi, \pi]$. In this subsection, we extend the results to the range of $x_1 \in [-\pi \ \pi]$ except for a thin strip near $\pm \pi/2$. Balancing the pendulum for the angle range of $\pi/2 < |x_1| \leq \pi$ is referred to as swing-up control of the pendulum. Recall that for $x_1 = \pm \pi/2$ the system is uncontrollable. We add two more rules (Rules 3 and 4) to the fuzzy model.

Rule 1:

IF $x_1(t)$ is about 0,

THEN $\dot{x}(t) = A_1 x(t) + B_1 u(t)$.

Rule 2:

IF $x_1(t)$ is about $\pm \pi/2$ ($|x_1(t)| < \pi/2$),

THEN $\dot{x}(t) = A_2 x(t) + B_2 u(t)$.

APPLICATION: INVERTED PENDULUM ON A CART

Rule 3:

IF $x_1(t)$ is about $\pm \pi/2$ $(|x_1(t)| > \pi/2)$,

THEN $\dot{x}(t) = A_3 x(t) + B_3 u(t)$.

Rule 4:

IF $x_1(t)$ is about π,

THEN $\dot{x}(t) = A_4 x(t) + B_4 u(t)$.

Here A_1, B_1, A_2, B_2 are the same as above and

$$A_3 = \begin{bmatrix} 0 & 1 \\ \dfrac{2g}{\pi(4l/3 - aml\beta^2)} & 0 \end{bmatrix}, \quad B_3 = \begin{bmatrix} 0 \\ \dfrac{a\beta}{4l/3 - aml\beta^2} \end{bmatrix},$$

$$A_4 = \begin{bmatrix} 0 & 1 \\ 0 & 0 \end{bmatrix}, \quad B_4 = \begin{bmatrix} 0 \\ \dfrac{a}{4l/3 - aml} \end{bmatrix}.$$

The membership functions of this four-rule fuzzy model are shown in Figure 2.19.

Again choose the closed-loop eigenvalues $[-2, -2]$ for $A_3 - B_3 F_3$ and $A_4 - B_4 F_4$. We have

$$F_3 = [2551.6 \quad 764.0],$$

$$F_4 = [22.6667 \quad 22.6667].$$

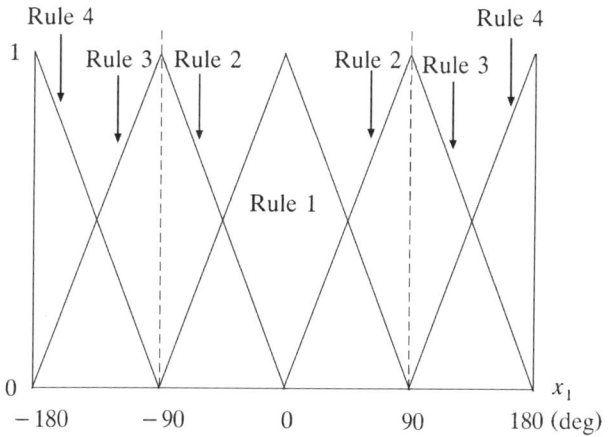

Fig. 2.19 Membership functions of four-rule model.

It follows that
$$A_3 - B_3F_3 = A_4 - B_4F_4 = G$$
and
$$G_{34} = \begin{bmatrix} 0 & 1 \\ -220.5230 & -67.4675 \end{bmatrix}.$$

Note that G_{34} is Hurwitz.

It can be shown that the P of (2.42) satisfies the additional stability conditions
$$\{A_i - B_iF_i\}^T P + P\{A_i - B_iF_i\} < 0, \quad i = 3, 4, \tag{2.47}$$
$$G_{34}^T P + PG_{34} < 0. \tag{2.48}$$

There is no overlap between membership values h_1 and h_3, h_1 and h_4, h_2 and h_3, and h_2 and h_4. Hence only G_{12} and G_{34} are needed in stability check.

The PDC controller is given as follows:

Rule 1:

    IF $x_1(t)$ is about 0,

        THEN $u(t) = -F_1 x(t)$.

Rule 2:

    IF $x_1(t)$ is about $\pm \pi/2$ ($|x_1(t)| < \pi/2$),

        THEN $u(t) = -F_2 x(t)$.

Rule 3:

    IF $x_1(t)$ is about $\pm \pi/2$ ($|x_1(t)| > \pi/2$),

        THEN $u(t) = -F_3 x(t)$.

Rule 4:

    IF $x_1(t)$ is about π,

        THEN $u(t) = -F_4 x(t)$.

That is,
$$u(t) = -h_1(x_1(t))F_1 x(t) - h_2(x_1(t))F_2 x(t) \\ - h_3(x_1(t))F_3 x(t) - h_4(x_1(t))F_4 x(t). \tag{2.49}$$

APPLICATION: INVERTED PENDULUM ON A CART

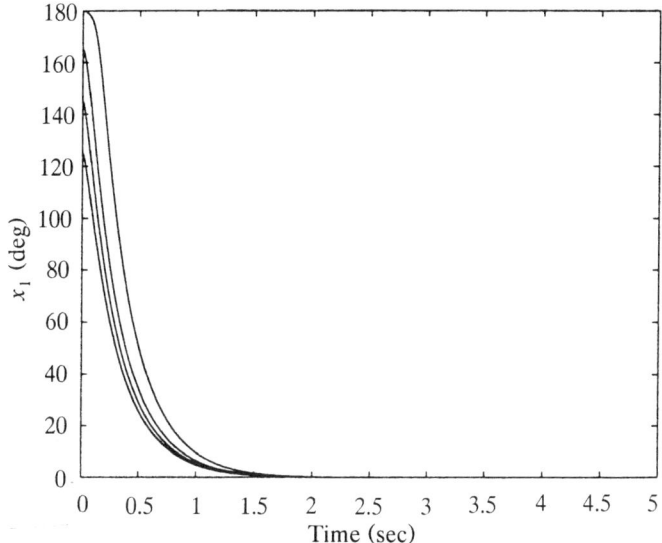

Fig. 2.20 Angle response using four-rule fuzzy control.

This control law guarantees stability of the fuzzy control system (four-rule fuzzy model + PDC control). This controller is applied to the original system (2.41) for evaluation of its performance. Simulation results demonstrate that the controller (2.49) is able to balance the pendulum for all initial angles except when $x_1(t)$ is in a thin strip $88° < |x_1(t)| < 94°$. The size of this thin strip can be reduced by adding more rules to the model and controller. Figure 2.20 illustrates the response of the closed-loop system for initial conditions $x_1 = 125°, 145°, 165°, 180°$ and $x_2 = 0$.

Note that the nonlinear controller (2.46) does not apply for $\pi/2 \leq |x_1(t)| \leq \pi$.

Some comparisons between the linear, nonlinear, and fuzzy control designs are summarized loosely in Table 2.1.

To test the robustness of this controller, the following simulations are conducted: (1) m is changed from 2.0 to 4.0 kg, (2) M is changed from 8.0 to 4.0 kg, and (3) $2l$ is changed from 1.0 to 0.5 m. For each case, we simulate

TABLE 2.1 Comparisons of Different Control Designs

	Work range	Simple?	Stability
Linear	$(-\pi/4 \ \pi/4)$	Yes	Local
Nonlinear	$(-\pi/2 \ \pi/2)$	No	Nonlocal
Fuzzy PDC	$[-\pi \ \pi]$	Yes	Nonlocal

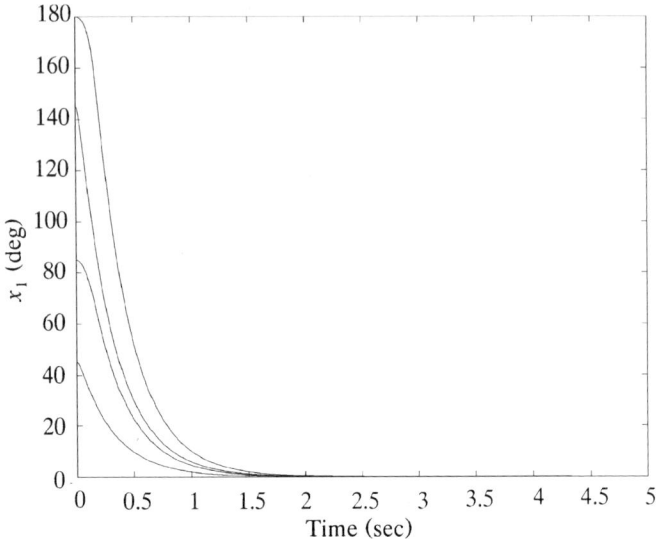

Fig. 2.21 Closed-loop angle response with m changed.

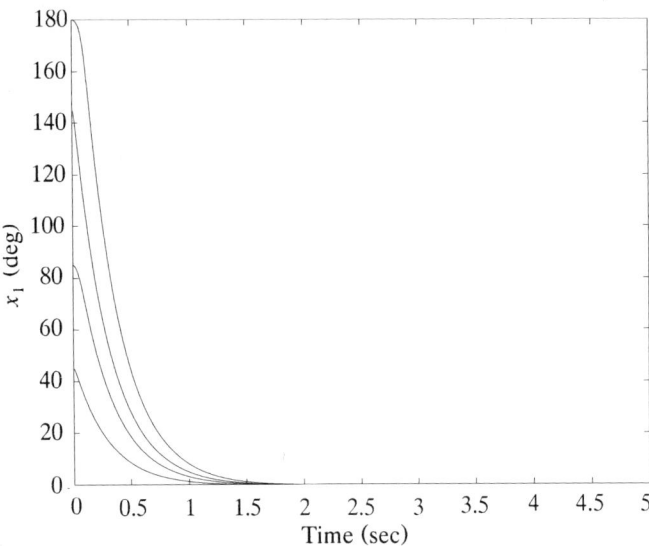

Fig. 2.22 Closed-loop angle response with M changed.

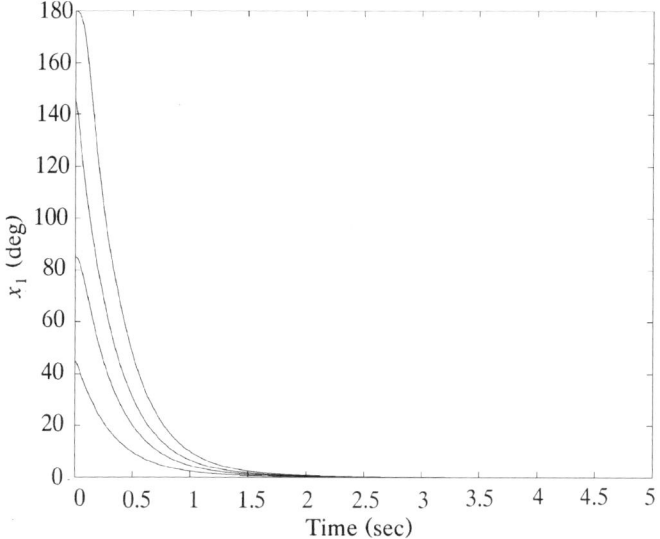

Fig. 2.23 Closed-loop angle response with l changed.

the closed-loop system for the following initial conditions $x_1 = 45°, 85°, 145°, 180°$ and $x_2(t) = 0$. The results are shown in Figures 2.21, 2.22, and 2.23, respectively, for cases 1, 2, and 3.

Robustness is not considered in this design. Robust fuzzy control design in Chapter 5 is applicable to this system.

REFERENCES

1. K. Tanaka and M. Sugeno, "Stability Analysis of Fuzzy Systems Using Lyapunov's Direct Method," *Proc. NAFIPS'90*, pp. 133–136, 1990.
2. K. Tanaka and M. Sugeno, "Stability Analysis and Design of Fuzzy Control Systems," *Fuzzy Sets Syst.*, Vol. 45, No. 2, pp. 135–156 (1992).
3. R. Langari and M. Tomizuka, "Analysis and Synthesis of Fuzzy Linguistic Control Systems," *Proc. 1990 ASME Winter Annual Meet.*, pp. 35–42, 1990.
4. L. X. Wang, *Adaptive Fuzzy Systems and Control: Design and Stability Analysis*, Prentice-Hall, Englewood Cliffs, NJ, 1994.
5. G. Chen and H. Ying, "On the Stability of Fuzzy Control Systems," *Proc. 3rd IFIS*, Houston, 1993.
6. S. S. Farinwata and G. Vachtsevanos, "Stability Analysis of the Fuzzy Logic Controller," *Proc. IEEE CDC*, San Antonio, 1993.
7. T. Takagi and M. Sugeno, "Fuzzy Identification of Systems and Its Applications to Modeling and Control," *IEEE Trans. Syst. Man. Cyber.*, Vol. 15, pp. 116–132, (1985).

8. M. Sugeno and G. T. Kang, "Structure Identification of Fuzzy Model," *Fuzzy Sets Syst.*, Vol. 28, pp. 329–346 (1986).
9. M. Sugeno, *Fuzzy Control*, Nikkan Kougyou Shinbunsha Publisher, Tokyo, 1988.
10. S. Kawamoto et al., "An Approach to Stability Analysis of Second Order Fuzzy Systems," *Proceedings of First IEEE International Conference on Fuzzy Systems*, Vol. 1, 1992, pp. 1427–1434.
11. K. Tanaka and M. Sano, "A Robust Stabilization Problem of Fuzzy Control Systems and Its Application to Backing Up Control of a Truck-Trailer," *IEEE Trans. Fuzzy Syst.*, Vol. 2, No. 2, pp. 119–134, (1994).
12. K. Tanaka, *A Theory of Advanced Fuzzy Control*, in Japanese, KYOURITSU Publishing Company, Tokyo, Japan, 1994.
13. S. Kawamoto et al, "An Approach to Stability Analysis of Second Order Fuzzy Systems," *Proc. FUZZ-IEEE'92*, pp. 1427–1434, 1992.
14. H. O. Wang, K. Tanaka, and M. F. Griffin, "Parallel Distributed Compensation of Nonlinear Systems by Takagi-Sugeno Fuzzy Model," *Proc. FUZZ-IEEE/IFES'95*, pp. 531–538, 1995.
15. H. O. Wang, K. Tanaka, and M. F. Griffin, "An Analytical Framework of Fuzzy Modeling and Control of Nonlinear Systems: Stability and Design Issues," *Proc. 1995 American Control Conference*, Seattle, 1995, pp. 2272–2276.
16. M. Sugeno and G. T. Kang, "Fuzzy Modeling and Control of Multilayer Incinerator," *Fuzzy Sets Syst.*, No. 18, pp. 329–346, (1986).
17. J-J E. Slotine and W. Li, *Applied Nonlinear Control*, Prentice Hall, Englewood Cliffs, NJ, 1991.
18. S. Boyd et al., *Linear Matrix Inequalities in Systems and Control Theory*, SIAM, Philadelphia, PA, 1994.
19. Yu. Nesterov and A. Nemirovsky, *Interior-Point Polynomial Methods in Convex Programming*, SIAM, Philadelphia, PA, 1994.
20. W. T. Baumann and W. J. Rugh, "Feedback Control of Nonlinear Systems by Extended Linearization," *IEEE Trans. Automatic Control*, Vol. AC-31, No. 1, pp. 40–46, (1986).
21. R. H. Cannon, *Dynamics of Physical Systems*, McGraw-Hill, New York, 1967.
22. H. E. Nusse and J. A. Yorke, *Numerical Investigations of Chaotic Systems: A Handbook for JAY's Dynamics*, Draft, Institute for Physical Science and Technology, University of Maryland, College Park, MD, 1992.
23. K. Tanaka and M. Sano, "Fuzzy Stability Criterion of a Class of Nonlinear Systems," *Inform. Sci.*, Vol. 71, No. 1 & 2, pp. 3–26, (1993).
24. H. O. Wang, K. Tanaka, and M. Griffin, "An Approach to Fuzzy Control of Nonlinear Systems: Stability and Design Issues," *IEEE Trans. Fuzzy Syst.*, Vol. 4, No. 1, pp.14–23 (1996).

CHAPTER 3

LMI CONTROL PERFORMANCE CONDITIONS AND DESIGNS

The preceding chapter introduced the concept and basic procedure of parallel distributed compensation and LMI-based designs. The goal of this chapter is to present the details of analysis and design via LMIs. This chapter forms a basic and important component of this book. To this end, it will be shown that various kinds of control performance specifications can be represented in terms of LMIs. The control performance specifications include stability conditions, relaxed stability conditions, decay rate conditions, constrains on control input and output, and disturbance rejection for both continuous and discrete fuzzy control systems [1–3]. Other more advanced control performance considerations utilizing LMI conditions will be presented in later chapters.

3.1 STABILITY CONDITIONS

In the 1990's, the issue of stability of fuzzy control systems has been investigated extensively in the framework of nonlinear system stability [1–18]. Today, there exist a large number of papers on stability analysis of fuzzy control in the literature. This section discusses some basic results on the stability of fuzzy control systems.

In the following, Theorems 5 and 6 deal with stability conditions for the open-loop systems. Theorem 5 can be readily obtained via Lyapunov stability theory. The proof of Theorem 6 is given in [4, 7].

THEOREM 5 [CFS] *The equilibrium of the continuous fuzzy system (2.3) with $u(t) = 0$ is globally asymptotically stable if there exists a common positive definite matrix **P** such that*

$$A_i^T P + P A_i < 0, \quad i = 1, 2, \ldots, r, \qquad (3.1)$$

*that is, a common **P** has to exist for all subsystems.*

THEOREM 6 [DFS] *The equilibrium of the discrete fuzzy system (2.5) with $u(t) = 0$ is globally asymptotically stable if there exists a common positive definite matrix **P** such that*

$$A_i^T P A_i - P < 0, \quad i = 1, 2, \ldots, r, \qquad (3.2)$$

*that is, a common **P** has to exist for all subsystems.*

Next, let us consider the stability of the closed-loop system. By substituting (2.23) into (2.3) and (2.5), we obtain (3.3) and (3.4), respectively.

CFS

$$\dot{x}(t) = \sum_{i=1}^{r} \sum_{j=1}^{r} h_i(z(t)) h_j(z(t)) \{A_i - B_i F_j\} x(t). \qquad (3.3)$$

DFS

$$x(t+1) = \sum_{i=1}^{r} \sum_{j=1}^{r} h_i(z(t)) h_j(z(t)) \{A_i - B_i F_j\} x(t). \qquad (3.4)$$

Denote

$$G_{ij} = A_i - B_i F_j.$$

Equations (3.3) and (3.4) can be rewritten as (3.5) and (3.6), respectively.

CFS

$$\dot{x}(t) = \sum_{i=1}^{r} h_i(z(t)) h_i(z(t)) G_{ii} x(t)$$

$$+ 2 \sum_{i=1}^{r} \sum_{i<j} h_i(z(t)) h_j(z(t)) \left\{ \frac{G_{ij} + G_{ji}}{2} \right\} x(t). \qquad (3.5)$$

DFS

$$x(t+1) = \sum_{i=1}^{r} h_i(z(t)) h_i(z(t)) G_{ii} x(t)$$

$$+ 2 \sum_{i=1}^{r} \sum_{i<j} h_i(z(t)) h_j(z(t)) \left\{ \frac{G_{ij} + G_{ji}}{2} \right\} x(t). \qquad (3.6)$$

STABILITY CONDITIONS

By applying the stability conditions for the open-loop system (Theorems 5 and 6) to (3.5) and (3.6), we can derive stability conditions for the CFS and the DFS, respectively.

THEOREM 7 [CFS] *The equilibrium of the continuous fuzzy control system described by (3.5) is globally asymptotically stable if there exists a common positive definite matrix P such that*

$$G_{ii}^T P + P G_{ii} < 0, \tag{3.7}$$

$$\left(\frac{G_{ij} + G_{ji}}{2}\right)^T P + P\left(\frac{G_{ij} + G_{ji}}{2}\right) \leq 0,$$

$$i < j \text{ s.t. } h_i \cap h_j \neq \phi. \tag{3.8}$$

Proof. It follows directly from Theorem 5.

For the explanation of the notation $i < j$ s.t. $h_i \cap h_j \neq \phi$, refer to Chapter 1.

THEOREM 8 [DFS] *The equilibrium of the discrete fuzzy control system described by (3.6) is globally asymptotically stable if there exists a common positive definite matrix P such that*

$$G_{ii}^T P G_{ii} - P < 0, \tag{3.9}$$

$$\left(\frac{G_{ij} + G_{ji}}{2}\right)^T P \left(\frac{G_{ij} + G_{ji}}{2}\right) - P \leq 0,$$

$$i < j \text{ s.t. } h_i \cap h_j \neq \phi. \tag{3.10}$$

Proof. It follows directly from Theorem 6.

The fuzzy control design problem is to determine F_j's ($j = 1, 2, \ldots, r$) which satisfy the conditions of Theorem 7 or 8 with a common positive definite matrix P.

Consider the common B matrix case, that is, $B_1 = B_2 = \cdots = B_r$. In this case, the stability conditions of Theorems 7 and 8 can be simplified as follows.

COROLLARY 1 *Assume that $B_1 = B_2 = \cdots = B_r$. The equilibrium of the fuzzy control system (3.5) is globally asymptotically stable if there exists a common positive definite matrix P satisfying (3.7).*

COROLLARY 2 *Assume that $B_1 = B_2 = \cdots = B_r$. The equilibrium of the fuzzy control system (3.6) is globally asymptotically stable if there exists a common positive definite matrix P satisfying (3.9).*

In other words, the corollaries state that in the common B case, $G_{ii}^T P + PG_{ii} < 0$ implies

$$\left(\frac{G_{ij}+G_{ji}}{2}\right)^T P + P\left(\frac{G_{ij}+G_{ji}}{2}\right) \leq 0$$

and $G_{ii}^T PG_{ii} - P < 0$ implies

$$\left(\frac{G_{ij}+G_{ji}}{2}\right)^T P\left(\frac{G_{ij}+G_{ji}}{2}\right) - P \leq 0$$

To check stability of the fuzzy control system, it has long been considered difficult to find a common positive definite matrix P satisfying the conditions of Theorems 5–8. A trial-and-error type of procedure was first used [4, 7, 9]. In [19], a procedure to construct a common P is given for second-order fuzzy systems, that is, the dimension of the state is 2. It was first stated in [11, 12, 17] that the common P problem for fuzzy controller design can be solved numerically, that is, the stability conditions of Theorems 5–8 can be expressed in LMIs. For example, to check the stability conditions of Theorem 7, we need to find P satisfying the LMIs

$$P > 0, \qquad G_{ii}^T P + PG_{ii} < 0,$$

$$\left(\frac{G_{ij}+G_{ji}}{2}\right)^T P + P\left(\frac{G_{ij}+G_{ji}}{2}\right) \leq 0, \quad i < j \text{ s.t. } h_i \cap h_j \neq \phi,$$

or determine that no such P exists. This is a convex feasibility problem. As shown in Chapter 2, this feasibility problem can be numerically solved very efficiently by means of the most powerful tools available to date in the mathematical programming literature.

3.2 RELAXED STABILITY CONDITIONS

We have shown that the stability analysis of the fuzzy control system is reduced to a problem of finding a common P. If r, that is the number of IF-THEN rules, is large, it might be difficult to find a common P satisfying the conditions of Theorem 7 (or Theorem 8). This section presents new stability conditions by relaxing the conditions of Theorems 7 and 8. Theorems 9 and 10 provide relaxed stability conditions [1–3]. First, we need the following corollaries to prove Theorems 9 and 10.

COROLLARY 3

$$\sum_{i=1}^{r} h_i^2(z(t)) - \frac{1}{r-1} \sum_{i=1}^{r}\sum_{i<j} 2h_i(z(t))h_j(z(t)) \geq 0,$$

where

$$\sum_{i=1}^{r} h_i(z(t)) = 1, \quad h_i(z(t)) \geq 0$$

for all i.

Proof. It holds since

$$\sum_{i=1}^{r} h_i^2(z(t)) - \frac{1}{r-1} \sum_{i=1}^{r} \sum_{i<j} 2h_i(z(t))h_j(z(t))$$

$$= \frac{1}{r-1} \sum_{i=1}^{r} \sum_{i<j} \{h_i(z(t)) - h_j(z(t))\}^2 \geq 0. \qquad \text{Q.E.D.}$$

COROLLARY 4 *If the number of rules that fire for all t is less than or equal to s, where $1 < s \leq r$, then*

$$\sum_{i=1}^{r} h_i^2(z(t)) - \frac{1}{s-1} \sum_{i=1}^{r} \sum_{i<j} 2h_i(z(t))h_j(z(t)) \geq 0,$$

where

$$\sum_{i=1}^{r} h_i(z(t)) = 1, \quad h_i(z(t)) \geq 0$$

for all i.

Proof. It follows directly from Corollary 3.

THEOREM 9 [CFS] *Assume that the number of rules that fire for all t is less than or equal to s, where $1 < s \leq r$. The equilibrium of the continuous fuzzy control system described by (3.5) is globally asymptotically stable if there exist a common positive definite matrix P and a common positive semidefinite matrix Q such that*

$$G_{ii}^T P + P G_{ii} + (s-1)Q < 0 \tag{3.11}$$

$$\left(\frac{G_{ij} + G_{ji}}{2}\right)^T P + P \left(\frac{G_{ij} + G_{ji}}{2}\right) - Q \leq 0,$$

$$i < j \text{ s.t. } h_i \cap h_j \neq \phi \tag{3.12}$$

where $s > 1$.

Proof. Consider a candidate of Lyapunov function $V(x(t)) = x^T(t)Px(t)$, where $P > 0$. Then,

$$\dot{V}(x(t)) = \sum_{i=1}^{r} \sum_{j=1}^{r} h_i(z(t))h_j(z(t))x^T(t)$$
$$\times \left[(A_i - B_iF_j)^T P + P(A_i - B_iF_j)\right]x(t)$$
$$= \sum_{i=1}^{r} h_i^2(z(t))x^T(t)\left[G_{ii}^T P + PG_{ii}\right]x(t)$$
$$+ \sum_{i=1}^{r}\sum_{i<j} 2h_i(z(t))h_j(z(t))x^T(t)$$
$$\times \left[\left(\frac{G_{ij} + G_{ji}}{2}\right)^T P + P\left(\frac{G_{ij} + G_{ji}}{2}\right)\right]x(t),$$

where
$$G_{ij} = A_i - B_iF_j.$$

From condition (3.12) and Corollary 4, we have

$$\dot{V}(x(t)) \leq \sum_{i=1}^{r} h_i^2(z(t))x^T(t)\left[G_{ii}^T P + PG_{ii}\right]x(t)$$
$$+ \sum_{i=1}^{r}\sum_{i<j} 2h_i(z(t))h_j(z(t))x^T(t)Qx(t)$$
$$\leq \sum_{i=1}^{r} h_i^2(z(t))x^T(t)\left[G_{ii}^T P + PG_{ii}\right]x(t)$$
$$+ (s-1)\sum_{i=1}^{r} h_i^2(z(t))x^T(t)Qx(t)$$
$$= \sum_{i=1}^{r} h_i^2(z(t))x^T(t)\left[G_{ii}^T P + PG_{ii} + (s-1)Q\right]x(t).$$

If condition (3.11) holds, $\dot{V}(x(t)) < 0$ at $x(t) \neq 0$. Q.E.D.

THEOREM 10 [DFS] *Assume that the number of rules that fire for all t is less than or equal to s, where $1 < s \leq r$. The equilibrium of the discrete fuzzy control system described by (3.6) is globally asymptotically stable if there exist a common positive definite matrix P and a common positive semidefinite matrix Q such that*

$$G_{ii}^T PG_{ii} - P + (s-1)Q < 0, \tag{3.13}$$

$$\left(\frac{G_{ij} + G_{ji}}{2}\right)^T P \left(\frac{G_{ij} + G_{ji}}{2}\right) - P - Q \leq 0,$$

$$i < j \text{ s.t. } h_i \cap h_j \neq \phi, \tag{3.14}$$

where $s > 1$.

Proof. Consider a candidate of Lyapunov function $V(x(t)) = x^T(t)Px(t)$, where $P > 0$. Then,

$$\Delta V(x(t)) = V(x(t+1)) - V(x(t))$$

$$= \sum_{i=1}^{r}\sum_{j=1}^{r}\sum_{k=1}^{r}\sum_{l=1}^{r} h_i(z(t))h_j(z(t))h_k(z(t))h_l(z(t))$$

$$\times x^T(t)\left[G_{ij}^T P G_{kl} - P\right]x(t)$$

$$= \frac{1}{4}\sum_{i=1}^{r}\sum_{j=1}^{r}\sum_{k=1}^{r}\sum_{l=1}^{r} h_i(z(t))h_j(z(t))h_k(z(t))h_l(z(t))$$

$$\times x^T(t)\left[(G_{ij} + G_{ji})^T P (G_{kl} + G_{lk}) - 4P\right]x(t)$$

$$\leq \frac{1}{4}\sum_{i=1}^{r}\sum_{j=1}^{r} h_i(z(t))h_j(z(t)) x^T(t)\left[H_{ij}^T P H_{ij} - 4P\right]x(t)$$

$$= \sum_{i=1}^{r}\sum_{j=1}^{r} h_i(z(t))h_j(z(t)) x^T(t)\left[\frac{H_{ij}^T}{2} P \frac{H_{ij}}{2} - P\right]x(t)$$

$$= \sum_{i=1}^{r} h_i^2(z(t)) x^T(t)\left[G_{ii}^T P G_{ii} - P\right]x(t)$$

$$+ 2\sum_{i=1}^{r}\sum_{i<j} h_i(z(t))h_j(z(t)) x^T(t)\left[\frac{H_{ij}^T}{2} P \frac{H_{ij}}{2} - P\right]x(t),$$

where $H_{ij} = G_{ij} + G_{ji}$.

From condition (3.14) and Corollary 4, the right side of the above inequality becomes

$$\leq \sum_{i=1}^{r} h_i^2(z(t)) x^T(t)\left[G_{ii}^T P G_{ii} - P\right]x(t)$$

$$+ 2\sum_{i=1}^{r}\sum_{i<j} h_i(z(t))h_j(z(t)) x^T(t) Q x(t)$$

$$\leq \sum_{i=1}^{r} h_i^2(z(t)) x^T(t)\left[G_{ii}^T P G_{ii} - P\right]x(t)$$

$$+ (s-1)\sum_{i=1}^{r} h_i^2(z(t)) x^T(t) Q x(t)$$

$$= \sum_{i=1}^{r} h_i^2(z(t)) x^T(t)\left[G_{ii}^T P G_{ii} - P + (s-1)Q\right]x(t).$$

If condition (3.13) holds, $\Delta V(x(t)) < 0$ at $x(t) \neq 0$. Q.E.D.

56 LMI CONTROL PERFORMANCE CONDITIONS AND DESIGNS

Corollary 4 is used in the proofs of Theorems 9 and 10. The use of Corollary 3 would lead to conservative results because $s \leq r$.

Remark 11 It is assumed in the derivations of Theorems 7–10 that the weight $h_i(z(t))$ of each rule in the fuzzy controller is equal to that of each rule in the fuzzy model for all t. Note that Theorems 7–10 cannot be used if the assumption does not hold. This fact will show up again in a case (case B) of fuzzy observer design given in Chapter 4. If the assumption does not hold, the following stability conditions should be used instead of Theorems 7–10:

$$G_{ij}^T P + P G_{ji} < 0$$

in the CFS case and

$$G_{ij}^T P G_{ji} - P < 0$$

in the DFS case. These conditions imply those of Theorems 7–10. These conditions may be regarded as robust stability conditions for premise part uncertainty [18].

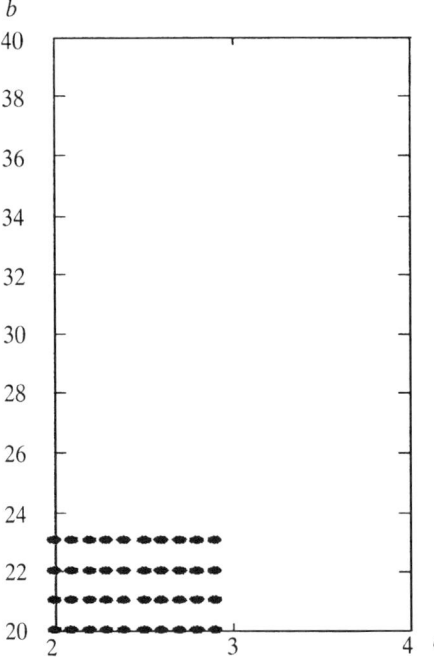

Fig. 3.1 Feasible area for the stability conditions of Theorem 7.

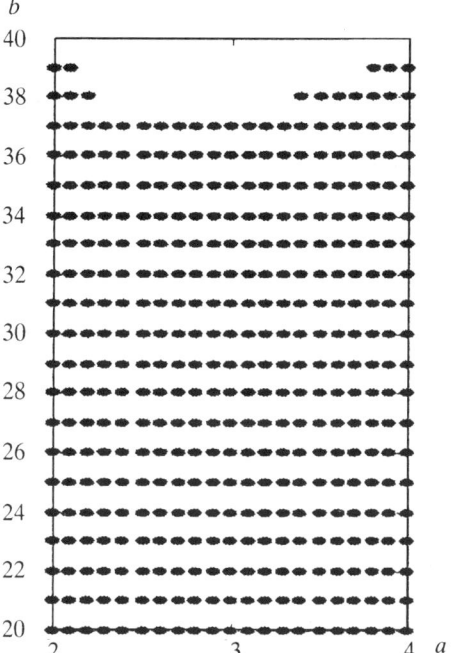

Fig. 3.2 Feasible area for the stability conditions of Theorem 9.

The conditions of Theorems 9 and 10 reduce to those of Theorems 7 and 8, respectively, when $Q = 0$.

Example 9 This example demonstrates the utility of the relaxed conditions in the CFS case. Consider the CFS, where $r = s = 2$,

$$A_1 = \begin{bmatrix} 2 & -10 \\ 1 & 0 \end{bmatrix}, \quad B_1 = \begin{bmatrix} 1 \\ 0 \end{bmatrix},$$

$$A_2 = \begin{bmatrix} a & -10 \\ 1 & 0 \end{bmatrix}, \quad B_2 = \begin{bmatrix} b \\ 0 \end{bmatrix}.$$

The local feedback gains F_1 and F_2 are determined by selecting $[-2 \ -2]$ as the eigenvalues of the subsystems in the PDC. Figures 3.1 and 3.2 show the feasible areas satisfying the conditions of Theorems 7 and 9 for the variables a and b, respectively. In these figures, the feasible areas are plotted for $a > 2$ and $b > 20$. A common P (and a common Q) satisfying the conditions of Theorem 7 (Figure 3.1) and Theorem 9 (Figure 3.2) exists if and only if the system parameters a and b are located in the feasible areas under $a > 2$ and $b > 20$. It is found in these figures that the conditions of Theorem 7 lead to conservative results.

3.3 STABLE CONTROLLER DESIGN

This section presents stable fuzzy controller designs for CFS and DFS.

We first present a stable fuzzy controller design problem which is to determine the feedback gains F_i for the CFS using the stability conditions of Theorem 7. The conditions (3.7) and (3.8) are not jointly convex in F_i and P. Now multiplying the inequality on the left and right by P^{-1} and defining a new variable $X = P^{-1}$, we rewrite the conditions as

$$-XA_i^T - A_i X + XF_i^T B_i^T + B_i F_i X > 0,$$
$$-XA_i^T - A_i X - XA_j^T - A_j X$$
$$+ XF_j^T B_i^T + B_i F_j X + XF_i^T B_j^T + B_j F_i X \geq 0.$$

Define $M_i = F_i X$ so that for $X > 0$ we have $F_i = M_i X^{-1}$. Substituting into the above inequalities yields

$$-XA_i^T - A_i X + M_i^T B_i^T + B_i M_i > 0,$$
$$-XA_i^T - A_i X - XA_j^T - A_j X$$
$$+ M_j^T B_i^T + B_i M_j + M_i^T B_j^T + B_j M_i \geq 0.$$

Using these LMI conditions, we define a stable fuzzy controller design problem.

Stable Fuzzy Controller Design: CFS Find $X > 0$ and M_i ($i = 1, \ldots, r$) satisfying

$$-XA_i^T - A_i X + M_i^T B_i^T + B_i M_i > 0, \qquad (3.15)$$
$$-XA_i^T - A_i X - XA_j^T - A_j X$$
$$+ M_j^T B_i^T + B_i M_j + M_i^T B_j^T + B_j M_i \geq 0,$$
$$i < j \text{ s.t } h_i \cap h_j \neq \phi \qquad (3.16)$$

where

$$X = P^{-1}, \qquad M_i = F_i X. \qquad (3.17)$$

The above conditions are LMIs with respect to variables X and M_i. We can find a positive definite matrix X and M_i satisfying the LMIs or determination that no such X and M_i exist. The feedback gains F_i and a common P can be obtained as

$$P = X^{-1}, \qquad F_i = M_i X^{-1} \qquad (3.18)$$

from the solutions X and M_i.

A stable fuzzy controller design problem for the DFS can be defined from the conditions of Theorem 8 as well:

$$X(A_i - B_i F_i)^T X^{-1}(A_i - B_i F_i)X - X < 0,$$

$$X\left\{\frac{A_i - B_i F_j + A_j - B_j F_i}{2}\right\}^T X^{-1}$$

$$\times \left\{\frac{A_i - B_i F_j + A_j - B_j F_i}{2}\right\}X - X \le 0.$$

Define $M_i = F_i X$ so that for $X > 0$ we have $F_i = M_i X^{-1}$. Substituting into the above inequalities yields

$$X - (A_i X - B_i M_i)^T X^{-1}(A_i X - B_i M_i) > 0,$$

$$X - X\left\{\frac{A_i X - B_i M_j + A_j X - B_j M_i}{2}\right\}^T X^{-1}$$

$$\times \left\{\frac{A_i X - B_i M_j + A_j X - B_j M_i}{2}\right\}X \ge 0.$$

These nonlinear (convex) inequalities can now be converted to LMIs using the Schur complement. The resulting LMIs are

$$\begin{bmatrix} X & XA_i^T - M_i^T B_i^T \\ A_i X - B_i M_i & X \end{bmatrix} > 0,$$

$$\begin{bmatrix} X & \left\{\dfrac{A_i X + A_j X - B_i M_j - B_j M_i}{2}\right\}^T \\ \left\{\dfrac{A_i X + A_j X - B_i M_j - B_j M_i}{2}\right\} & X \end{bmatrix} \ge 0$$

in X and F_i.

Stable Fuzzy Controller Design: DFS Find $X > 0$ and M_i ($i = 1, \ldots, r$) satisfying

$$\begin{bmatrix} X & XA_i^T - M_i^T B_i^T \\ A_i X - B_i M_i & X \end{bmatrix} > 0. \tag{3.19}$$

$$\begin{bmatrix} X & \left\{\dfrac{A_i X + A_j X - B_i M_j - B_j M_i}{2}\right\}^T \\ \left\{\dfrac{A_i X + A_j X - B_i M_j - B_j M_i}{2}\right\} & X \end{bmatrix} \ge 0,$$

$$i < j \text{ s.t. } h_i \cap h_j \ne \phi, \tag{3.20}$$

where
$$X = P^{-1}, \quad M_i = F_i X. \qquad (3.21)$$

The feedback gain F_i and a common P can be obtained as
$$P = X^{-1}, \quad F_i = M_i X^{-1} \qquad (3.22)$$
from the solutions X and M_i.

From the relaxed stability conditions of Theorem 9, the design problem to determine the feedback gains F_i for CFS can be defined as well.

Fuzzy Controller Design Using Relaxed Stability Conditions: CFS Find $X > 0$, $Y \geq 0$, and M_i ($i = 1, \ldots, r$) satisfying

$$-XA_i^T - A_i X + M_i^T B_i^T + B_i M_i - (s-1)Y > 0, \qquad (3.23)$$

$$2Y - XA_i^T - A_i X - XA_j^T - A_j X$$
$$+ M_j^T B_i^T + B_i M_j + M_i^T B_j^T + B_j M_i \geq 0,$$
$$i < j \text{ s.t. } h_i \cap h_j \neq \phi, \qquad (3.24)$$

where
$$X = P^{-1}, \quad M_i = F_i X, \quad Y = XQX. \qquad (3.25)$$

The above conditions are LMIs with respect to variables X, Y, and M_i. We can find a positive definite matrix X, a positive semidefinite matrix Y, and M_i satisfying the LMIs or determine that no such X, Y, and M_i exist. The feedback gains F_i, a common P, and a common Q can be obtained as

$$P = X^{-1}, \quad F_i = M_i X^{-1}, \quad Q = PYP \qquad (3.26)$$

from the solutions X, Y, and M_i.

From the relaxed conditions of Theorem 10, the design problem for DFS can be defined as well.

Fuzzy Controller Design Using Relaxed Stability Conditions: DFS Find $X > 0$, $Y \geq 0$, and M_i ($i = 1, \ldots, r$) satisfying

$$\begin{bmatrix} X - (s-1)Y & XA_i^T - M_i^T B_i^T \\ A_i X - B_i M_i & X \end{bmatrix} > 0, \qquad (3.27)$$

$$\begin{bmatrix} X + Y & \frac{1}{2}\{A_i X + A_j X - B_i M_j - B_j M_i\}^T \\ \frac{1}{2}\{A_i X + A_j X - B_i M_j - B_j M_i\} & X \end{bmatrix} \geq 0,$$
$$i < j \text{ s.t. } h_i \cap h_j \neq \phi, \qquad (3.28)$$

where
$$X = P^{-1}, \qquad M_i = F_i X, \qquad Y = XQX.$$

The feedback gain F_i, a common P, and a common Q can be obtained as
$$P = X^{-1}, \qquad F_i = M_i X^{-1}, \qquad Q = PYP$$

from the solutions X, Y, and M_i.

The conditions (3.27) and (3.28) can be obtained as follows: Multiplying both sides of (3.13) by P^{-1} gives
$$P^{-1} G_{ii}^T P G_{ii} P^{-1} - P^{-1} + (s-1) P^{-1} Q P^{-1} < 0.$$

Therefore,
$$P^{-1} - (s-1) P^{-1} Q P^{-1}$$
$$- \left(A_i P^{-1} - B_i F_i P^{-1} \right)^T P \left(A_i P^{-1} - B_i F_i P^{-1} \right) > 0.$$

Since $P^{-1} = X$, we have
$$X - (s-1) XQX - (A_i X - B_i F_i X)^T X^{-1} (A_i X - B_i F_i X) > 0.$$

Define $M_i = F_i X$ and $Y = XQX$. By substituting into the above inequality, we obtain
$$X - (s-1) Y - (A_i X - B_i M_i)^T X^{-1} (A_i X - B_i M_i) > 0.$$

It easily follows that the above inequality can be transformed into (3.27) by the Schur complement procedure.

Similarly, from (3.14), we have
$$P^{-1} \left(\frac{G_{ij} + G_{ji}}{2} \right)^T P \left(\frac{G_{ij} + G_{ji}}{2} \right) P^{-1} - P^{-1} - P^{-1} Q P^{-1} \leq 0.$$

Therefore,
$$P^{-1} + P^{-1} Q P^{-1}$$
$$- \tfrac{1}{4} \left(G_{ij} P^{-1} + G_{ji} P^{-1} \right)^T P \left(G_{ij} P^{-1} + G_{ji} P^{-1} \right) \geq 0.$$

Since $P^{-1} = X$, we have
$$X + XQX - \left\{ \tfrac{1}{2} (G_{ij} X + G_{ji} X)^T \right\} X^{-1} \left\{ \tfrac{1}{2} (G_{ij} X + G_{ji} X) \right\}$$
$$= X + XQX - \left\{ \tfrac{1}{2} (A_i X + A_j X - B_i F_j X - B_j F_i X)^T \right\} X^{-1}$$
$$\times \left\{ \tfrac{1}{2} (A_i X + A_j X - B_i F_j X - B_j F_i X) \right\} \geq 0.$$

By substituting $M_i = F_i X$ and $Y = XQX$ into the above inequality, we obtain

$$X + Y - \left\{\tfrac{1}{2}(A_i X + A_j X - B_i M_j - B_j M_i)^T\right\} X^{-1}$$
$$\times \left\{\tfrac{1}{2}(A_i X + A_j X - B_i M_j - B_j M_i)\right\} \geq 0.$$

Equation (3.28) is obtained by applying the Schur complement.

3.4 DECAY RATE

The speed of response is related to decay rate, that is, the largest Lyapunov exponent. This section deals with the decay rate fuzzy controller design [1–3].

Decay Rate Controller Design: CFS The condition that $\dot{V}(x(t)) \leq -2\alpha V(x(t))$ [20] for all trajectories is equivalent to

$$G_{ii}^T P + P G_{ii} + 2\alpha P < 0 \tag{3.29}$$

for all i and

$$\left(\frac{G_{ij} + G_{ji}}{2}\right)^T P + P \left(\frac{G_{ij} + G_{ji}}{2}\right) + 2\alpha P \leq 0 \tag{3.30}$$

for $i < j$ excepting the pairs (i, j) such that $h_i(z(t))h_j(z(t)) = 0, \forall\, t$, where $\alpha > 0$. Therefore, the largest lower bound on the decay rate that we can find using a quadratic Lyapunov function can be found by solving the following GEVP (generalized eigenvalue minimization problem) in X and α:

maximize α
$X, M_1, \ldots, M_r$

subject to

$$X > 0,$$
$$-XA_i^T - A_i X + M_i^T B_i^T + B_i M_i - 2\alpha X > 0, \tag{3.31}$$
$$-XA_i^T - A_i X - XA_j^T - A_j X$$
$$+ M_j^T B_i^T + B_i M_j + M_i^T B_j^T + B_j M_i - 4\alpha X \geq 0,$$
$$i < j \text{ s.t. } h_i \cap h_j \neq \phi, \tag{3.32}$$

where

$$X = P^{-1}, \quad M_i = F_i X. \tag{3.33}$$

Decay Rate Fuzzy Controller Design: DFS The condition that $\Delta V(x(t)) \leq (\alpha^2 - 1)V(x(t))$ [20] for all trajectories is equivalent to

$$G_{ii}^T P G_{ii} - \alpha^2 P < 0, \quad (3.34)$$

$$\left(\frac{G_{ij} + G_{ji}}{2}\right)^T P \left(\frac{G_{ij} + G_{ji}}{2}\right) - \alpha^2 P \leq 0,$$

$$i < j \text{ s.t. } h_i \cap h_j \neq \phi, \quad (3.35)$$

where $\alpha < 1$. Therefore, we define the following GEVP in X and β, where $\beta = \alpha^2$:

$$\underset{X, M_1, \ldots, M_r}{\text{minimize}} \ \beta$$

subject to

$X > 0$,

$$\begin{bmatrix} \beta X & X A_i^T - M_i^T B_i^T \\ A_i X - B_i M_i & X \end{bmatrix} > 0, \quad (3.36)$$

$$\begin{bmatrix} \beta X & \left\{\dfrac{A_i X + A_j X - B_i M_j - B_j M_i}{2}\right\}^T \\ \left\{\dfrac{A_i X + A_j X - B_i M_j - B_j M_i}{2}\right\} & X \end{bmatrix} \geq 0,$$

$$i < j \text{ s.t. } h_i \cap h_j \neq \phi, \quad (3.37)$$

where

$$X = P^{-1}, \quad M_i = F_i X. \quad (3.38)$$

It should be noted that $0 \leq \beta < 1$.

Remark 12 The decay rate fuzzy controller designs reduce to the stable fuzzy controller designs when $\alpha = 0$ and $\beta = 1$. Therefore, a fuzzy controller that satisfies the LMI conditions of (3.31) and (3.32) or (3.36) and (3.37) is a stable fuzzy controller. In other words, the LMI conditions of (3.15) and (3.16) or (3.19) and (3.20) are special cases of those of (3.31) and (3.32) or (3.36) and (3.37), respectively.

Decay Rate Controller Design Using Relaxed Stability Conditions: CFS

The condition that $\dot{V}(x(t)) \leq -2\alpha V(x(t))$ for all trajectories is equivalent to

$$G_{ii}^T P + P G_{ii} + (s-1)Q + 2\alpha P < 0,$$

$$\left(\frac{G_{ij} + G_{ji}}{2}\right)^T P + P\left(\frac{G_{ij} + G_{ji}}{2}\right) - Q + 2\alpha P \leq 0,$$

$$i < j \text{ s.t. } h_i \cap h_j \neq \phi,$$

where $\alpha > 0$. Therefore, the largest lower bound on the decay rate that we can find using a quadratic Lyapunov function can be found by solving the following GEVP in X and α:

$$\underset{X, Y, M_1, \ldots, M_r}{\text{maximize}} \quad \alpha$$

subject to

$$X > 0, \, Y \geq 0,$$

$$\begin{aligned}-XA_i^T - A_i X + M_i^T B_i^T + B_i M_i \\ -(s-1)Y - 2\alpha X > 0,\end{aligned} \quad (3.39)$$

$$\begin{aligned}2Y - XA_i^T - A_i X - XA_j^T - A_j X \\ + M_j^T B_i^T + B_i M_j + M_i^T B_j^T + B_j M_i \\ - 4\alpha X \geq 0,\end{aligned}$$

$$i < j \text{ s.t. } h_i \cap h_j \neq \phi, \quad (3.40)$$

where

$$X = P^{-1}, \quad M_i = F_i X, \quad Y = XQX.$$

Decay Rate Controller Design Using Relaxed Stability Conditions: DFS

The condition that $\Delta V(x(t)) \leq (\alpha^2 - 1)V(x(t))$ for all trajectories is equivalent to

$$G_{ii}^T P G_{ii} - \alpha^2 P + (s-1)Q < 0, \quad (3.41)$$

$$\left(\frac{G_{ij} + G_{ji}}{2}\right)^T P \left(\frac{G_{ij} + G_{ji}}{2}\right) - \alpha^2 P - Q \leq 0,$$

$$i < j \text{ s.t. } h_i \cap h_j \neq \phi, \quad (3.42)$$

where $\alpha < 1$. Therefore, we define the following GEVP in X and β where $\beta = \alpha^2$.

$$\underset{X, Y, M_1, \ldots, M_r}{\text{minimize}} \ \beta$$

subject to

$X > 0, Y \geq 0$,

$$\begin{bmatrix} \beta X - (s-1)Y & XA_i^T - M_i^T B_i^T \\ A_i X - B_i M_i & X \end{bmatrix} > 0, \quad (3.43)$$

$$\begin{bmatrix} \beta X + Y & \frac{1}{2}\{A_i X + A_j X - B_i M_j - B_j M_i\}^T \\ \frac{1}{2}\{A_i X + A_j X - B_i M_j - B_j M_i\} & X \end{bmatrix} \geq 0,$$

$$i < j \text{ s.t. } h_i \cap h_j \neq \phi, \quad (3.44)$$

where

$$X = P^{-1}, \quad M_i = F_i X, \quad Y = XQX.$$

It should be noted that $0 \leq \beta < 1$.

The condition (3.43) is derived as follows. Multiplying both sides of (3.41) by P^{-1} gives

$$P^{-1} G_{ii}^T P G_{ii} P^{-1} - \alpha^2 P^{-1} + (s-1) P^{-1} Q P^{-1} < 0.$$

Therefore, from $G_{ii} = A_i - B_i F_i$,

$$(A_i P^{-1} - B_i F_i P^{-1})^T P (A_i P^{-1} - B_i F_i P^{-1}) - \alpha^2 P^{-1} + (s-1) P^{-1} Q P^{-1} < 0.$$

From $X = P^{-1}$, $Y = XQX$, $M_i = F_i X$, and $\beta = \alpha^2$, we have

$$\beta X - (s-1) Y - (A_i X - B_i M_i)^T X^{-1} (A_i X - B_i M_i) > 0.$$

Therefore, the Schur complement procedure yields

$$\begin{bmatrix} \beta X - (s-1)Y & XA_i^T - M_i^T B_i^T \\ A_i X - B_i M_i & X \end{bmatrix} > 0. \quad (3.45)$$

Inequality (3.44) can be obtained from (3.42) in the same fashion.

Remark 13 A fuzzy controller that satisfies the LMI conditions of (3.39) and (3.40) [or (3.43) and (3.44)] is a stable fuzzy controller. In other words, it also satisfies the LMI conditions of (3.23) and (3.24) [or (3.27) and (3.28)].

Remark 14 As illustrated in Example 9, the conditions of Theorems 9 and 10 lead to less conservative results for the stability of a given fuzzy control system. For the design of stabilizing fuzzy controllers, it is recommended to use the conditions of these theorems together with other control performance considerations such as pole placement LMI conditions.

3.5 CONSTRAINTS ON CONTROL INPUT AND OUTPUT

3.5.1 Constraint on the Control Input

THEOREM 11 *Assume that the initial condition $x(0)$ is known. The constraint $\|u(t)\|_2 \leq \mu$ is enforced at all times $t \geq 0$ if the LMIs*

$$\begin{bmatrix} 1 & x(0)^T \\ x(0) & X \end{bmatrix} \geq \mathbf{0}, \qquad (3.46)$$

$$\begin{bmatrix} X & M_i^T \\ M_i & \mu^2 I \end{bmatrix} \geq 0 \qquad (3.47)$$

hold, where $X = P^{-1}$ and $M_i = F_i X$.

Proof. Assume that $V(x(t)) = x^T(t) P x(t)$ is a Lyapunov function and

$$x^T(0) P x(0) \leq 1.$$

Then,

$$1 - x^T(0) X^{-1} x(0) \geq 0, \qquad (3.48)$$

where $X = P^{-1}$. The inequality (3.48) is transformed into (3.46) by the Schur complement procedure.

The derivation of (3.47) is as follows: From $\|u(t)\|_2 \leq \mu$,

$$u^T(t) u(t) = \sum_{i=1}^{r} \sum_{j=1}^{r} h_i(z(t)) h_j(z(t)) x^T(t) F_i^T F_j x(t) \leq \mu^2.$$

Therefore,

$$\frac{1}{\mu^2} \sum_{i=1}^{r} \sum_{j=1}^{r} h_i(z(t)) h_j(z(t)) x^T(t) F_i^T F_j x(t) \leq 1. \qquad (3.49)$$

CONSTRAINTS ON CONTROL INPUT AND OUTPUT   **67**

Since $x^T(t)X^{-1}x(t) < x^T(0)X^{-1}x(0) \leq 1$ for $t > 0$, if

$$\frac{1}{\mu^2} \sum_{i=1}^{r} \sum_{j=1}^{r} h_i(z(t))h_j(z(t))x^T(t)F_i^T F_j x(t) \leq x^T(t)X^{-1}x(t), \quad (3.50)$$

then (3.49) holds. Therefore, we have

$$\sum_{i=1}^{r} \sum_{j=1}^{r} h_i(z(t))h_j(z(t))x^T(t)\left(\frac{1}{\mu^2}F_i^T F_j - X^{-1}\right)x(t) \leq 0. \quad (3.51)$$

From the left side of (3.51),

$$\frac{1}{2} \sum_{i=1}^{r} \sum_{j=1}^{r} h_i(z(t))h_j(z(t))x^T(t)\left(\frac{1}{\mu^2}F_i^T F_j + \frac{1}{\mu^2}F_j^T F_i - 2X^{-1}\right)x(t)$$

$$= \frac{1}{2} \sum_{i=1}^{r} \sum_{j=1}^{r} h_i(z(t))h_j(z(t))x^T(t)$$

$$\times \left[\frac{1}{\mu^2}(F_i^T F_i + F_j^T F_j) - \frac{1}{\mu^2}(F_i^T - F_j^T)(F_i - F_j) - 2X^{-1}\right]x(t)$$

$$\leq \frac{1}{2} \sum_{i=1}^{r} \sum_{j=1}^{r} h_i(z(t))h_j(z(t))x^T(t)\left[\frac{1}{\mu^2}(F_i^T F_i + F_j^T F_j) - 2X^{-1}\right]x(t)$$

$$= \sum_{i=1}^{r} h_i(z(t))x^T(t)\left(\frac{1}{\mu^2}F_i^T F_i - X^{-1}\right)x(t).$$

If

$$\frac{1}{\mu^2}F_i^T F_i - X^{-1} \leq \mathbf{0}, \quad (3.52)$$

(3.51) holds. By defining $M_i = F_i X$ for (3.52), we obtain

$$\frac{1}{\mu^2}M_i^T M_i - X \leq \mathbf{0}.$$

Inequality (3.47) can be obtained from the above inequality by the Schur Complement procedure.

Another solution to obtain (3.47) is as follows. From (3.51), we have

$$\sum_{i=1}^{r} h_i(z(t))\begin{bmatrix} X^{-1} & F_i^T \\ F_i & \mu^2 I \end{bmatrix} \geq \mathbf{0}.$$

Multiplying both side of the above inequality by block-diag $[X\ I]$ gives

$$\sum_{i=1}^{r} h_i(z(t)) \begin{bmatrix} X & M_i^T \\ M_i & \mu^2 I \end{bmatrix} \geq \mathbf{0}.$$

Hence we arrive at the condition (3.47). This derivation is more direct and compact. Q.E.D.

The LMIs are available for both CFSs and DFSs. A design problem of stable fuzzy controllers satisfying the input constraint can be defined as follows: Find $X > \mathbf{0}$, $Y \geq \mathbf{0}$, and M_i ($i = 1, \ldots, r$) satisfying (3.23) and (3.24) [or (3.27) and (3.28)] and (3.46) and (3.47).

3.5.2 Constraint on the Output

THEOREM 12 *Assume that the initial condition $x(0)$ is known. The constraint $\|y(t)\|_2 \leq \lambda$ is enforced at all times $t \geq 0$ if the LMIs*

$$\begin{bmatrix} 1 & x(0)^T \\ x(0) & X \end{bmatrix} \geq \mathbf{0}, \tag{3.53}$$

$$\begin{bmatrix} X & XC_i^T \\ C_i X & \lambda^2 I \end{bmatrix} \geq \mathbf{0} \tag{3.54}$$

hold, where $X = P^{-1}$.

Proof. The proof can be completed in the same procedure as in Theorem 11.

The LMIs are available for both CFSs and DFSs. A design problem of stable fuzzy controllers satisfying the output constraint can be defined as follows: Find $X > \mathbf{0}$, $Y \geq \mathbf{0}$, and M_i ($i = 1, \ldots, r$) satisfying (3.23) and (3.24) [or (3.27) and (3.28)] and (3.53) and (3.54).

3.6 INITIAL STATE INDEPENDENT CONDITION

The above LMI design conditions for input and output constraints depend on the initial states of the system. This means that the feedback gains F_i must be again determined using the above LMIs if the initial states $x(0)$ change. This is a disadvantage of using the LMIs on the control input and output. We modify the LMI constraints on the control input and output, where $x(0)$ is unknown but the upper bound ϕ of $\|x(0)\|$ is known, that is, $\|x(0)\| \leq \phi$. To encompass a large set of initial states, we can set ϕ to be a large quantity even if $x(0)$ is unknown. Of course, a large ϕ could lead to conservative designs.

The modified LMI is accomplished by the following results.

THEOREM 13 *Assume that* $\|x(0)\| \leq \phi$, *where* $x(0)$ *is unknown but the upper bound* ϕ *is known. Then,*

$$x^T(0) X^{-1} x(0) \leq 1 \tag{3.55}$$

if

$$\phi^2 I \leq X, \tag{3.56}$$

where $X = P^{-1}$.

Proof. From (3.56),

$$X^{-1} \leq \frac{1}{\phi^2} I.$$

Therefore,

$$x^T(0) X^{-1} x(0) \leq \frac{1}{\phi^2} x^T(0) x(0) \leq 1. \qquad \text{Q.E.D.}$$

Note that (3.55) is equivalent to (3.46) and (3.53). The condition (3.56) can be used instead of (3.55). A design example using the initial state independent condition will be presented in Chapter 8.

3.7 DISTURBANCE REJECTION

This section presents a disturbance rejection fuzzy controller design for the Takagi-Sugeno fuzzy models. Consider the following CFS with disturbance [1]:

$$\dot{x}(t) = \sum_{i=1}^{r} h_i(z(t))\{A_i x(t) + B_i u(t) + E_i v(t)\}, \tag{3.57}$$

$$y(t) = \sum_{i=1}^{r} h_i(z(t)) C_i x(t), \tag{3.58}$$

where $v(t)$ is the disturbance. The disturbance rejection can be realized by minimizing γ subject to

$$\sup_{\|v(t)\|_2 \neq 0} \frac{\|y(t)\|_2}{\|v(t)\|_2} \leq \gamma. \tag{3.59}$$

THEOREM 14 [CFS] *The feedback gains* F_i *that stabilize the fuzzy model and minimize* γ *in* (3.59) *can be obtained by solving the following minimization problem based on LMIs.*

70 LMI CONTROL PERFORMANCE CONDITIONS AND DESIGNS

$$\underset{x, M_1, \ldots, M_r}{\text{minimize }} \gamma^2$$

subject to

$X > 0$,

$$\begin{bmatrix} -\frac{1}{2}\{XA_i^T - M_j^T B_i^T + A_i X - B_i M_j \\ +XA_j^T - M_i^T B_j^T + A_j X - B_j M_i\} & -\frac{1}{2}(E_i + E_j) & \frac{1}{2}X(C_i + C_j)^T \\ -\frac{1}{2}(E_i + E_j)^T & \gamma^2 I & 0 \\ \frac{1}{2}(C_i + C_j)X & 0 & I \end{bmatrix} \geq 0,$$

$$i \leq j \text{ s.t. } h_i(z(t)) \cap h_j(z(t)) \neq \phi, \qquad (3.61)$$

where

$$M_i = F_i X.$$

Proof. Suppose there exists a quadratic function $V(x(t)) = x^T(t) P x(t)$, $P > 0$, and $\gamma \geq 0$ such that, for all t,

$$\dot{V}(x(t)) + y^T(t) y(t) - \gamma^2 v^T(t) v(t) \leq 0 \qquad (3.62)$$

for (3.57) and (3.58). By integrating (3.62) from 0 to T, we obtain

$$\int_0^T \left(\dot{V}(x(t)) + y^T(t) y(t) - \gamma^2 v^T(t) v(t) \right) dt \leq 0.$$

By assuming that initial condition $x(0) = 0$, we have

$$V(x(T)) + \int_0^T \left(y^T(t) y(t) - \gamma^2 v^T(t) v(t) \right) dt \leq 0. \qquad (3.63)$$

Since $V(x(T)) \geq 0$, this implies

$$\frac{\|y(t)\|_2}{\|v(t)\|_2} \leq \gamma.$$

Therefore the L_2 gain of the fuzzy model is less than γ if (3.62) holds. We derive an LMI condition from (3.62). From (3.62),

$$\dot{x}^T(t)Px(t) + x^T(t)P\dot{x}(t)$$

$$+ \sum_{i=1}^{r}\sum_{j=1}^{r} h_i(z(t))h_j(z(t))x^T C_i^T C_j x(t) - \gamma^2 v^T(t)v(t)$$

$$= \sum_{i=1}^{r}\sum_{j=1}^{r} h_i(z(t))h_j(z(t))x^T(t)(A_i - B_i F_j)^T Px(t)$$

$$+ \sum_{i=1}^{r}\sum_{j=1}^{r} h_i(z(t))h_j(z(t))x^T(t)P(A_i - B_i F_j)x(t)$$

$$+ \sum_{i=1}^{r}\sum_{j=1}^{r} h_i(z(t))h_j(z(t))x^T C_i^T C_j x(t) - \gamma^2 v^T(t)v(t)$$

$$+ \sum_{i=1}^{r} h_i(z(t))v^T(t)E_i^T Px(t) + \sum_{i=1}^{r} h_i(z(t))x^T(t)PE_i v(t)$$

$$= \sum_{i=1}^{r}\sum_{j=1}^{r} h_i(z(t))h_j(z(t))[x^T(t) \; v^T(t)]$$

$$\times \begin{bmatrix} \begin{pmatrix} (A_i - B_i F_j)^T P \\ +P(A_i - B_i F_j) \\ +C_i^T C_j \end{pmatrix} & PE_i \\ E_i^T P & -\gamma^2 I \end{bmatrix} \begin{bmatrix} x(t) \\ v(t) \end{bmatrix} \leq 0. \quad (3.64)$$

From (3.64), we have the following conditions:

$$\begin{bmatrix} \begin{pmatrix} -\sum_{i=1}^{r}\sum_{j=1}^{r} h_i(z(t))h_j(z(t))\{(A_i - B_i F_j)^T P \\ +P(A_i - B_i F_j) + C_i^T C_j\} \\ -\sum_{i=1}^{r} h_i(z(t))E_i^T P \end{pmatrix} & -P\sum_{i=1}^{r} h_i(z(t))E_i \\ & \gamma^2 I \end{bmatrix} \geq 0.$$

$$(3.65)$$

The left-hand side of (3.65) can be decomposed as follows:

$$
\begin{bmatrix} \left(\begin{array}{c} -\sum_{i=1}^{r}\sum_{j=1}^{r} h_i(z(t))h_j(z(t)) \\ \times\{(A_i - B_iF_j)^T P + P(A_i - B_iF_j)\} \end{array}\right) & -P\sum_{i=1}^{r} h_i(z(t))E_i \\ -\sum_{i=1}^{r} h_i(z(t))E_i^T P & \gamma^2 I \end{bmatrix}
$$

$$
-\begin{bmatrix} \sum_{i=1}^{r}\sum_{j=1}^{r} h_i(z(t))h_j(z(t))\, C_i^T C_j & 0 \\ 0 & 0 \end{bmatrix}
$$

$$
= \begin{bmatrix} \left(\begin{array}{c} -\sum_{i=1}^{r}\sum_{j=1}^{r} h_i(z(t))h_j(z(t)) \\ \times\{(A_i - B_iF_j)^T P + P(A_i - B_iF_j)\} \end{array}\right) & -P\sum_{i=1}^{r} h_i(z(t))E_i \\ -\sum_{i=1}^{r} h_i(z(t))E_i^T P & \gamma^2 I \end{bmatrix}
$$

$$
-\begin{bmatrix} \sum_{i=1}^{r} h_i(z(t))C_i^T \\ 0 \end{bmatrix}\begin{bmatrix} \sum_{i=1}^{r} h_i(z(t))C_i & 0 \end{bmatrix} \geq \mathbf{0}. \qquad (3.66)
$$

Inequality (3.67) is equivalent to

$$
\begin{bmatrix} \left(\begin{array}{c} -\sum_{i=1}^{r}\sum_{j=1}^{r} h_i(z(t))h_j(z(t)) \\ \times\{(A_i - B_iF_j)^T P \\ + P(A_i - B_iF_j)\} \end{array}\right) & -P\sum_{i=1}^{r} h_i(z(t))E_i & \sum_{i=1}^{r} h_i(z(t))C_i^T \\ -\sum_{i=1}^{r} h_i(z(t))E_i^T P & \gamma^2 I & 0 \\ \sum_{i=1}^{r} h_i(z(t))C_i & 0 & I \end{bmatrix} \geq \mathbf{0}. \quad (3.67)
$$

Inequality (3.67) can be rewritten as

$$
\sum_{i=1}^{r}\sum_{j=1}^{r} h_i(z(t))h_j(z(t))
$$

$$
\begin{bmatrix} \left(\begin{array}{c} -\frac{1}{2}\{(A_i - B_iF_j)^T P + P(A_i - B_iF_j) \\ + (A_j - B_jF_i)^T P + P(A_j - B_jF_i)\} \end{array}\right) & -\frac{1}{2}P(E_i + E_j) & \frac{1}{2}(C_i + C_j)^T \\ -\frac{1}{2}(E_i + E_j)^T P & \gamma^2 I & 0 \\ \frac{1}{2}(C_i + C_j) & 0 & I \end{bmatrix} \geq \mathbf{0}.
$$

Therefore, we have

$$\begin{bmatrix} \begin{pmatrix} -\frac{1}{2}\{(A_i - B_i F_j)^T P + P(A_i - B_i F_j) \\ + (A_j - B_j F_i)^T P + P(A_j - B_j F_i)\} \end{pmatrix} & -\frac{1}{2} P(E_i + E_j) & \frac{1}{2}(C_i + C_j)^T \\ -\frac{1}{2}(E_i + E_j)^T P & \gamma^2 I & 0 \\ \frac{1}{2}(C_i + C_j) & 0 & I \end{bmatrix} \geq 0. \quad (3.68)$$

By multiplying both side of (3.68) by block-diag $\{X\ I\ I\}$, (3.61) is obtained, where $X = P^{-1}$. Q.E.D.

Next, consider the following DFS with disturbance [21]:

$$x(t+1) = \sum_{i=1}^{r} h_i(z(t))\{A_i x(t) + B_i u(t) + E_i v(t)\}, \quad (3.69)$$

$$y(t) = \sum_{i=1}^{r} h_i(z(t)) C_i x(t), \quad (3.70)$$

where $v(t)$ is the disturbance. The disturbance rejection can be realized by minimizing γ subject to

$$\sup_{\|v(t)\|_2 \neq 0} \frac{\|y(t)\|_2}{\|v(t)\|_2} \leq \gamma. \quad (3.71)$$

THEOREM 15 [DFS] *The feedback gains F_i that stabilize the fuzzy model and minimize γ in (3.71) can be obtained by solving the following LMIs:*

$$\underset{X, M_1, \ldots, M_r}{\text{minimize}} \ \gamma^2$$

subject to

$X > 0$,

$$\begin{bmatrix} X & 0 & \frac{1}{2}\begin{pmatrix} (A_i X - B_i M_j) \\ + A_j X - B_j M_i)^T \end{pmatrix} & \frac{1}{2} X(C_i + C_j)^T \\ 0 & \gamma^2 I & \frac{1}{2}(E_i + E_j)^T & 0 \\ \frac{1}{2}\begin{pmatrix} (A_i X - B_i M_j) \\ + A_j X - B_j M_i) \end{pmatrix} & \frac{1}{2}(E_i + E_j) & X & 0 \\ \frac{1}{2}(C_i + C_j) X & 0 & 0 & I \end{bmatrix} \geq 0,$$

$$i \leq j \text{ s.t. } h_i \cap h_j \neq \phi, \quad (3.72)$$

where $X > 0$ and $M_i = F_i X$.

Proof. Suppose there exists a quadratic function $V(x(t)) = x^T(t)Px(t)$, $P > 0$, and $\gamma \geq 0$ such that, for all t,

$$\Delta V(x(t)) + y^T(t)y(t) - \gamma^2 v^T(t)v(t) \leq 0 \qquad (3.73)$$

for (3.69) and (3.70). From (3.73), we obtain

$$\sum_{t=0}^{T} \{\Delta V(x(t)) + y^T(t)y(t) - \gamma^2 v^T(t)v(t)\} \leq 0.$$

By assuming that initial condition $x(0) = 0$, we obtain

$$V(x(T)) + \sum_{t=0}^{T} (y^T(t)y(t) - \gamma^2 v^T(t)v(t)) \leq 0. \qquad (3.74)$$

Since $V(x(T)) \geq 0$, this implies

$$\frac{\|y(t)\|_2}{\|v(t)\|_2} \leq \gamma.$$

Therefore the L_2 gain of the fuzzy model is less than γ if (3.73) holds. We derive an LMI condition from (3.73):

$$\gamma^2 v^T(t)v(t) - y^T(t)y(t) - \Delta V(x(t))$$

$$= \gamma^2 v^T(t)v(t) - x^T(t)\left(\sum_{i=1}^{r} h_i(z(t))C_i\right)^T \left(\sum_{i=1}^{r} h_i(z(t))C_i\right)x(t)$$

$$- \left\{\sum_{i=1}^{r}\sum_{j=1}^{r} h_i(z(t))h_j(z(t))(A_i - B_iF_j)x(t) + \sum_{i=1}^{r} h_i(z(t))E_i v(t)\right\}^T$$

$$\times P\left\{\sum_{i=1}^{r}\sum_{j=1}^{r} h_i(z(t))h_j(z(t))(A_i - B_iF_j)x(t) + \sum_{i=1}^{r} h_i(z(t))E_i v(t)\right\}$$

$$+ x^T(t)Px(t)$$

$$= \begin{bmatrix} x^T(t) & v^T(t) \end{bmatrix} \begin{bmatrix} P & 0 \\ 0 & \gamma^2 I \end{bmatrix} \begin{bmatrix} x(t) \\ v(t) \end{bmatrix}$$

$$- \begin{bmatrix} x^T(t) & v^T(t) \end{bmatrix} \left\{\sum_{i=1}^{r}\sum_{j=1}^{r} h_i(z(t))h_j(z(t))[A_i - B_iF_j \quad E_i]\right\}^T$$

$$\times P\left\{\sum_{i=1}^{r}\sum_{j=1}^{r} h_i(z(t))h_j(z(t))[A_i - B_iF_j \quad E_i]\right\}\begin{bmatrix} x(t) \\ v(t) \end{bmatrix}$$

DISTURBANCE REJECTION 75

$$-x^T(t)\left(\sum_{i=1}^{r}h_i(z(t))C_i\right)^T\left(\sum_{i=1}^{r}h_i(z(t))C_i\right)x(t)$$

$$=\begin{bmatrix}x^T(t) & v^T(t)\end{bmatrix}$$

$$\times\begin{bmatrix}P-\left(\sum_{i=1}^{r}h_i(z(t))C_i\right)^T\left(\sum_{i=1}^{r}h_i(z(t))C_i\right) & 0 \\ 0 & \gamma^2 I\end{bmatrix}\begin{bmatrix}x(t)\\v(t)\end{bmatrix}$$

$$-\begin{bmatrix}x^T(t) & v^T(t)\end{bmatrix}\left\{\sum_{i=1}^{r}\sum_{j=1}^{r}h_i(z(t))h_j(z(t))\begin{bmatrix}A_i-B_iF_j & E_i\end{bmatrix}\right\}^T$$

$$\times P\left\{\sum_{i=1}^{r}\sum_{j=1}^{r}h_i(z(t))h_j(z(t))\begin{bmatrix}A_i-B_iF_j & E_i\end{bmatrix}\right\}\begin{bmatrix}x(t)\\v(t)\end{bmatrix}\geq \mathbf{0}.$$

From the Schur complement, we obtain the LMI condition:

$$\begin{bmatrix} \left(P-\left(\sum_{i=1}^{r}h_i(z(t))C_i\right)^T \times \left(\sum_{i=1}^{r}h_i(z(t))C_i\right)\right) & 0 & \left(\sum_{i=1}^{r}\sum_{j=1}^{r}h_i(z(t))h_j(z(t)) \times \{A_i-B_iF_j\}^T\right) \\ 0 & \gamma^2 I & \sum_{i=1}^{r}h_i(z(t))E_i^T \\ \left(\sum_{i=1}^{r}\sum_{j=1}^{r}h_i(z(t))h_j(z(t)) \times \{A_i-B_iF_j\}\right) & \sum_{i=1}^{r}h_i(z(t))E_i & P^{-1} \end{bmatrix}$$

$$=\begin{bmatrix} P & 0 & \left(\sum_{i=1}^{r}\sum_{j=1}^{r}h_i(z(t))h_j(z(t)) \times \{A_i-B_iF_j\}^T\right) \\ 0 & \gamma^2 I & \sum_{i=1}^{r}h_i(z(t))E_i^T \\ \left(\sum_{i=1}^{r}\sum_{j=1}^{r}h_i(z(t))h_j(z(t)) \times \{A_i-B_iF_j\}\right) & \sum_{i=1}^{r}h_i(z(t))E_i & P^{-1} \end{bmatrix}$$

$$-\begin{bmatrix}\left(\sum_{i=1}^{r}h_i(z(t))C_i\right)^T \\ 0 \\ 0\end{bmatrix}\begin{bmatrix}\sum_{i=1}^{r}h_i(z(t))C_i & 0 & 0\end{bmatrix}\geq \mathbf{0}. \quad (3.75)$$

Inequality (3.75) is equivalent to

$$\begin{bmatrix} P & 0 & \left(\sum_{i=1}^{r}\sum_{j=1}^{r} h_i(z(t))h_j(z(t))\right) \times \{A_i - B_i F_j\}^T & \left(\sum_{i=1}^{r} h_i(z(t))C_i\right)^T \\ 0 & \gamma^2 I & \sum_{i=1}^{r} h_i(z(t))E_i^T & 0 \\ \left(\sum_{i=1}^{r}\sum_{j=1}^{r} h_i(z(t))h_j(z(t))\right) \times \{A_i - B_i F_j\} & \sum_{i=1}^{r} h_i(z(t))E_i & P^{-1} & 0 \\ \sum_{i=1}^{r} h_i(z(t))C_i & 0 & 0 & I \end{bmatrix}$$

$$= \sum_{i=1}^{r}\sum_{j=1}^{r} h_i(z(t))h_j(z(t))$$

$$\times \begin{bmatrix} P & 0 & \frac{1}{2}(A_i - B_i F_j + A_j - B_j F_i)^T & \frac{1}{2}(C_i + C_j)^T \\ 0 & \gamma^2 I & \frac{1}{2}(E_i + E_j)^T & 0 \\ \frac{1}{2}(A_i - B_i F_j + A_j - B_j F_i) & \frac{1}{2}(E_i + E_j) & P^{-1} & 0 \\ \frac{1}{2}(C_i + C_j) & 0 & 0 & I \end{bmatrix} \geq 0.$$

(3.76)

Therefore,

$$\begin{bmatrix} P & 0 & \frac{1}{2}(A_i - B_i F_j + A_j - B_j F_i)^T & \frac{1}{2}(C_i + C_j)^T \\ 0 & \gamma^2 I & \frac{1}{2}(E_i + E_j)^T & 0 \\ \frac{1}{2}(A_i - B_i F_j + A_j - B_j F_i) & \frac{1}{2}(E_i + E_j) & P^{-1} & 0 \\ \frac{1}{2}(C_i + C_j) & 0 & 0 & I \end{bmatrix} \geq 0,$$

$$i \leq j \text{ s.t. } h_i \cap h_j \neq \phi. \quad (3.77)$$

By multiplying both sides of (3.77) by block-diag[$X\ I\ I\ I$], (3.72) is obtained, where $X = P^{-1}$. Q.E.D.

A design example for disturbance rejection will be discussed in Chapter 8.

3.8 DESIGN EXAMPLE: A SIMPLE MECHANICAL SYSTEM

Let us consider an example of dc motor controlling an inverted pendulum via a gear train [22]. Fuzzy modeling for the nonlinear system was done in [3],

[23] and [24]. The fuzzy model is as follows:

Plant Rule 1

IF $x_1(t)$ is M_1,

$$\text{THEN} \quad \begin{cases} \dot{x}(t) = A_1 x(t) + B_1 u(t), \\ y(t) = C_1 x(t). \end{cases} \quad (3.78)$$

Plant Rule 2

IF $x_1(t)$ is M_2,

$$\text{THEN} \quad \begin{cases} \dot{x}(t) = A_2 x(t) + B_2 u(t), \\ y(t) = C_2 x(t). \end{cases} \quad (3.79)$$

Here,

$$x(t) = \begin{bmatrix} x_1(t) & x_2(t) & x_3(t) \end{bmatrix}^T,$$

$$A_1 = \begin{bmatrix} 0 & 1 & 0 \\ 9.8 & 0 & 1 \\ 0 & -10 & -10 \end{bmatrix}, \quad B_1 = \begin{bmatrix} 0 \\ 0 \\ 10 \end{bmatrix},$$

$$C_1 = \begin{bmatrix} 1 & 0 & 0 \end{bmatrix}$$

$$A_2 = \begin{bmatrix} 0 & 1 & 0 \\ 0 & 0 & 1 \\ 0 & -10 & -10 \end{bmatrix}, \quad B_2 = \begin{bmatrix} 0 \\ 0 \\ 10 \end{bmatrix},$$

$$C_2 = \begin{bmatrix} 1 & 0 & 0 \end{bmatrix}.$$

The angle of the pendulum is $x_1(t)$, $x_2(t) = \dot{x}_1(t)$, and $x_3(t)$ is current of the motor. The M_1 and M_2 are fuzzy sets defined as

$$M_1(x_1(t)) = \begin{cases} \dfrac{\sin x_1(t)}{x_1(t)}, & x_1(t) \neq 0, \\ 1, & x_1(t) = 0, \end{cases}$$

$$M_2(x_1(t)) = 1 - M_1(x_1(t)).$$

This fuzzy model exactly represents the dynamics of the nonlinear mechanical system under $-\pi \leq x_1(t) \leq \pi$. Note that the fuzzy model has a common B matrix, that is, $B_1 = B_2$. The fuzzy controller design of the common B matrix cases is simple in general. To show the effect of the LMI-based designs, we consider a more difficult case, that is, we change B_2 as follows:

$$B_2 = \begin{bmatrix} 0 \\ 0 \\ 20 \end{bmatrix}.$$

3.8.1 Design Case 1: Decay Rate

We first design a stable fuzzy controller by considering the decay rate. The design problem of the CFS is defined as follows:

$$\underset{X, Y, M_1, \ldots, M_r}{\text{maximize}} \quad \alpha$$

subject to $X > 0$, $Y \geq 0$, (3.39) and (3.40).

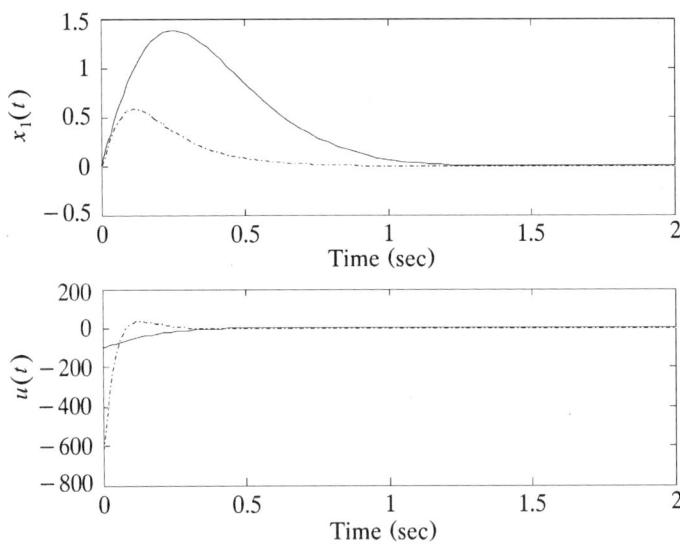

Fig. 3.3 Design examples 1 and 2.

We obtain

$$\alpha = 5.0,$$
$$F_1 = [282.3129 \quad 62.4176 \quad 3.2238],$$
$$F_2 = [110.4644 \quad 24.9381 \quad 1.2716],$$
$$P = X^{-1} = \begin{bmatrix} 105.108 & 20.4393 & 1.05294 \\ 20.4393 & 4.29985 & 0.23680 \\ 1.05294 & 0.23680 & 0.01567 \end{bmatrix} > 0,$$
$$Q = X^{-1}YX^{-1} = \begin{bmatrix} 1432.034 & 299.8039 & 16.26773 \\ 299.8039 & 63.19188 & 3.449801 \\ 16.26773 & 3.449801 & 0.190786 \end{bmatrix} \geq 0.$$

The dotted line in Figure 3.3 shows the responses of $y(t)$ $[= x_1(t)]$ and $u(t)$.

3.8.2 Design Case 2: Decay Rate + Constraint on the Control Input

It can be seen in the design example 1 that $\max_t \|u(t)\|_2 = 624$. In practical design, there is a limitation of control input. It is important to consider not only the decay rate but also the constraint on the control input. The design problem that considers the decay rate and the constraint on the control input is defined as follows, where $\mu = 100$ and $x(0) = [0 \quad 10 \quad 0]^T$:

$$\underset{X,Y,M_1,\ldots,M_r}{\text{maximize}} \quad \alpha$$

subject to $X > 0$, $Y \geq 0$ (3.39), (3.40), (3.46), and (3.47).

The solution is obtained as

$$\alpha = 4.23,$$
$$F_1 = [38.3637 \quad 9.9338 \quad 0.7203],$$
$$F_2 = [18.2429 \quad 6.4771 \quad 0.5118],$$
$$P = \begin{bmatrix} 0.1578 & 0.03847 & 0.002738 \\ 0.03847 & 0.009995 & 0.000742 \\ 0.002738 & 0.000742 & 5.831 \times 10^{-5} \end{bmatrix} > 0,$$
$$Q = \begin{bmatrix} 0.001250 & 0.000281 & 4.275 \times 10^{-5} \\ 0.000281 & 0.0001215 & 6.332 \times 10^{-6} \\ 4.275 \times 10^{-5} & 6.332 \times 10^{-6} & 1.976 \times 10^{-6} \end{bmatrix} \geq 0.$$

80   LMI CONTROL PERFORMANCE CONDITIONS AND DESIGNS

The real line in Figure 3.3 shows the responses of $y(t)(=x_1(t))$ and $u(t)$. The designed controller realizes the input constraint $\max_t \|u(t)\|_2 = 99.3 < \mu$.

3.8.3 Design Case 3: Stability + Constraint on the Control Input

It is also possible to design a stable fuzzy controller satisfying the constraint on the control input, where $\mu = 100$.

Find $X > 0$, $Y \geq 0$, and M_i ($i = 1, \ldots, r$) satisfying (3.23), (3.24), (3.53), and (3.54).

The solution is obtained as

$$F_1 = [13.0065 \quad 3.6948 \quad 0.1786],$$

$$F_2 = [7.7309 \quad 2.7900 \quad 0.1163],$$

$$P = \begin{bmatrix} 0.0335 & 0.0106 & 0.0015 \\ 0.0106 & 0.0036 & 0.0005 \\ 0.0015 & 0.0005 & 0.0001 \end{bmatrix} > \mathbf{0},$$

$$Q = \begin{bmatrix} 0.0522 & 0.0203 & 0.0040 \\ 0.0203 & 0.0082 & 0.0016 \\ 0.0040 & 0.0016 & 0.0003 \end{bmatrix} \geq \mathbf{0}.$$

The dotted line in Figure 3.4 shows the responses of $y(t)$ $[= x_1(t)]$ and $u(t)$. It can be found that $\max_t \|u(t)\|_2 = 38.1 < \mu$.

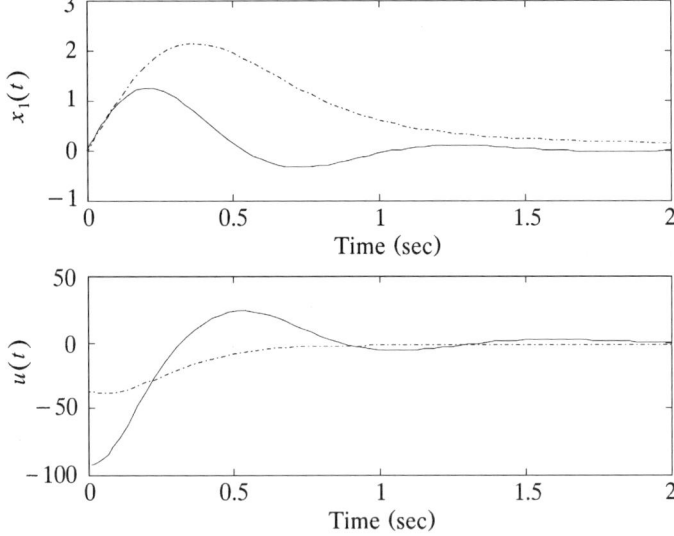

Fig. 3.4 Design examples 3 and 4.

3.8.4 Design Case 4: Stability + Constraint on the Control Input + Constraint on the Output

The response of the control system in the design example 3 has a large output error $(\max_t \|y(t)\|_2 = 2.16)$ since the constraint on the output is not considered in the fuzzy controller design. To improve the response, we can design a fuzzy controller by adding the constraint on the output.

Find $X > 0$, $Y \geq 0$, and M_i $(i = 1, \ldots, r)$ satisfying (3.23), (3.24), (3.46), (3.47), and (3.54) where $\mu = 100$ and $\lambda = 2$.

The solution is obtained as

$$F_1 = [59.2819 \quad 9.3038 \quad 0.5580],$$

$$F_2 = [33.7254 \quad 7.4115 \quad 0.4122],$$

$$P = \begin{bmatrix} 0.5478 & 0.0519 & 0.0034 \\ 0.0519 & 0.0098 & 0.0006 \\ 0.0034 & 0.0006 & 0.0001 \end{bmatrix} > 0,$$

$$Q = \begin{bmatrix} 0.9936 & 0.0334 & 0.0075 \\ 0.0334 & 0.0118 & 0.0008 \\ 0.0075 & 0.0008 & 0.0001 \end{bmatrix} \geq 0.$$

The real line in Figure 3.4 shows the responses of $y(t)$ $[= x_1(t)]$ and $u(t)$. The response of the control system satisfies the constraints $\max_t \|u(t)\|_2 = 93 < \mu$ and $\max_t \|y(t)\|_2 = 1.25 < \lambda$.

REFERENCES

1. K. Tanaka, T. Taniguchi, and H. O. Wang, "Model-Based Fuzzy Control of TORA System: Fuzzy Regulator and Fuzzy Observer Design via LMIs that Represent Decay Rate, Disturbance Rejection, Robustness, Optimality," Seventh IEEE International Conference on Fuzzy Systems, Alaska, 1998, pp. 313–318.
2. K. Tanaka, T. Ikeda, and H. O. Wang, "Design of Fuzzy Control Systems Based on Relaxed LMI Stability Conditions," 35th IEEE Conference on Decision and Control, Kobe, Vol. 1, 1996, pp. 598–603.
3. K. Tanaka, T. Ikeda, and H. O. Wang, "Fuzzy Regulators and Fuzzy Observers," *IEEE Trans. Fuzzy Syst.*, Vol. 6, No. 2, pp. 250–265 (1998).
4. K. Tanaka and M. Sugeno, "Stability Analysis of Fuzzy Systems Using Lyapunov's Direct Method," *Proc. of NAFIPS'90*, pp. 133–136, 1990.
5. R. Langari and M. Tomizuka, "Analysis and Synthesis of Fuzzy Linguistic Control Systems," 1990 ASME Winter Annual Meeting, 1990, pp. 35–42.
6. S. Kitamura and T. Kurozumi, "Extended Circle Criterion and Stability Analysis of Fuzzy Control Systems," in *Proc. of the International Fuzzy Eng. Symp.'91*, Vol. 2, 1991, pp. 634–643.

7. K. Tanaka and M. Sugeno, "Stability Analysis and Design of Fuzzy Control Systems," *Fuzzy Sets Systs.* Vol. 45, No. 2, pp. 135–156 (1992).
8. S. S. Farinwata et al., "Stability Analysis of The Fuzzy Logic Controller Designed by The Phase Portrait Assignment Algorithm," *Proc. of 2nd IEEE International Conference on Fuzzy Systems*, 1993, pp. 1377–1382.
9. K. Tanaka and M. Sano, "Fuzzy Stability Criterion of a Class of Nonlinear Systems," *Inform. Sci.*, Vol. 71, Nos. 1 & 2, pp. 3–26 (1993).
10. K. Tanaka and M. Sugeno, "Concept of Stability Margin or Fuzzy Systems and Design of Robust Fuzzy Controllers," in *Proceedings of 2nd IEEE International Conference on Fuzzy Systems*, Vol. 1, 1993, pp. 29–34.
11. H. O. Wang, K. Tanaka, and M. Griffin, "Parallel Distributed Compensation of Nonlinear Systems by Takagi and Sugeno's Fuzzy Model.," *Proceedings of FUZZ-IEEE'95*, 1995, pp. 531–538.
12. H. O. Wang, K. Tanaka, and M. Griffin, "An Analytical Framework of Fuzzy Modeling and Control of Nonlinear Systems," 1995 American Control Conference, Vol 3, Seattle, 1995, pp. 2272–2276.
13. S. Singh, "Stability Analysis of Discrete Fuzzy Control Systems," *Proceedings of First IEEE International Conference on Fuzzy Systems*, 1992, pp. 527–534.
14. R. Katoh et al., "Graphical Stability Analysis of a Fuzzy Control System," *Proceedings of IEEE International Conference on IECON'93*, Vol. 1, 1993, pp. 248–253.
15. C.-L. Chen et al., "Analysis and Design of Fuzzy Control Systems," *Fuzzy Sets and Syst.*, Vol. 57, pp. 125–140 (1993).
16. F. Hara and M. Ishibe, "Simulation Study on the Existence of Limit Cycle Oscillation in a Fuzzy Control System," *Proceedings of the Korea-Japan Joint Conference on Fuzzy Systems and Engineering*, 1992, pp. 25–28.
17. H. O. Wang, K. Tanaka, and M. Griffin, "An Approach to Fuzzy Control of Nonlinear Systems: Stability and Design Issues," *IEEE Trans. Fuzzy Syst.*, Vol. 4, No. 1, pp. 14–23 (1996).
18. K. Tanaka and M. Sano, "A Robust Stabilization Problem of Fuzzy Controller Systems and Its Applications to Backing up Control of a Truck-Trailer," *IEEE Trans. Fuzzy Syst.*, Vol. 2, No. 2, pp. 119–134 (1994).
19. S. Kawamoto et al. "An Approach to Stability Analysis of Second Order Fuzzy Systems," *Proceedings of First IEEE International Conference on Fuzzy Systems*, Vol. 1, 1992, pp. 1427–1434.
20. A. Ichikawa et al., *Control Hand Book*, Ohmu Publisher, 1993, Tokyo in Japanese.
21. K. Tanaka , T. Taniguchi, and H. O. Wang, "Trajectory Control of an Articulated Vehicle with Triple Trailers," 1999 IEEE International Conference on Control Applications, Vol. 2, Hawaii, August 1999.
22. J. G. Kushewski et. al., "Application of Feedforward Neural Networks to Dynamical System Identification and Control," *IEEE Trans. Control Sys. Technol.*, Vol. 1, No. 1, pp. 37–49 (1993).
23. K. Tanaka and M. Sano, "On Design of Fuzzy Regulators and Fuzzy Observers," *Proc. 10th Fuzzy System Symposium*, 1994, pp. 411–414 in Japanese.
24. S. Kawamoto, et. al., "Nonlinear Control and Rigorous Stability Analysis Based on Fuzzy System for Inverted Pendulum," *Proc. of FUZZ-IEEE'96*, Vol. 2, 1996, pp. 1427–1432.

CHAPTER 4

FUZZY OBSERVER DESIGN

In practical applications, the state of a system is often not readily available. Under such circumstances, the question arises whether it is possible to determine the state from the system response to some input over some period of time. For linear systems, a linear observer [1] provides an affirmative answer if the system is observable. Likewise, a systematic design method of fuzzy regulators and fuzzy observers plays an important role for fuzzy control systems. This chapter presents the concept of fuzzy observers and two design procedures for fuzzy observer-based control [2, 3]. In linear system theory, one of the most important results on observer design is the so-called separation principle, that is, the controller and observer design can be carried out separately without compromising the stability of the overall closed-loop system. In this chapter, it is shown that a similar separation principle also holds for a large class of fuzzy control systems.

4.1 FUZZY OBSERVER

Up to this point we have mainly dealt with LMI-based fuzzy control designs involving state feedback. In real-world control problems, however, it is often the case that the complete information of the states of a system is not always available. In such cases, one need to resort to output feedback design methods such as observer-based designs. This chapter presents fuzzy observer design methodologies involving state estimation for T-S fuzzy models. Alternatively, output feedback design can be treated in the framework of dynamic feedback, which is the subject of Chapter 12.

As in all observer designs, fuzzy observers [4] are required to satisfy

$$x(t) - \hat{x}(t) \to 0 \quad \text{as } t \to \infty,$$

where $\hat{x}(t)$ denotes the state vector estimated by a fuzzy observer. This condition guarantees that the steady-state error between $x(t)$ and $\hat{x}(t)$ converges to 0. As in the case of controller design, the PDC concept is employed to arrive at the following fuzzy observer structures:

CFS

Observer Rule i

    IF $z_1(t)$ is M_{i1} and $\cdots$ and $z_p(t)$ is M_{ip}

        THEN

$$\dot{\hat{x}}(t) = A_i \hat{x}(t) + B_i u(t) + K_i(y(t) - \hat{y}(t)),$$
$$\hat{y}(t) = C_i \hat{x}(t), \quad i = 1, 2, \ldots, r. \tag{4.1}$$

DFS

Observer Rule i

    IF $z_1(t)$ is M_{i1} and $\cdots$ and $z_p(t)$ is M_{ip}

        THEN

$$\hat{x}(t+1) = A_i \hat{x}(t) + B_i u(t) + K_i(y(t) - \hat{y}(t)),$$
$$\hat{y}(t) = C_i \hat{x}(t), \quad i = 1, 2, \ldots, r. \tag{4.2}$$

The fuzzy observer has the linear state observer's laws in its consequent parts. The steady-state error between $x(t)$ and $\hat{x}(t)$ will be discussed in the next section.

4.2 DESIGN OF AUGMENTED SYSTEMS

This section presents LMI-based designs for an augmented system containing both the fuzzy controller and observer.

The dependence of the premise variables on the state variables makes it necessary to consider two cases for fuzzy observer design:

    Case A $z_1(t), \ldots, z_p(t)$ do not depend on the state variables estimated by a fuzzy observer.

DESIGN OF AUGMENTED SYSTEMS

Case B $z_1(t), \ldots, z_p(t)$ depend on the state variables estimated by a fuzzy observer.

Obviously the stability analysis and design of the augmented system for Case A are more straightforward, whereas the stability analysis and design for Case B are complicated since the premise variables depend on the state variables, which have to estimated by a fuzzy observer. This fact leads to significant difference between $z(t)$ (Case A) and $\hat{z}(t)$ (Case B) in the design of fuzzy observer and controller.

4.2.1 Case A

The fuzzy observer for Case A is represented as follows:

CFS

$$\hat{x}(t) = \frac{\sum_{i=1}^{r} w_i(z(t))\{A_i\hat{x}(t) + B_i u(t) + K_i(y(t) - \hat{y}(t))\}}{\sum_{i=1}^{r} w_i(z(t))}$$

$$= \sum_{i=1}^{r} h_i(z(t))\{A_i\hat{x}(t) + B_i u(t) + K_i(y(t) - \hat{y}(t))\}, \quad (4.3)$$

$$\hat{y}(t) = \sum_{i=1}^{r} h_i(z(t))C_i\hat{x}(t). \quad (4.4)$$

DFS

$$\hat{x}(t+1) = \frac{\sum_{i=1}^{r} w_i(z(t))\{A_i\hat{x}(t) + B_i u(t) + K_i(y(t) - \hat{y}(t))\}}{\sum_{i=1}^{r} w_i(z(t))}$$

$$= \sum_{i=1}^{r} h_i(z(t))\{A_i\hat{x}(t) + B_i u(t) + K_i(y(t) - \hat{y}(t))\}, \quad (4.5)$$

$$\hat{y}(t) = \sum_{i=1}^{r} h_i(z(t))C_i\hat{x}(t). \quad (4.6)$$

We use the same weight $w_i(z(t))$ as that of the ith rule of the fuzzy models (2.3) and (2.4), and (2.5) and (2.6). The fuzzy observer design is to determine the local gains K_i in the consequent parts.

In the presence of the fuzzy observer for Case A, the PDC fuzzy controller takes on the following form, instead of (2.23):

$$u(t) = -\frac{\sum_{i=1}^{r} w_i(z(t)) F_i \hat{x}(t)}{\sum_{i=1}^{r} w_i(z(t))} = -\sum_{i=1}^{r} h_i(z(t)) F_i \hat{x}(t). \quad (4.7)$$

Combining the fuzzy controller (4.7) and the fuzzy observers (4.3)–(4.6) and denoting $e(t) = x(t) - \hat{x}(t)$, we obtain the following system representations:

CFS

$$\dot{x}(t) = \sum_{i=1}^{r} \sum_{j=1}^{r} h_i(z(t)) h_j(z(t)) \{(A_i - B_i F_j) x(t) + B_i F_j e(t)\},$$

$$\dot{e}(t) = \sum_{i=1}^{r} \sum_{j=1}^{r} h_i(z(t)) h_j(z(t)) \{A_i - K_i C_j\} e(t).$$

DFS

$$x(t+1) = \sum_{i=1}^{r} \sum_{j=1}^{r} h_i(z(t)) h_j(z(t)) \{(A_i - B_i F_j) x(t) + B_i F_j e(t)\},$$

$$e(t+1) = \sum_{i=1}^{r} \sum_{j=1}^{r} h_i(z(t)) h_j(z(t)) \{A_i - K_i C_j\} e(t).$$

Therefore, the augmented systems are represented as follows:

CFS

$$\dot{x}_a(t) = \sum_{i=1}^{r} \sum_{j=1}^{r} h_i(z(t)) h_j(z(t)) G_{ij} x_a(t)$$

$$= \sum_{i=1}^{r} h_i(z(t)) h_i(z(t)) G_{ii} x_a(t)$$

$$+ 2 \sum_{i=1}^{r} \sum_{i<j} h_i(z(t)) h_j(z(t)) \frac{G_{ij} + G_{ji}}{2} x_a(t), \quad (4.8)$$

DFS

$$x_a(t+1) = \sum_{i=1}^{r} \sum_{j=1}^{r} h_i(z(t)) h_j(z(t)) G_{ij} x_a(t)$$

$$= \sum_{i=1}^{r} h_i(z(t)) h_i(z(t)) G_{ii} x_a(t)$$

$$+ 2 \sum_{i=1}^{r} \sum_{i<j} h_i(z(t)) h_j(z(t)) \frac{G_{ij} + G_{ji}}{2} x_a(t), \quad (4.9)$$

where

$$x_a(t) = \begin{bmatrix} x(t) \\ e(t) \end{bmatrix},$$

$$G_{ij} = \begin{bmatrix} A_i - B_i F_j & B_i F_j \\ 0 & A_i - K_i C_j \end{bmatrix}. \quad (4.10)$$

By applying Theorems 7 and 8 to the augmented system (4.8) and (4.9), respectively, we arrive at the following theorems.

THEOREM 16 [CFS] *The equilibrium of the augmented system described by (4.8) is globally asymptotically stable if there exists a common positive definite matrix P such that*

$$G_{ii}^T P + P G_{ii} < 0, \quad (4.11)$$

$$\left(\frac{G_{ij} + G_{ji}}{2}\right)^T P + P \left(\frac{G_{ij} + G_{ji}}{2}\right) < 0,$$

$$i < j \text{ s.t. } h_i \cap h_j \neq \phi. \quad (4.12)$$

Proof. It follows directly from Theorem 7.

THEOREM 17 [DFS] *The equilibrium of the augmented system described by (4.9) is globally asymptotically stable if there exists a common positive definite matrix P such that*

$$G_{ii}^T P G_{ii} - P < 0, \quad (4.13)$$

$$\left(\frac{G_{ij} + G_{ji}}{2}\right)^T P \left(\frac{G_{ij} + G_{ji}}{2}\right) - P < 0,$$

$$i < j \text{ s.t. } h_i \cap h_j \neq \phi. \quad (4.14)$$

Proof. It follows directly from Theorem 8.

Recall that Theorems 9 and 10 represent less conservative conditions than those of Theorems 7 and 8. Therefore, by applying Theorems 9 and 10 to (4.8) and (4.9), respectively, we can obtain the following less conservative conditions:

THEOREM 18 [CFS] *The equilibrium of the augmented system described by (4.8) is globally asymptotically stable if there exist a common positive definite matrix P and a common positive semidefinite matrix Q such that*

$$G_{ii}^T P + P G_{ii} + (s-1)Q < 0, \tag{4.15}$$

$$\left(\frac{G_{ij} + G_{ji}}{2}\right)^T P + P\left(\frac{G_{ij} + G_{ji}}{2}\right) - Q \leq 0,$$

$$i < j \text{ s.t. } h_i \cap h_j \neq \phi, \tag{4.16}$$

where $s > 1$.

Proof. It follows directly from Theorem 9.

THEOREM 19 [DFS] *The equilibrium of the augmented system described by (4.9) is globally asymptotically stable if there exist a common positive definite matrix P and a common positive semidefinite matrix Q such that*

$$G_{ii}^T P G_{ii} - P + (s-1)Q < 0, \tag{4.17}$$

$$\left(\frac{G_{ij} + G_{ji}}{2}\right)^T P \left(\frac{G_{ij} + G_{ji}}{2}\right) - P - Q \leq 0,$$

$$i < j \text{ s.t. } h_i \cap h_j \neq \phi, \tag{4.18}$$

where $s > 1$.

Proof. It follows directly from Theorem 10.

As a further refinement, we can incorporate the decay rate condition into the augmented systems as follows:

CFS: The condition that $\dot{V}(x_a(t)) \leq -2\alpha V(x_a(t))$ for all trajectories is equivalent to

$$G_{ii}^T P + P G_{ii} + (s-1)Q + 2\alpha P < 0, \tag{4.19}$$

$$\left(\frac{G_{ij} + G_{ji}}{2}\right)^T P + P\left(\frac{G_{ij} + G_{ji}}{2}\right) - Q + 2\alpha P \leq 0,$$

$$i < j \text{ s.t. } h_i \cap h_j \neq \phi, \tag{4.20}$$

where $\alpha > 0$.

DFS: The condition that $\Delta V(x_a(t)) \leq (\alpha^2 - 1)V(x_a(t))$ for all trajectories is equivalent to

$$G_{ii}^T P G_{ii} - \alpha^2 P + (s-1)Q < 0, \tag{4.21}$$

$$\left(\frac{G_{ij} + G_{ji}}{2}\right)^T P \left(\frac{G_{ij} + G_{ji}}{2}\right) - \alpha^2 P - Q \leq 0,$$

$$i < j \text{ s.t. } h_i \cap h_j \neq \phi, \tag{4.22}$$

where $\alpha < 1$.

DESIGN OF AUGMENTED SYSTEMS

Next we consider the controller and observer design problem. The approach is to transform the conditions above for CFS and DFS into LMI ones so as to directly determine the feedback gains F_i and the observer gains K_i. The transformation procedure can be similarly applied to all theorems in this section. In the following, we present some representative results. Other cases are left as exercises for the readers.

Design Procedure for Case A: CFS Assume that the number of rules that fire for all t is less than or equal to s, where $1 < s \leq r$. The largest bound on the decay rate that we can find using a quadratic Lyapunov function can be found by solving the GEVP.

$$\underset{P_1, P_2, Y, Q_{22}, M_{1i}, N_{2i}}{\text{maximize}} \quad \alpha$$

subject to $\alpha > 0$,

$$P_1, P_2 > 0, \quad Y \geq 0, \quad Q_{22} \geq 0,$$

$$P_1 A_i^T - M_{1i}^T B_i^T + A_i P_1 - B_i M_{1i} + (s-1)Y + 2\alpha P_1 < 0,$$

$$A_i^T P_2 - C_i^T N_{2i}^T + P_2 A_i - N_{2i} C_i + (s-1)Q_{22} + 2\alpha P_2 < 0,$$

$$P_1 A_i^T - M_{1j}^T B_i^T + A_i P_1 - B_i M_{1j} - 2Y + 4\alpha P_1$$
$$+ P_1 A_j^T - M_{1i}^T B_j^T + A_j P_1 - B_j M_{1i} < 0,$$

$$i < j \text{ s.t. } h_i \cap h_j \neq \phi,$$

$$A_i^T P_2 - C_j^T N_{2i}^T + P_2 A_i - N_{2i} C_j - 2Q_{22} + 4\alpha P_2$$
$$+ A_j^T P_2 - C_i^T N_{2j}^T + P_2 A_j - N_{2j} C_i < 0,$$

$$i < j \text{ s.t. } h_i \cap h_j \neq \phi,$$

where $s > 1$, $M_{1i} = F_i P_1$, $N_{2i} = P_2 K_i$, and $Y = P_1 Q_{11} P_1$.

The matrices P_1, P_2, Q_{22}, M_{1i}, N_{2i}, and Y can be found by using convex optimization techniques involving LMIs if they exist. The feedback gains and the observer gains can then be obtained as $F_i = M_{1i} P_1^{-1}$ and $K_i = P_2^{-1} N_{2i}$. The design conditions above address decay rate and relaxed stability conditions and are reduced to the stable controller design problem if we set $\alpha = 0$, $Y = 0$, and $Q_{22} = 0$.

The design problem for discrete systems can be handled similarly.

Design Procedure for Case A: DFS

$$P_1, P_2 > 0,$$

$$\begin{bmatrix} P_1 & P_1 A_i^T - M_{1i}^T B_i^T \\ A_i P_1 - B_i M_{1i} & P_1 \end{bmatrix} > 0, \quad (4.23)$$

$$\begin{bmatrix} P_2 & A_i^T P_2 - C_i^T N_{2i}^T \\ P_2 A_i - N_{2i}^T C_i & P_2 \end{bmatrix} > 0, \qquad (4.24)$$

$$\begin{bmatrix} 4P_1 & \begin{pmatrix} P_1 A_i^T - M_{1j}^T B_i^T \\ + P_1 A_j^T - M_{1i}^T B_j^T \end{pmatrix} \\ \begin{pmatrix} A_i P_1 - B_i M_{1j} \\ + A_j P_1 - B_j M_{1i} \end{pmatrix} & P_1 \end{bmatrix} > 0, \qquad (4.25)$$

$$\begin{bmatrix} 4P_2 & \begin{pmatrix} A_i^T P_2 - C_j^T N_{2i} \\ + A_j^T P_2 - C_i^T N_{2i}^T \end{pmatrix} \\ \begin{pmatrix} P_2 A_i - N_{2i}^T C_j \\ + P_2 A_i - N_{2i}^T C_j \end{pmatrix} & P_2 \end{bmatrix} > 0. \qquad (4.26)$$

Remark 15 Note that in the designs above the controller gains and the observer gains can be determined separately. This powerful result is similar to the well-known separation principle for linear systems. Unfortunately, such a separation principle only holds for Case A and does not hold for Case B [3].

Finally, we would like to point out, as in Chapter 3, that a variety of control performance specifications can be incorporated into the LMI-based observer and controller design.

4.2.2 Case B

In Case B we deal with the situation when the premise variables $z(t)$ are unknown since they depend on the state variables to be estimated by fuzzy observers. As a result, we must use $w_i(\hat{z}(t))$ instead of $w_i(z(t))$. In other words, in Case B, $h_i(z(t)) \neq h_i(\hat{z}(t))$ because of $z(t) \neq \hat{z}(t)$ in general.

The fuzzy observers for Case B are of the following forms, instead of (4.3) or (4.5):

CFS

$$\dot{\hat{x}}(t) = \sum_{i=1}^{r} h_i(\hat{z}(t))\{A_i \hat{x}(t) + B_i u(t) + K_i(y(t) - \hat{y}(t))\}, \quad (4.27)$$

$$\hat{y}(t) = \sum_{i=1}^{r} h_i(\hat{z}(t)) C_i \hat{x}(t).$$

DFS

$$\hat{x}(t+1) = \sum_{i=1}^{r} h_i(\hat{z}(t))\{A_i \hat{x}(t) + B_i u(t) + K_i(y(t) - \hat{y}(t))\}, \quad (4.28)$$

$$\hat{y}(t) = \sum_{i=1}^{r} h_i(\hat{z}(t)) C_i \hat{x}(t).$$

Accordingly, instead of (4.7), the PDC fuzzy controller becomes

$$u(t) = -\frac{\sum_{i=1}^{r} w_i(\hat{z}(t)) F_i \hat{x}(t)}{\sum_{i=1}^{r} w_i(\hat{z}(t))} = -\sum_{i=1}^{r} h_i(\hat{z}(t)) F_i \hat{x}(t). \quad (4.29)$$

Then the augmented systems are obtained as follows:

CFS

$$\dot{x}_a(t) = \sum_{i=1}^{r} \sum_{j=1}^{r} \sum_{k=1}^{r} h_i(z(t)) h_j(\hat{z}(t)) h_k(\hat{z}(t)) G_{ijk} x_a(t)$$

$$= \sum_{i=1}^{r} \sum_{j=1}^{r} h_i(z(t)) h_j(\hat{z}(t)) h_j(\hat{z}(t)) G_{ijj} x_a(t)$$

$$+ 2 \sum_{i=1}^{r} \sum_{j<k}^{r} h_i(z(t)) h_j(\hat{z}(t)) h_k(\hat{z}(t)) \frac{G_{ijk} + G_{ikj}}{2} x_a(t). \quad (4.30)$$

DFS

$$x_a(t+1) = \sum_{i=1}^{r} \sum_{j=1}^{r} \sum_{k=1}^{r} h_i(z(t)) h_j(\hat{z}(t)) h_k(\hat{z}(t)) G_{ijk} x_a(t)$$

$$= \sum_{i=1}^{r} \sum_{j=1}^{r} h_i(z(t)) h_j(\hat{z}(t)) h_j(\hat{z}(t)) G_{ijj} x_a(t)$$

$$+ 2 \sum_{i=1}^{r} \sum_{j<k}^{r} h_i(z(t)) h_j(\hat{z}(t)) h_k(\hat{z}(t)) \frac{G_{ijk} + G_{ikj}}{2} x_a(t),$$

$$(4.31)$$

where

$$x_a(t) = \begin{bmatrix} x(t) \\ e(t) \end{bmatrix},$$

$$e(t) = x(t) - \hat{x}(t),$$

$$G_{ijk} = \begin{bmatrix} A_i - B_i F_k & B_i F_k \\ S^1_{ijk} & S^2_{ijk} \end{bmatrix},$$

$$S^1_{ijk} = (A_i - A_j) - (B_i - B_j) F_k + K_j (C_k - C_i),$$

$$S^2_{ijk} = A_j - K_j C_k + (B_i - B_j) F_k. \quad (4.32)$$

The following stability theorem for the augmented system (4.30) can be derived from Theorem 7.

THEOREM 20 [CFS] *The equilibrium of the augmented system described by (4.30) is globally asymptotically stable if there exists a common positive definite matrix P such that*

$$G_{ijj}^T P + P G_{ijj} < 0, \qquad (4.33)$$

$$\left(\frac{G_{ijk} + G_{ikj}}{2}\right)^T P + P\left(\frac{G_{ijk} + G_{ikj}}{2}\right) < 0,$$

$$\forall i, j < k \text{ s.t. } h_i \cap h_j \cap h_k \neq \phi. \qquad (4.34)$$

Proof. It follows directly from Theorem 7.

The following stability theorem for the augmented system (4.31) can be derived from Theorem 8.

THEOREM 21 [DFS] *The equilibrium of the augmented system described by (4.31) is globally asymptotically stable if there exists a common positive definite matrix P such that*

$$G_{ijj}^T P G_{ijj} - P < 0, \qquad (4.35)$$

$$\left(\frac{G_{ijk} + G_{ikj}}{2}\right)^T P \left(\frac{G_{ijk} + G_{ikj}}{2}\right) - P < 0,$$

$$\forall i, j < k \text{ s.t. } h_i \cap h_j \cap h_k \neq \phi. \qquad (4.36)$$

Proof. It follows directly from Theorem 8.

Remark 16 Consider the common C matrix case, that is, $C_1 = C_2 = \cdots = C_r = C$. In this case,

$$S_{ijk}^1 = (A_i - A_j) - (B_i - B_j) F_k,$$

$$S_{ijk}^2 = A_j - K_j C + (B_i - B_j) F_k.$$

The conditions of Theorems 20 and 21 imply those of Theorems 18 and 19, respectively.

Remark 17 We can no longer apply the relaxed conditions (Theorems 9 and 10) to Case B because of $h_i(z(t)) \neq h_i(\hat{z}(t))$ in general.

4.3 DESIGN EXAMPLE

Consider the following nonlinear system:

$$\dot{x}_1(t) = x_2(t) + \sin x_3(t) + (x_1^2(t) + 1)u(t),$$
$$\dot{x}_2(t) = x_1(t) + 2x_2(t),$$
$$\dot{x}_3(t) = x_1^2(t)x_2(t) + x_1(t),$$
$$\dot{x}_4(t) = \sin x_3(t),$$
$$y_1(t) = (x_1^2(t) + 1)x_4(t) + x_2(t),$$
$$y_2(t) = x_2(t) + x_3(t).$$

Assume that $x_1(t)$ and $x_3(t)$ are observable. In other words, $x_2(t)$ and $x_4(t)$ are estimated using a fuzzy observer. It is also assumed that

$$x_1(t) \in [-a, a], \quad x_3(t) \in [-b, b],$$

where a and b are positive values. The nonlinear terms are $x_1^2(t)$ and $\sin x_3(t)$. The nonlinear terms can be represented as

$$x_1^2(t) = M_1^1(x_1(t)) \cdot a^2 + M_1^2(x_1(t)) \cdot 0,$$
$$\sin x_3(t) = M_2^1(x_3(t)) \cdot 1 \cdot x_3(t) + M_2^2(x_3(t)) \cdot \frac{\sin b}{b} \cdot x_3(t),$$

where

$$M_1^1(x_1(t)), \quad M_1^2(x_1(t)), \quad M_2^1(x_3(t)), \quad M_2^2(x_3(t)) \in [0, 1],$$
$$M_1^1(x_1(t)) + M_1^2(x_1(t)) = 1, \quad M_2^1(x_3(t)) + M_2^2(x_3(t)) = 1.$$

By solving the equations, they are obtained as follows:

$$M_1^1(x_1(t)) = \frac{x_1^2}{a^2},$$

$$M_1^2(x_1(t)) = 1 - M_1^1(x_1(t)) = 1 - \frac{x^2(t)}{a^2},$$

$$M_2^1(x_3(t)) = \begin{cases} \dfrac{b \cdot \sin x_3(t) - \sin b \cdot x_3(t)}{x_3(t) \cdot (b - \sin b)}, & x_3(t) \neq 0, \\ 1, & x_3(t) = 0, \end{cases}$$

$$M_2^2(x_3(t)) = 1 - M_2^1(x_3(t))$$

$$= \begin{cases} \dfrac{b \cdot (x_3(t) - \sin x_3(t))}{x_3(t) \cdot (b - \sin b)}, & x_3(t) \neq 0, \\ 0, & x_3(t) = 0, \end{cases}$$

where

$$x_1(t) \in [-a, a], \quad x_3(t) \in [-b, b].$$

The terms M_1^1, M_2^1, M_1^2, and M_2^2 can be interpreted as membership functions of fuzzy sets. By using these fuzzy sets, the nonlinear system can be represented by the following T-S fuzzy model:

Model Rule 1

IF $x_1(t)$ is M_1^1 and $x_3(t)$ is M_2^1,

$$\textbf{THEN} \begin{cases} \dot{x}(t) = A_1 x(t) + B_1 u(t), \\ y(t) = C_1 x(t). \end{cases} \quad (4.37)$$

Model Rule 2

IF $x_1(t)$ is M_1^1 and $x_3(t)$ is M_2^2,

$$\textbf{THEN} \begin{cases} \dot{x}(t) = A_2 x(t) + B_2 u(t), \\ y(t) = C_2 x(t). \end{cases} \quad (4.38)$$

Model Rule 3

IF $x_1(t)$ is M_1^2 and $x_3(t)$ is M_2^1,

$$\textbf{THEN} \begin{cases} \dot{x}(t) = A_3 x(t) + B_3 u(t), \\ y(t) = C_3 x(t). \end{cases} \quad (4.39)$$

Model Rule 4

IF $x_1(t)$ is M_1^2 and $x_3(t)$ is M_2^2,

$$\textbf{THEN} \begin{cases} \dot{x}(t) = A_4 x(t) + B_4 u(t), \\ y(t) = C_4 x(t). \end{cases} \quad (4.40)$$

Here,
$$x(t) = \begin{bmatrix} x_1(t) & x_2(t) & x_3(t) & x_4(t) \end{bmatrix}^T,$$

$$A_1 = \begin{bmatrix} 0 & 1 & 1 & 0 \\ 1 & 2 & 0 & 0 \\ 1 & a^2 & 0 & 0 \\ 0 & 0 & 1 & 0 \end{bmatrix}, \quad B_1 = \begin{bmatrix} 1+a^2 \\ 0 \\ 0 \\ 0 \end{bmatrix},$$

$$C_1 = \begin{bmatrix} 0 & 1 & 0 & 1+a^2 \\ 0 & 1 & 1 & 0 \end{bmatrix},$$

$$A_2 = \begin{bmatrix} 0 & 1 & \sin b/b & 0 \\ 1 & 2 & 0 & 0 \\ 1 & a^2 & 0 & 0 \\ 0 & 0 & \sin b/b & 0 \end{bmatrix}, \quad B_2 = \begin{bmatrix} 1+a^2 \\ 0 \\ 0 \\ 0 \end{bmatrix},$$

$$C_2 = \begin{bmatrix} 0 & 1 & 0 & 1+a^2 \\ 0 & 1 & 1 & 0 \end{bmatrix},$$

$$A_3 = \begin{bmatrix} 0 & 1 & 1 & 0 \\ 1 & 2 & 0 & 0 \\ 1 & 0 & 0 & 0 \\ 0 & 0 & 1 & 0 \end{bmatrix}, \quad B_3 = \begin{bmatrix} 1 \\ 0 \\ 0 \\ 0 \end{bmatrix},$$

$$C_3 = \begin{bmatrix} 0 & 1 & 0 & 1 \\ 0 & 1 & 1 & 0 \end{bmatrix},$$

$$A_4 = \begin{bmatrix} 0 & 1 & \sin b/b & 0 \\ 1 & 2 & 0 & 0 \\ 1 & 0 & 0 & 0 \\ 0 & 0 & \sin b/b & 0 \end{bmatrix}, \quad B_4 = \begin{bmatrix} 1 \\ 0 \\ 0 \\ 0 \end{bmatrix},$$

$$C_4 = \begin{bmatrix} 0 & 1 & 0 & 1 \\ 0 & 1 & 1 & 0 \end{bmatrix}.$$

Note that it exactly represents the nonlinear system under the condition

$$x_1(t) \in [-a, a], \quad x_3(t) \in [-b, b].$$

In this simulation, we use $a = 0.8$ and $b = 0.6$. The design procedure for Case A is adapted since the premise variables are independent of the variables $x_2(t)$ and $x_4(t)$ to be estimated.

Figure 4.1 shows a simulation result, where the dotted lines denote the state variables estimated by the fuzzy observer. We found F_i and K_i satisfying the LMI conditions by the *Design Procedure for Case A* using a convex optimization technique involving LMIs.

The designed fuzzy controller stabilizes the overall control system. The fuzzy observer estimates the states of the nonlinear system without steady-

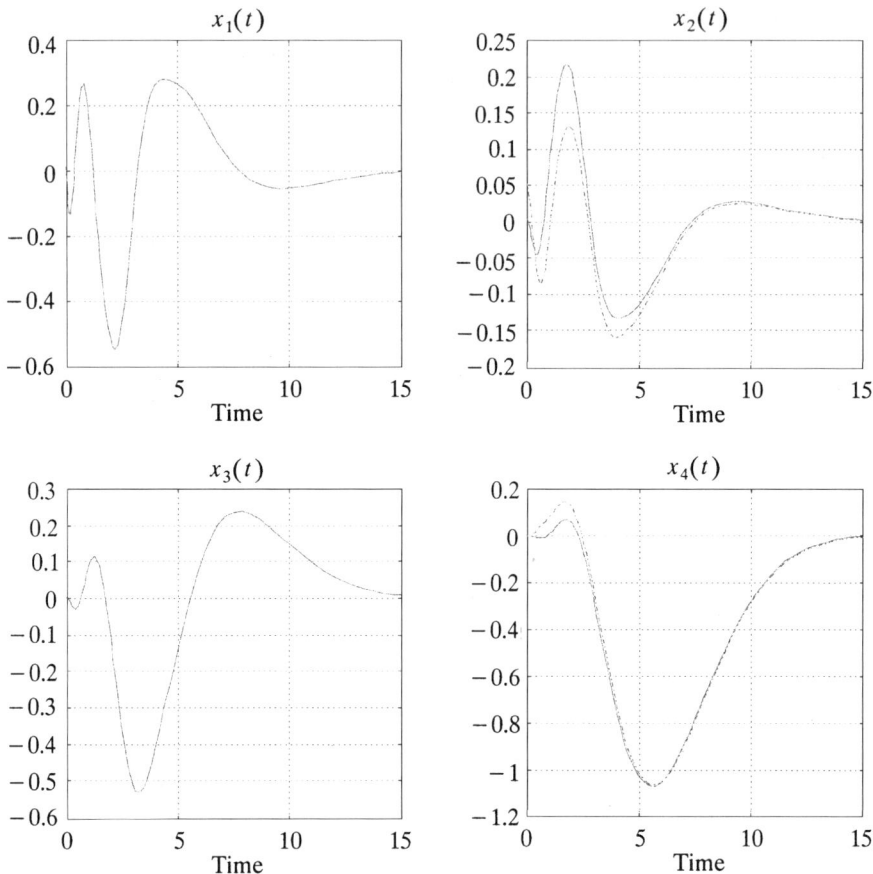

Fig. 4.1 Simulation result.

state errors for the range

$$x_1(t) \in [-0.8, 0.8], \quad x_3(t) \in [-0.6, 0.6].$$

REFERENCES

1. R. E. Kalman, "On the General Theory of Control Systems," in *Proc. IFAC*, Vol. 1, Butterworths, London, 1961, pp. 481–492.
2. K. Tanaka and H. O. Wang, "Fuzzy Regulators and Fuzzy Observers: A Linear Matrix Inequality Approach," 36th IEEE Conference on Decision and Control, Vol. 2, San Diego, 1997, pp. 1315–1320.
3. K. Tanaka, T. Ikeda, and H. O. Wang, "Fuzzy Regulators and Fuzzy Observers," *IEEE Trans. Fuzzy Syst.*, Vol. 6, No. 2, pp. 250–265 (1998).
4. K. Tanaka and M. Sano, "On the Concept of Fuzzy Regulators and Fuzzy Observers," *Proceedings of Third IEEE International Conference on Fuzzy Systems*, Vol. 2, June 1994, pp. 767–772.

CHAPTER 5

ROBUST FUZZY CONTROL

This chapter deals with the issue of robust fuzzy control [1–3]. In general, there exist an infinite number of stabilizing controllers if the plant is stabilizable. The selection of a particular controller among this group of available controllers is often decided by certain specifications of control performance. Fuzzy control designs which guarantee a number of control performance considerations were presented in Chapter 3. The LMI-based techniques ensure not only stabilization but also, for example, good speed of response, avoidance of actuator saturation, and output error constraint. In this and next chapters, a systematic treatment is given for two advanced and important issues of control performance, namely, robustness and optimality, in fuzzy control system designs. The robustness issue is dictated by practical control applications in which there are always uncertainties associated with, for example, the plant, actuators, and sensors in a control system. Robust control addresses these uncertainties and aims to derive the best design possible under the circumstances. This chapter presents such a robust fuzzy control methodology, whereas optimal fuzzy control based on quadratic performance functions will be treated in the next chapter.

This chapter defines a class of Takagi-Sugeno fuzzy systems with uncertainty. Robust stability conditions for this class of systems are derived by applying the relaxed stability conditions described in Chapter 3. This chapter also gives a design method that selects the robust fuzzy controller so as to maximize the norm of the uncertain blocks out of the class of stabilizing PDC controllers. This chapter focuses on robust fuzzy control for CFS. For the design of robust fuzzy control for DFS, refer to [4, 5].

5.1 FUZZY MODEL WITH UNCERTAINTY

To address the robustness of fuzzy control systems, a first and necessary step is to introduce a class of fuzzy systems with uncertainty. For this purpose, we introduce uncertainty blocks to the Takagi-Sugeno fuzzy model to arrive at the following fuzzy model with uncertainty:

Plant Rule i

IF $z_1(t)$ is M_{i1} and $\cdots$ and $z_p(t)$ is M_{ip}

THEN $\dot{x}(t) = (A_i + D_{ai}\Delta_{ai}(t)E_{ai})x(t)$
$$+ (B_i + D_{bi}\Delta_{bi}(t)E_{bi})u(t), \quad i = 1, 2, \ldots, r, \quad (5.1)$$

where the uncertain blocks satisfy

$$\|\Delta_{ai}(t)\| \leq \frac{1}{\gamma_{ai}}, \quad (5.2)$$

$$\Delta_{ai}(t) = \Delta_{ai}^T(t), \quad (5.3)$$

$$\|\Delta_{bi}(t)\| \leq \frac{1}{\gamma_{bi}}, \quad (5.4)$$

$$\Delta_{bi}(t) = \Delta_{bi}^T(t) \quad (5.5)$$

for all i. The fuzzy model is represented as

$$\dot{x}(t) = \sum_{i=1}^{r} h_i(z(t))\{(A_i + D_{ai}\Delta_{ai}(t)E_{ai})x(t)$$
$$+ (B_i + D_{bi}\Delta_{bi}(t)E_{bi})u(t)\}. \quad (5.6)$$

The fuzzy model (5.1) [or (5.6)] contains uncertainty in the consequent parts. The robust stability for the fuzzy model with *premise uncertainty* was first discussed in [6] and [7]. This chapter will focus on the consequent uncertainty.

5.2 ROBUST STABILITY CONDITION

To begin with, this section presents a stability condition for the uncertain fuzzy model (5.1) [i.e., (5.6)]. By substituting the PDC controller (2.23) into

(5.6), we have

$$\dot{x}(t) = \sum_{i=1}^{r}\sum_{j=1}^{r} h_i(z(t))h_j(z(t))$$

$$\times \left\{ A_i - B_i F_j + \begin{bmatrix} D_{ai} & D_{bi} \end{bmatrix} \begin{bmatrix} \Delta_{ai} & 0 \\ 0 & \Delta_{bi} \end{bmatrix} \begin{bmatrix} E_{ai} \\ -E_{bi} F_j \end{bmatrix} \right\} x(t)$$

$$= \sum_{i=1}^{r} h_i^2(z(t)) \left\{ A_i - B_i F_i + \begin{bmatrix} D_{ai} & D_{bi} \end{bmatrix} \begin{bmatrix} \Delta_{ai} & 0 \\ 0 & \Delta_{bi} \end{bmatrix} \begin{bmatrix} E_{ai} \\ -E_{bi} F_i \end{bmatrix} \right\} x(t)$$

$$+ \sum_{i=1}^{r}\sum_{i<j} h_i(z(t)) h_j(z(t))$$

$$\times \left\{ A_i - B_i F_j + A_j - B_j F_i + \begin{bmatrix} D_{ai} & D_{bi} \end{bmatrix} \begin{bmatrix} \Delta_{ai} & 0 \\ 0 & \Delta_{bi} \end{bmatrix} \begin{bmatrix} E_{ai} \\ -E_{bi} F_j \end{bmatrix} \right.$$

$$\left. + \begin{bmatrix} D_{aj} & D_{bj} \end{bmatrix} \begin{bmatrix} \Delta_{aj} & 0 \\ 0 & \Delta_{bj} \end{bmatrix} \begin{bmatrix} E_{aj} \\ -E_{bj} F_i \end{bmatrix} \right\} x(t). \tag{5.7}$$

The following theorem presents robust stability conditions for the fuzzy model (5.1) [i.e., (5.6)] with a given PDC fuzzy controller (2.23). This theorem provides a basis for the robust stabilization problem which is considered in the next section.

THEOREM 22 *The fuzzy system* (5.1) [*i.e.*, (5.6)] *is stabilized via the PDC controller* (2.23) *if there exist a common positive definite matrix* P *and a common positive semidefinite matrix* Q_0 *satisfying*

$$S_{ii} + (s-1)Q_1 < 0, \tag{5.8}$$

$$T_{ij} - 2Q_2 < 0, \quad i < j \text{ s.t. } h_i \cap h_j \neq \phi, \tag{5.9}$$

where $s > 1$,

$$S_{ii} = \begin{bmatrix} (A_i - B_i F_i)^T P + P(A_i - B_i F_i) & P D_{ai} & P D_{bi} & E_{ai}^T & -F_i^T E_{bi}^T \\ D_{ai}^T P & -I & 0 & 0 & 0 \\ D_{bi}^T P & 0 & -I & 0 & 0 \\ E_{ai} & 0 & 0 & -\gamma_{ai}^2 I & 0 \\ -E_{bi} F_i & 0 & 0 & 0 & -\gamma_{bi}^2 I \end{bmatrix},$$

$$T_{ij} = \begin{bmatrix} \begin{pmatrix} (A_i - B_iF_j)^T P \\ +P(A_i - B_iF_j) \\ +(A_j - B_jF_i)^T P \\ +P(A_j - B_jF_i) \end{pmatrix} & PD_{ai} & PD_{bi} & PD_{aj} & PD_{bj} & E_{ai}^T & -F_j^T E_{bi}^T & E_{aj}^T & -F_i^T E_{bj}^T \\ D_{ai}^T P & -I & 0 & 0 & 0 & 0 & 0 & 0 & 0 \\ D_{bi}^T P & 0 & -I & 0 & 0 & 0 & 0 & 0 & 0 \\ D_{aj}^T P & 0 & 0 & -I & 0 & 0 & 0 & 0 & 0 \\ D_{bj}^T P & 0 & 0 & 0 & -I & 0 & 0 & 0 & 0 \\ E_{ai} & 0 & 0 & 0 & 0 & -\gamma_{ai}^2 I & 0 & 0 & 0 \\ -E_{bi}F_j & 0 & 0 & 0 & 0 & 0 & -\gamma_{bi}^2 I & 0 & 0 \\ E_{aj} & 0 & 0 & 0 & 0 & 0 & 0 & -\gamma_{aj}^2 I & 0 \\ -E_{bj}F_i & 0 & 0 & 0 & 0 & 0 & 0 & 0 & -\gamma_{bj}^2 I \end{bmatrix},$$

$$Q_1 = \text{block-diag}\begin{pmatrix} Q_0 & 0 & 0 & 0 & 0 \end{pmatrix},$$

$$Q_2 = \text{block-diag}\begin{pmatrix} Q_0 & 0 & 0 & 0 & 0 & 0 & 0 & 0 & 0 \end{pmatrix}.$$

Proof. Consider the T-S fuzzy control system with uncertainty (5.1), where $\Delta_{ai}(t)$ and $\Delta_{bi}(t)$ are the uncertain blocks satisfying

$$\|\Delta_{ai}(t)\| \leq \frac{1}{\gamma_{ai}}, \qquad \Delta_{ai}(t) = \Delta_{ai}^T(t),$$

$$\|\Delta_{bi}(t)\| \leq \frac{1}{\gamma_{bi}}, \qquad \Delta_{bi}(t) = \Delta_{bi}^T(t).$$

Consider a candidate of Lyapunov functions $x^T(t)Px(t)$. Then,

$$\frac{d}{dt} x^T(t) P x(t)$$

$$= \dot{x}^T(t) P x(t) + x^T(t) P \dot{x}(t)$$

$$= \sum_{i=1}^r h_i^2(z(t)) x^T(t) \left\{ \left(A_i - B_iF_i + \begin{bmatrix} D_{ai} & D_{bi} \end{bmatrix} \begin{bmatrix} \Delta_{ai} & 0 \\ 0 & \Delta_{bi} \end{bmatrix} \begin{bmatrix} E_{ai} \\ -E_{bi}F_i \end{bmatrix} \right)^T P \right.$$

$$\left. + P \left(A_i - B_iF_i + \begin{bmatrix} D_{ai} & D_{bi} \end{bmatrix} \begin{bmatrix} \Delta_{ai} & 0 \\ 0 & \Delta_{bi} \end{bmatrix} \begin{bmatrix} E_{ai} \\ -E_{bi}F_i \end{bmatrix} \right) \right\} x(t)$$

$$+ \sum_{i=1}^{r} \sum_{i<j} h_i(z(t)) h_j(z(t)) x^T(t)$$

$$\times \left\{ \left(A_i - B_i F_j + \begin{bmatrix} D_{ai} & D_{bi} \end{bmatrix} \begin{bmatrix} \Delta_{ai} & 0 \\ 0 & \Delta_{bi} \end{bmatrix} \begin{bmatrix} E_{ai} \\ -E_{bi} F_j \end{bmatrix} \right)^T P \right.$$

$$+ P \left(A_i - B_i F_j + \begin{bmatrix} D_{ai} & D_{bi} \end{bmatrix} \begin{bmatrix} \Delta_{ai} & 0 \\ 0 & \Delta_{bi} \end{bmatrix} \begin{bmatrix} E_{ai} \\ -E_{bi} F_j \end{bmatrix} \right)$$

$$+ \left(A_j - B_j F_i + \begin{bmatrix} D_{aj} & D_{bj} \end{bmatrix} \begin{bmatrix} \Delta_{aj} & 0 \\ 0 & \Delta_{bj} \end{bmatrix} \begin{bmatrix} E_{aj} \\ -E_{bj} F_i \end{bmatrix} \right)^T P$$

$$+ P \left(A_j - B_j F_i + \begin{bmatrix} D_{aj} & D_{bj} \end{bmatrix} \begin{bmatrix} \Delta_{aj} & 0 \\ 0 & \Delta_{bj} \end{bmatrix} \begin{bmatrix} E_{aj} \\ -E_{bj} F_i \end{bmatrix} \right) \right\} x(t)$$

$$= \sum_{i=1}^{r} h_i^2(z(t)) x^T(t)$$

$$\times \left\{ (A_i - B_i F_i)^T P + P(A_i - B_i F_i) + P \begin{bmatrix} D_{ai} & D_{bi} \end{bmatrix} \begin{bmatrix} D_{ai}^T \\ D_{bi}^T \end{bmatrix} P \right.$$

$$+ \begin{bmatrix} E_{ai}^T & -(E_{bi} F_i)^T \end{bmatrix} \begin{bmatrix} \Delta_{ai} & 0 \\ 0 & \Delta_{bi} \end{bmatrix}^T \begin{bmatrix} \Delta_{ai} & 0 \\ 0 & \Delta_{bi} \end{bmatrix} \begin{bmatrix} E_{ai} \\ -E_{bi} F_i \end{bmatrix}$$

$$- \left(\begin{bmatrix} D_{ai}^T \\ D_{bi}^T \end{bmatrix} P - \begin{bmatrix} \Delta_{ai} & 0 \\ 0 & \Delta_{bi} \end{bmatrix} \begin{bmatrix} E_{ai} \\ -E_{bi} F_i \end{bmatrix} \right)^T$$

$$\times \left(\begin{bmatrix} D_{ai}^T \\ D_{bi}^T \end{bmatrix} P - \begin{bmatrix} \Delta_{ai} & 0 \\ 0 & \Delta_{bi} \end{bmatrix} \begin{bmatrix} E_{ai} \\ -E_{bi} F_i \end{bmatrix} \right) \right\} x(t)$$

$$+ \sum_{i=1}^{r} \sum_{i<j} h_i(z(t)) h_j(z(t)) x^T(t)$$

$$\times \left\{ (A_i - B_i F_j)^T P + P(A_i - B_i F_j) + P \begin{bmatrix} D_{ai} & D_{bi} \end{bmatrix} \begin{bmatrix} D_{ai}^T \\ D_{bi}^T \end{bmatrix} P \right.$$

$$+ \begin{bmatrix} E_{ai}^T & -(E_{bi} F_j)^T \end{bmatrix} \begin{bmatrix} \Delta_{ai} & 0 \\ 0 & \Delta_{bi} \end{bmatrix}^T \begin{bmatrix} \Delta_{ai} & 0 \\ 0 & \Delta_{bi} \end{bmatrix} \begin{bmatrix} E_{ai} \\ -E_{bi} F_j \end{bmatrix}$$

$$- \left(\begin{bmatrix} D_{ai}^T \\ D_{bi}^T \end{bmatrix} P - \begin{bmatrix} \Delta_{ai} & 0 \\ 0 & \Delta_{bi} \end{bmatrix} \begin{bmatrix} E_{ai} \\ -E_{bi} F_j \end{bmatrix} \right)^T \left(\begin{bmatrix} D_{ai}^T \\ D_{bi}^T \end{bmatrix} P - \begin{bmatrix} \Delta_{ai} & 0 \\ 0 & \Delta_{bi} \end{bmatrix} \begin{bmatrix} E_{ai} \\ -E_{bi} F_j \end{bmatrix} \right)$$

$$+ (A_j - B_j F_i)^T P + P(A_j - B_j F_i) + P \begin{bmatrix} D_{aj} & D_{bj} \end{bmatrix} \begin{bmatrix} D_{aj}^T \\ D_{bj}^T \end{bmatrix} P$$

$$+ \begin{bmatrix} E_{aj}^T & -(E_{bj} F_i)^T \end{bmatrix} \begin{bmatrix} \Delta_{aj} & 0 \\ 0 & \Delta_{bj} \end{bmatrix}^T \begin{bmatrix} \Delta_{aj} & 0 \\ 0 & \Delta_{bj} \end{bmatrix} \begin{bmatrix} E_{aj} \\ -E_{bj} F_i \end{bmatrix}$$

$$- \left(\begin{bmatrix} D_{aj}^T \\ D_{bj}^T \end{bmatrix} P - \begin{bmatrix} \Delta_{aj} & 0 \\ 0 & \Delta_{bj} \end{bmatrix} \begin{bmatrix} E_{aj} \\ -E_{bj} F_i \end{bmatrix} \right)^T$$

$$\left. \times \left(\begin{bmatrix} D_{aj}^T \\ D_{bj}^T \end{bmatrix} P - \begin{bmatrix} \Delta_{aj} & 0 \\ 0 & \Delta_{bj} \end{bmatrix} \begin{bmatrix} E_{aj} \\ -E_{bj} F_i \end{bmatrix} \right) \right\} x(t). \quad (5.10)$$

If

$$(A_i - B_i F_j)^T P + P(A_i - B_i F_j) + P \begin{bmatrix} D_{ai} & D_{bi} \end{bmatrix} \begin{bmatrix} D_{ai}^T \\ D_{bi}^T \end{bmatrix} P$$

$$+ \begin{bmatrix} E_{ai}^T & -(E_{bi} F_j)^T \end{bmatrix} \begin{bmatrix} \frac{1}{\gamma_{ai}^2} I & 0 \\ 0 & \frac{1}{\gamma_{bi}^2} I \end{bmatrix} \begin{bmatrix} E_{ai} \\ -E_{bi} F_j \end{bmatrix}$$

$$+ (A_j - B_j F_i)^T P + P(A_j - B_j F_i) + P \begin{bmatrix} D_{aj} & D_{bj} \end{bmatrix} \begin{bmatrix} D_{aj}^T \\ D_{bj}^T \end{bmatrix} P$$

$$+ \begin{bmatrix} E_{aj}^T & -(E_{bj} F_i)^T \end{bmatrix} \begin{bmatrix} \frac{1}{\gamma_{aj}^2} I & 0 \\ 0 & \frac{1}{\gamma_{bj}^2} I \end{bmatrix} \begin{bmatrix} E_{aj} \\ -E_{bj} F_i \end{bmatrix} - 2 Q_0 < 0, \quad (5.11)$$

then

$$\frac{d}{dt} x^T(t) P x(t) < \sum_{i=1}^{r} h_i^2(z(t)) x^T(t)$$

$$\times \left\{ (A_i - B_i F_i)^T P + P(A_i - B_i F_i) + P \begin{bmatrix} D_{ai} & D_{bi} \end{bmatrix} \begin{bmatrix} D_{ai}^T \\ D_{bi}^T \end{bmatrix} P \right.$$

$$+ \begin{bmatrix} E_{ai}^T & -(E_{bi} F_i)^T \end{bmatrix} \begin{bmatrix} \Delta_{ai} & 0 \\ 0 & \Delta_{bi} \end{bmatrix}^T \begin{bmatrix} \Delta_{ai} & 0 \\ 0 & \Delta_{bi} \end{bmatrix} \begin{bmatrix} E_{ai} \\ -E_{bi} F_i \end{bmatrix}$$

$$- \left(\begin{bmatrix} D_{ai}^T \\ D_{bi}^T \end{bmatrix} P - \begin{bmatrix} \Delta_{ai} & 0 \\ 0 & \Delta_{bi} \end{bmatrix} \begin{bmatrix} E_{ai} \\ -E_{bi} F_i \end{bmatrix} \right)^T$$

$$\left. \times \left(\begin{bmatrix} D_{ai}^T \\ D_{bi}^T \end{bmatrix} P - \begin{bmatrix} \Delta_{ai} & 0 \\ 0 & \Delta_{bi} \end{bmatrix} \begin{bmatrix} E_{ai} \\ -E_{bi} F_i \end{bmatrix} \right) \right\} x(t)$$

$$+ 2 \sum_{i=1}^{r} \sum_{i<j} h_i(z(t)) h_j(z(t)) x^T(t) Q_0 x^T(t)$$

$$\leq \sum_{i=1}^{r} h_i^2(z(t)) x^T(t)$$

$$\times \left\{ (A_i - B_i F_i)^T P + P(A_i - B_i F_i) + P \begin{bmatrix} D_{ai} & D_{bi} \end{bmatrix} \begin{bmatrix} D_{ai}^T \\ D_{bi}^T \end{bmatrix} P \right.$$

$$+ \begin{bmatrix} E_{ai}^T & -(E_{bi} F_i)^T \end{bmatrix} \begin{bmatrix} \Delta_{ai} & 0 \\ 0 & \Delta_{bi} \end{bmatrix}^T \begin{bmatrix} \Delta_{ai} & 0 \\ 0 & \Delta_{bi} \end{bmatrix} \begin{bmatrix} E_{ai} \\ -E_{bi} F_i \end{bmatrix}$$

$$- \left(\begin{bmatrix} D_{ai}^T \\ D_{bi}^T \end{bmatrix} P - \begin{bmatrix} \Delta_{ai} & 0 \\ 0 & \Delta_{bi} \end{bmatrix} \begin{bmatrix} E_{ai} \\ -E_{bi} F_i \end{bmatrix} \right)^T$$

$$\left. \times \left(\begin{bmatrix} D_{ai}^T \\ D_{bi}^T \end{bmatrix} P - \begin{bmatrix} \Delta_{ai} & 0 \\ 0 & \Delta_{bi} \end{bmatrix} \begin{bmatrix} E_{ai} \\ -E_{bi} F_i \end{bmatrix} \right) \right\} x(t)$$

$$+ (s - 1) \sum_{i=1}^{r} h_i^2(z(t)) x^T(t) Q_0 x(t)$$

$$= \sum_{i=1}^{r} h_i^2(z(t)) x^T(t)$$

$$\times \left\{ (A_i - B_i F_i)^T P + P(A_i - B_i F_i) + (s-1)Q_0 \right.$$

$$+ P \begin{bmatrix} D_{ai} & D_{bi} \end{bmatrix} \begin{bmatrix} D_{ai}^T \\ D_{bi}^T \end{bmatrix} P$$

$$+ \begin{bmatrix} E_{ai}^T & -(E_{bi}F_i)^T \end{bmatrix} \begin{bmatrix} \Delta_{ai} & 0 \\ 0 & \Delta_{bi} \end{bmatrix}^T \begin{bmatrix} \Delta_{ai} & 0 \\ 0 & \Delta_{bi} \end{bmatrix} \begin{bmatrix} E_{ai} \\ -E_{bi}F_i \end{bmatrix}$$

$$- \left(\begin{bmatrix} D_{ai}^T \\ D_{bi}^T \end{bmatrix} P - \begin{bmatrix} \Delta_{ai} & 0 \\ 0 & \Delta_{bi} \end{bmatrix} \begin{bmatrix} E_{ai} \\ -E_{bi}F_i \end{bmatrix} \right)^T$$

$$\times \left(\begin{bmatrix} D_{ai}^T \\ D_{bi}^T \end{bmatrix} P - \begin{bmatrix} \Delta_{ai} & 0 \\ 0 & \Delta_{bi} \end{bmatrix} \begin{bmatrix} E_{ai} \\ -E_{bi}F_i \end{bmatrix} \right) \right\} x(t).$$

If

$$(A_i - B_i F_i)^T P + P(A_i - B_i F_i) + (s-1)Q_0$$

$$+ P \begin{bmatrix} D_{ai} & D_{bi} \end{bmatrix} \begin{bmatrix} D_{ai}^T \\ D_{bi}^T \end{bmatrix} P$$

$$+ \begin{bmatrix} E_{ai}^T & -(E_{bi}F_i)^T \end{bmatrix} \begin{bmatrix} \frac{1}{\gamma_{ai}^2} I & 0 \\ 0 & \frac{1}{\gamma_{bi}^2} I \end{bmatrix} \begin{bmatrix} E_{ai} \\ -E_{bi}F_i \end{bmatrix} < 0, \qquad (5.12)$$

then

$$\frac{d}{dt} x^T(t) P x(t) < 0$$

at $x(t) \neq 0$. Since

$$\Delta_{ai}^T(t)\Delta_{ai}(t) \leq \frac{1}{\gamma_{ai}^2}I, \quad \Delta_{bi}^T(t)\Delta_{bi}(t) \leq \frac{1}{\gamma_{bi}^2}I,$$

$$-\left(\begin{bmatrix} D_{ai}^T \\ D_{bi}^T \end{bmatrix} P - \begin{bmatrix} \Delta_{ai} & 0 \\ 0 & \Delta_{bi} \end{bmatrix} \begin{bmatrix} E_{ai} \\ -E_{bi}F_i \end{bmatrix}\right)^T$$

$$\times \left(\begin{bmatrix} D_{ai}^T \\ D_{bi}^T \end{bmatrix} P - \begin{bmatrix} \Delta_{ai} & 0 \\ 0 & \Delta_{bi} \end{bmatrix} \begin{bmatrix} E_{ai} \\ -E_{bi}F_i \end{bmatrix}\right) \leq 0.$$

By the Schur complement, (5.12) and (5.11) are rewritten as (5.8) and (5.9), respectively. Q.E.D.

When $Q_1 = 0$ and $Q_2 = 0$, that is, $Q_0 = 0$, the relaxed robust stability conditions are reduced to just the robust conditions:

$$P > 0, \quad S_{ii} < 0, \quad T_{ij} < 0, \quad i < j \text{ s.t. } h_i \cap h_j \neq \phi.$$

As a result, by utilizing the relaxed stability conditions, less conservative results can be obtained in the robust stability analysis.

5.3 ROBUST STABILIZATION

We define a robust stabilization problem so as to select a PDC fuzzy controller, in the class of PDC controllers (2.23) satisfying the robust stability conditions (5.8) and (5.9), to maximize the norm of the uncertainty blocks, or equivalently, to minimize γ_{ai} and γ_{bi} in (5.7). The following theorem provides a solution to the robust stabilization problem.

THEOREM 23 *The feedback gains F_i that stabilize the fuzzy model (5.1) and maximize the norms of the uncertain blocks (i.e., minimize γ_{ai} and γ_{bi}) can be obtained by solving the following LMIs, where α_i, $\beta_i > 0$ are design parameters:*

$$\underset{\gamma_{ai}^2, \gamma_{bi}^2, X, M_1, \ldots, M_r, Y_0}{\text{minimize}} \sum_{i=1}^{r} \{\alpha_i \gamma_{ai}^2 + \beta_i \gamma_{bi}^2\}$$

subject to

$$X > 0, \quad Y_0 \geq 0, \quad \hat{S}_{ii} + (s-1)Y_1 < 0, \quad (5.13)$$

$$\hat{T}_{ij} - 2Y_2 < 0, \quad i < j \text{ s.t. } h_i \cap h_j \neq \phi, \quad (5.14)$$

where $s > 1$,

$$\hat{S}_{ii} = \begin{bmatrix} \begin{pmatrix} XA_i^T + A_i X \\ -B_i M_i - M_i^T B_i^T \end{pmatrix} & * & * & * & * \\ D_{ai}^T & -I & 0 & 0 & 0 \\ D_{bi}^T & 0 & -I & 0 & 0 \\ E_{ai} X & 0 & 0 & -\gamma_{ai}^2 I & 0 \\ -E_{bi} M_i & 0 & 0 & 0 & -\gamma_{bi}^2 I \end{bmatrix},$$

$$\hat{T}_{ij} = \begin{bmatrix} \begin{pmatrix} XA_i^T + A_i X \\ -B_i M_j - M_j^T B_i^T \\ +XA_j^T + A_j X \\ -B_j M_i - M_i^T B_j^T \end{pmatrix} & D_{ai} & D_{bi} & D_{aj} & D_{bj} & XE_{ai}^T & -M_j^T E_{bi}^T & XE_{aj}^T & -M_i^T E_{bj}^T \\ D_{ai}^T & -I & 0 & 0 & 0 & 0 & 0 & 0 & 0 \\ D_{bi}^T & 0 & -I & 0 & 0 & 0 & 0 & 0 & 0 \\ D_{aj}^T & 0 & 0 & -I & 0 & 0 & 0 & 0 & 0 \\ D_{bj}^T & 0 & 0 & 0 & -I & 0 & 0 & 0 & 0 \\ E_{ai} X & 0 & 0 & 0 & 0 & -\gamma_{ai}^2 I & 0 & 0 & 0 \\ -E_{bi} M_j & 0 & 0 & 0 & 0 & 0 & -\gamma_{bi}^2 I & 0 & 0 \\ E_{aj} X & 0 & 0 & 0 & 0 & 0 & 0 & -\gamma_{ai}^2 I & 0 \\ -E_{bj} M_i & 0 & 0 & 0 & 0 & 0 & 0 & 0 & -\gamma_{bi}^2 I \end{bmatrix},$$

$$Y_1 = \text{block-diag}\begin{pmatrix} Y_0 & 0 & 0 & 0 & 0 \end{pmatrix},$$

$$Y_2 = \text{block-diag}\begin{pmatrix} Y_0 & 0 & 0 & 0 & 0 & 0 & 0 & 0 & 0 \end{pmatrix},$$

where

$$Y_0 = XQ_0 X$$

and the asterisk denotes the transposed elements (matrices) for symmetric positions.

Proof. The main idea is to transform the conditions of Theorem 22 into

LMIs:

$$\{\text{block-diag}\begin{bmatrix} X & I & I & I & I \end{bmatrix}\}\{S_{ii} + (s-1)Q_1\}$$
$$\times \{\text{block-diag}\begin{bmatrix} X & I & I & I & I \end{bmatrix}\}$$
$$= \begin{bmatrix} \begin{pmatrix} XA_i^T + A_i X \\ -B_i M_i - M_i^T B_i^T \end{pmatrix} & * & * & * & * \\ D_{ai}^T & -I & 0 & 0 & 0 \\ D_{bi}^T & 0 & -I & 0 & 0 \\ E_{ai} X & 0 & 0 & -\gamma_{ai}^2 I & 0 \\ -E_{bi} M_i & 0 & 0 & 0 & -\gamma_{bi}^2 I \end{bmatrix}$$
$$+ (s-1) \cdot \text{block-diag}\begin{pmatrix} XQ_0 X & 0 & 0 & 0 & 0 \end{pmatrix}$$
$$= \hat{S}_{ii} + (s-1)Y_1, \tag{5.15}$$

$$\{\text{block-diag}\begin{bmatrix} X & I & I & I & I & I & I & I & I \end{bmatrix}\} \cdot \{T_{ij} - 2Q_2\}$$
$$\times \{\text{block-diag}\begin{bmatrix} X & I & I & I & I & I & I & I & I \end{bmatrix}\}$$
$$= \begin{bmatrix} \begin{pmatrix} XA_i^T + A_i X \\ -B_i M_j - M_j^T B_i^T \\ +XA_j^T + A_j X \\ -B_j M_i - M_i^T B_j^T \end{pmatrix} & D_{ai} & D_{bi} & D_{aj} & D_{bj} & XE_{ai}^T & -M_j^T E_{bi}^T & XE_{aj}^T & -M_i^T E_{bj}^T \\ D_{ai}^T & -I & 0 & 0 & 0 & 0 & 0 & 0 & 0 \\ D_{bi}^T & 0 & -I & 0 & 0 & 0 & 0 & 0 & 0 \\ D_{aj}^T & 0 & 0 & -I & 0 & 0 & 0 & 0 & 0 \\ D_{bj}^T & 0 & 0 & 0 & -I & 0 & 0 & 0 & 0 \\ E_{ai} X & 0 & 0 & 0 & 0 & -\gamma_{ai}^2 I & 0 & 0 & 0 \\ -E_{bi} M_j & 0 & 0 & 0 & 0 & 0 & -\gamma_{bi}^2 I & 0 & 0 \\ E_{aj} X & 0 & 0 & 0 & 0 & 0 & 0 & -\gamma_{aj}^2 I & 0 \\ -E_{bj} M_i & 0 & 0 & 0 & 0 & 0 & 0 & 0 & -\gamma_{bj}^2 I \end{bmatrix}$$
$$- 2 \cdot \text{block-diag}\begin{pmatrix} XQ_0 X & 0 & 0 & 0 & 0 & 0 & 0 & 0 & 0 \end{pmatrix}$$
$$= \hat{T}_{ij} - 2Y_2, \tag{5.16}$$

where

$$X = P^{-1}, \quad M_i = F_i P^{-1}$$

for all i. Q.E.D.

The feedback gains can be obtained as

$$F_i = M_i X^{-1}$$

from the solutions X and M_i of the above LMIs.

A design example for robust fuzzy control will be presented in Chapter 7. H_∞ control for the fuzzy model (5.1) was first discussed in [8]. Since then, a number of papers considering H_∞ control for fuzzy control systems have appeared in the literature. Chapters 13 and 15 give an extensive treatment of H_∞ control for fuzzy control systems.

REFERENCES

1. K. Tanaka, T. Taniguchi, and H. O. Wang, "Robust and Optimal Fuzzy Control: A Linear Matrix Inequality Approach," 1999 International Federation of Automatic Control (IFAC) World Congress, Beijing, July 1999, pp. 213–218.
2. K. Tanaka, M. Nishimura, and H. O. Wang, "Multi-Objective Fuzzy Control of High Rise/High Speed Elevators Using LMIs," 1998 American Control Conference, 1998, pp. 3450–3454
3. K. Tanaka, T. Taniguchi, and H. Wang, "Model-Based Fuzzy Control of TORA System: Fuzzy Regulator and Fuzzy Observer Design via LMIs that Represent Decay Rate, Disturbance Rejection, Robustness, Optimality," Seventh IEEE International Conference on Fuzzy Systems, Alaska, 1998, pp.313–318.
4. K. Tanaka , T. Hori, K. Yamafuji, and H. O. Wang, "An Integrated Algorithm of Fuzzy Modeling and Controller Design for Nonlinear Systems," 1999 IEEE International Conference on Fuzzy Systems, Vol. 2, Seoul, August 1999, pp. 887–892.
5. K. Tanaka , T. Hori, K. Yamafuji, and H. O. Wang, "An Integrated Fuzzy Control System Design for Nonlinear Systems," 38th IEEE Conference on Decision and Control, Phoenix, Dec. 1999, pp. 4349–4354.
6. K. Tanaka and M. Sugeno, "Concept of Stability Margin of Fuzzy Systems and Design of Robust Fuzzy Controllers," in *Proceedings of 2nd IEEE International Conference on Fuzzy System*, Vol. 1, 1993, pp. 29–34.
7. K. Tanaka and M. Sano, "A Robust Stabilization Problem of Fuzzy Controller Systems and Its Applications to Backing up Control of a Truck-Trailer," *IEEE Trans. on Fuzzy Syst*. Vol. 2, No. 2, pp. 119–134, (1994).
8. K. Tanaka, T. Ikeda, and H. O. Wang, "Robust Stabilization of a Class of Uncertain Nonlinear System via Fuzzy Control: Quadratic Stabilizability, H^∞ control theory and linear matrix inequalities," *IEEE Trans. Fuzzy Syst*., Vol. 4, No. 1, pp. 1–13 (1996).

CHAPTER 6

OPTIMAL FUZZY CONTROL

In control design, it is often of interest to synthesize a controller to satisfy, in an optimal fashion, certain performance criteria and constraints in addition to stability. The subject of optimal control addresses this aspect of control system design. For linear systems, the problem of designing optimal controllers reduces to solving algebraic Riccati equations (AREs), which are usually easy to solve and detailed discussion of their solutions can be found in many textbooks [1]. However, for a general nonlinear system, the optimization problem reduces to the so-called Hamilton-Jacobi (HJ) equations, which are nonlinear partial differential equations (PDEs) [2]. Different from their counterparts for linear systems, HJ equations are usually hard to solve both numerically and analytically. Results have been given on the relationship between solution of the HJ equation and the invariant manifold for the Hamiltonian vector field. Progress has also been made on the numerical computation of the approximated solution of HJ equations [3]. But few results so far can provide an effective way of designing optimal controllers for general nonlinear systems.

In this chapter, we propose an alternative approach to nonlinear optimal control based on fuzzy logic. The optimal fuzzy control methodology presented in this chapter is based on a quadratic performance function [4-7] utilizing the relaxed stability conditions. The optimal fuzzy controller is designed by solving a minimization problem that minimizes the upper bound of a given quadratic performance function. In a strict sense, this approach is a suboptimal design. One of the advantages of this methodology is that the design conditions are represented in terms of LMIs. Refer to [8] for a more thorough treatment of optimal fuzzy control.

6.1 QUADRATIC PERFORMANCE FUNCTION AND STABILIZING CONTROL

The control objective of optimal fuzzy control is to minimize certain performance functions. In this chapter, we present a fuzzy controller design to minimize the upper bound of the following quadratic performance function (6.1):

$$J = \int_0^\infty \{y^T(t)Wy(t) + u^T(t)Ru(t)\}\,dt, \qquad (6.1)$$

where

$$y(t) = \sum_{i=1}^r h_i(z(t))C_i x(t).$$

The following theorem presents a basis to the optimal fuzzy control problem. The set of conditions given herein, however, are not in terms of LMIs. The LMI-based optimal fuzzy control design will be addressed in the next section.

THEOREM 24 *The fuzzy system (2.3) and (2.4) can be stabilized by the PDC fuzzy controller (2.23) if there exist a common positive definite matrix P and a common positive semidefinite matrix Q_0 satisfying*

$$U_{ii} + (s-1)Q_3 < 0 \qquad (6.2)$$

$$V_{ij} - 2Q_4 < 0, \quad i < j \text{ s.t. } h_i \cap h_j \neq \phi, \qquad (6.3)$$

where $s > 1$,

$$U_{ii} = \begin{bmatrix} (A_i - B_i F_i)^T P \\ +P(A_i - B_i F_i) & C_i^T & -F_i^T \\ C_i & -W^{-1} & 0 \\ -F_i & 0 & -R^{-1} \end{bmatrix}, \qquad (6.4)$$

$$V_{ij} = \begin{bmatrix} (A_i - B_i F_j)^T P \\ +P(A_i - B_i F_j) \\ +(A_j - B_j F_i)^T P & C_i^T & -F_j^T & C_j^T & -F_i^T \\ +P(A_j - B_j F_i) \\ C_i & -W^{-1} & 0 & 0 & 0 \\ -F_j & 0 & -R^{-1} & 0 & 0 \\ C_j & 0 & 0 & -W^{-1} & 0 \\ -F_i & 0 & 0 & 0 & -R^{-1} \end{bmatrix}, \qquad (6.5)$$

QUADRATIC PERFORMANCE FUNCTION AND STABILIZING CONTROL 111

$$Q_3 = \text{block-diag}\begin{pmatrix} Q_0 & 0 & 0 \end{pmatrix},$$

$$Q_4 = \text{block-diag}\begin{pmatrix} Q_0 & 0 & 0 & 0 & 0 \end{pmatrix}.$$

Then, the performance function satisfies

$$J < x^T(0)\, P x(0),$$

where $x^T(0) P x(0)$ acts as an upper bound of J.

Proof. Let us define the following new variable

$$\hat{y}(t) = \begin{bmatrix} y(t) \\ u(t) \end{bmatrix} = \sum_{i=1}^{r} h_i(z(t)) \begin{bmatrix} C_i \\ -F_i \end{bmatrix} x(t).$$

Equation (6.1) can be rewritten as

$$J = \int_0^\infty \hat{y}^T(t) \begin{bmatrix} W & 0 \\ 0 & R \end{bmatrix} \hat{y}(t)\, dt.$$

Assume that there exists a common positive definite matrix P and a common positive semidefinite matrix Q_0 satisfying (6.2) and (6.3). Then, from Schur complements, we have

$$(A_i - B_i F_i)^T P + P(A_i - B_i F_i) + (s-1)Q_0$$
$$+ \begin{bmatrix} C_i^T & -F_i^T \end{bmatrix} \begin{bmatrix} W & 0 \\ 0 & R \end{bmatrix} \begin{bmatrix} C_i \\ -F_i \end{bmatrix} < 0 \qquad (6.6)$$

and

$$(A_i - B_i F_j)^T P + P(A_i - B_i F_j)$$
$$+ (A_j - B_j F_i)^T P + P(A_j - B_j F_i) - 2Q_0$$
$$+ \begin{bmatrix} C_i^T & -F_j^T \end{bmatrix} \begin{bmatrix} W & 0 \\ 0 & R \end{bmatrix} \begin{bmatrix} C_i \\ -F_j \end{bmatrix}$$
$$+ \begin{bmatrix} C_j^T & -F_i^T \end{bmatrix} \begin{bmatrix} W & 0 \\ 0 & R \end{bmatrix} \begin{bmatrix} C_j \\ -F_i \end{bmatrix} < 0. \qquad (6.7)$$

From (6.6) and (6.7), we obtain

$$(A_i - B_i F_i)^T P + P(A_i - B_i F_i) + (s-1)Q_0 < 0 \qquad (6.8)$$

112 OPTIMAL FUZZY CONTROL

and

$$(A_i - B_i F_j)^T P + P(A_i - B_i F_j)$$
$$+ (A_j - B_j F_i)^T P + P(A_j - B_j F_i) - 2Q_0 < 0. \quad (6.9)$$

It is clear from Theorem 9 in Chapter 3 that the fuzzy control system is globally asymptotically stable if (6.2) and (6.3) hold.

Next, it will be proved that the quadratic performance function satisfies $J < x^T(0) P x(0)$. Consider a Lyapunov function candidate $x^T(t) P x(t)$. Then, from (6.6), (6.7), and the Appendix,

$$\frac{d}{dt} x^T(t) P x(t)$$

$$= \dot{x}^T(t) P x(t) + x^T(t) P \dot{x}(t)$$

$$= \sum_{i=1}^{r} \sum_{j=1}^{r} h_i(z(t)) h_j(z(t)) x^T(t) \{(A_i - B_i F_j)^T P + P(A_i - B_i F_j)\} x(t)$$

$$= \sum_{i=1}^{r} h_i^2(z(t)) x^T(t) \{(A_i - B_i F_i)^T P + P(A_i - B_i F_i)\} x(t)$$

$$+ \sum_{i=1}^{r} \sum_{i \neq j} h_i(z(t)) h_j(z(t)) x^T(t) \{(A_i - B_i F_j)^T P + P(A_i - B_i F_j)\} x(t)$$

$$< \sum_{i=1}^{r} h_i^2(z(t)) x^T(t) \{(A_i - B_i F_i)^T P + P(A_i - B_i F_i)\} x(t)$$

$$- x^T(t) \Bigg\{ \sum_{i=1}^{r} \sum_{i<j} h_i(z(t)) h_j(z(t)) [C_i^T \ -F_j^T] \begin{bmatrix} W & 0 \\ 0 & R \end{bmatrix} \begin{bmatrix} C_i \\ -F_j \end{bmatrix}$$

$$+ \sum_{i=1}^{r} \sum_{i<j} h_i(z(t)) h_j(z(t)) [C_j^T \ -F_i^T] \begin{bmatrix} W & 0 \\ 0 & R \end{bmatrix} \begin{bmatrix} C_j \\ -F_i \end{bmatrix} \Bigg\} x(t)$$

$$+ 2 \sum_{i=1}^{r} \sum_{i<j} h_i(z(t)) h_j(z(t)) x^T(t) Q_0 x(t)$$

$$< -x^T(t) \Bigg\{ \sum_{i=1}^{r} h_i^2(z(t)) [C_i^T \ -F_i^T] \begin{bmatrix} W & 0 \\ 0 & R \end{bmatrix} \begin{bmatrix} C_i \\ -F_i \end{bmatrix} \Bigg\} x(t)$$

$$- x^T(t) \Bigg\{ \sum_{i=1}^{r} \sum_{i<j} h_i(z(t)) h_j(z(t)) [C_i^T \ -F_j^T] \begin{bmatrix} W & 0 \\ 0 & R \end{bmatrix} \begin{bmatrix} C_i \\ -F_j \end{bmatrix}$$

$$+ \sum_{i=1}^{r} \sum_{i<j} h_i(z(t)) h_j(z(t)) \begin{bmatrix} C_j^T & -F_i^T \end{bmatrix} \begin{bmatrix} W & 0 \\ 0 & R \end{bmatrix} \begin{bmatrix} C_i \\ -F_i \end{bmatrix} \Big\} x(t)$$

$$- (s-1) \sum_{i=1}^{r} h_i^2(z(t)) x^T(t) Q_0 x(t)$$

$$+ 2 \sum_{i=1}^{r} \sum_{i<j} h_i(z(t)) h_j(z(t)) x^T(t) Q_0 x(t)$$

$$\leq -x^T(t) \Big\{ \sum_{i=1}^{r} h_i^2(z(t)) \begin{bmatrix} C_i^T & -F_i^T \end{bmatrix} \begin{bmatrix} W & 0 \\ 0 & R \end{bmatrix} \begin{bmatrix} C_i \\ -F_i \end{bmatrix} \Big\} x(t)$$

$$- x^T(t) \Big\{ \sum_{i=1}^{r} \sum_{i<j} h_i(z(t)) h_j(z(t)) \begin{bmatrix} C_i^T & -F_j^T \end{bmatrix} \begin{bmatrix} W & 0 \\ 0 & R \end{bmatrix} \begin{bmatrix} C_j \\ -F_i \end{bmatrix}$$

$$+ \sum_{i=1}^{r} \sum_{i<j} h_i(z(t)) h_j(z(t)) \begin{bmatrix} C_j^T & -F_i^T \end{bmatrix} \begin{bmatrix} W & 0 \\ 0 & R \end{bmatrix} \begin{bmatrix} C_i \\ -F_j \end{bmatrix} \Big\} x(t)$$

$$- (s-1) \sum_{i=1}^{r} h_i^2(z(t)) x^T(t) Q_0 x(t)$$

$$+ 2 \sum_{i=1}^{r} \sum_{i<j} h_i(z(t)) h_j(z(t)) x^T(t) Q_0 x(t)$$

$$= -x^T(t) \Big\{ \sum_{i=1}^{r} \sum_{j=1}^{r} h_i(z(t)) h_j(z(t)) \begin{bmatrix} C_i^T & -F_i^T \end{bmatrix} \begin{bmatrix} W & 0 \\ 0 & R \end{bmatrix} \begin{bmatrix} C_j \\ -F_j \end{bmatrix} \Big\} x(t)$$

$$- \Big((s-1) \sum_{i=1}^{r} h_i^2(z(t)) - 2 \sum_{i=1}^{r} \sum_{i<j} h_i(z(t)) h_j(z(t)) \Big) x^T(t) Q_0 x(t)$$

$$= -x^T(t) \Big\{ \Big(\sum_{i=1}^{r} h_i(z(t)) \begin{bmatrix} C_i^T & -F_i^T \end{bmatrix} \Big) \begin{bmatrix} W & 0 \\ 0 & R \end{bmatrix} \Big(\sum_{i=1}^{r} h_i(z(t)) \begin{bmatrix} C_i \\ -F_i \end{bmatrix} \Big) \Big\} x(t)$$

$$- \Big((s-1) \sum_{i=1}^{r} h_i^2(z(t)) - 2 \sum_{i=1}^{r} \sum_{i<j} h_i(z(t)) h_j(z(t)) \Big) x^T(t) Q_0 x(t)$$

$$= -y^T(t) \begin{bmatrix} W & 0 \\ 0 & R \end{bmatrix} y(t)$$

$$- \Big((s-1) \sum_{i=1}^{r} h_i^2(z(t)) - 2 \sum_{i=1}^{r} \sum_{i<j} h_i(z(t)) h_j(z(t)) \Big) x^T(t) Q_0 x(t)$$

$$\leq -y^T(t) \begin{bmatrix} W & 0 \\ 0 & R \end{bmatrix} y(t).$$

Therefore,

$$\frac{d}{dt}x^T(t)Px(t) < -\hat{y}^T(t)\begin{bmatrix} W & 0 \\ 0 & R \end{bmatrix}\hat{y}(t).$$

Integrating both side from 0 to ∞, we get

$$J = \int_0^\infty y^T(t)\begin{bmatrix} W & 0 \\ 0 & R \end{bmatrix}\hat{y}(t)\,dt < -x^T(t)Px(t)\big|_0^\infty.$$

Since the fuzzy control system is stable,

$$J = \int_0^\infty y^T(t)\begin{bmatrix} W & 0 \\ 0 & R \end{bmatrix}\hat{y}(t)\,dt < x^T(0)Px(0). \tag{6.10}$$

Q.E.D.

Remark 18 The above design procedure guarantees $J < x^T(0)Px(0)$ for all the values of $h_i(z(t)) \in [0, 1]$.

When $Q_3 = 0$ and $Q_4 = 0$, that is, $Q_0 = 0$, the relaxed conditions in Theorem 24 are reduced to the following conditions:

$$P > 0, \quad U_{ii} < 0, \quad V_{ij} < 0, \quad i < j \text{ s.t. } h_i \cap h_j \neq \phi.$$

Then, the performance function J' satisfies $J' < x^T(0)Px(0)$.

6.2 OPTIMAL FUZZY CONTROLLER DESIGN

We present a design problem to minimize the upper bound of the performance function based on the results derived in Theorem 24. As shown in the previous section, $x^T(0)Px(0)$ gives an upper bound of J under the conditions of Theorem 24. The optimal fuzzy controller to be introduced is in the strict sense a "sub-optimal" controller since $x^T(0)Px(0)$ will be minimized instead of J in the control design procedure. The following theorem summarizes the design conditions for such scheme.

THEOREM 25 *The feedback gains to minimize the upper bound of the performance function can be obtained by solving the following LMIs. From the solution of the LMIs, the feedback gains are obtained as*

$$F_i = M_i X^{-1}$$

for all i. Then, the performance function satisfies $J < x^T(0)Px(0) < \lambda$.

$$\underset{X, M_1, \ldots, M_r, Y_0}{\text{minimize}} \quad \lambda$$

subject to

$$X > 0, \quad Y_0 \geq 0,$$

$$\begin{bmatrix} \lambda & x^T(0) \\ x(0) & X \end{bmatrix} > 0, \tag{6.11}$$

$$\hat{U}_{ii} + (s-1)Y_3 < 0, \tag{6.12}$$

$$\hat{V}_{ij} - 2Y_4 < 0, \quad i < j \text{ s.t. } h_i \cap h_j \neq \phi, \tag{6.13}$$

where $s > 1$,

$$\hat{U}_{ii} = \begin{bmatrix} \begin{pmatrix} XA_i^T + A_i X \\ -B_i M_i - M_i^T B_i^T \end{pmatrix} & XC_i^T & -M_i^T \\ C_i X & -W^{-1} & 0 \\ -M_i & 0 & -R^{-1} \end{bmatrix},$$

$$\hat{V}_{ij} = \begin{bmatrix} \begin{pmatrix} XA_i^T + A_i X \\ -B_i M_j - M_j^T B_i^T \\ +XA_j^T + A_j X \\ -B_j M_i - M_i^T B_j^T \end{pmatrix} & XC_i^T & -M_j^T & XC_j^T & -M_i^T \\ C_i X & -W^{-1} & 0 & 0 & 0 \\ -M_j & 0 & -R^{-1} & 0 & 0 \\ C_j X & 0 & 0 & -W^{-1} & 0 \\ -M_i & 0 & 0 & 0 & -R^{-1} \end{bmatrix},$$

$$Y_3 = \text{block-diag}\begin{pmatrix} Y_0 & 0 & 0 \end{pmatrix},$$

$$Y_4 = \text{block-diag}\begin{pmatrix} Y_0 & 0 & 0 & 0 & 0 \end{pmatrix}.$$

Proof. The main idea here is to transform the inequality $J < x^T(0) P x(0) < \lambda$ and the conditions of Theorem 24 into LMIs:

$$\{\text{block-diag}\,[X\ I\ I]\}\cdot\{U_{ii}+(s-1)Q_3\}\cdot\{\text{block-diag}\,[X\ I\ I]\}$$

$$=\begin{bmatrix} \begin{pmatrix} XA_i^T+A_iX \\ -B_iM_i-M_i^TB_i^T \end{pmatrix} & XC_i^T & -M_i^T \\ C_iX & -W^{-1} & 0 \\ -M_i & 0 & -R^{-1} \end{bmatrix}$$

$$+(s-1)\cdot\text{block-diag}\,(XQ_0X\ 0\ 0)$$

$$=\hat{U}_{ii}+(s-1)Y_3,$$

where

$$Y_0 = XQ_0X.$$

We obtain the following condition as well:

$$\{\text{block-diag}\,[X\ I\ I]\}\cdot\{V_{ij}-2Q_4\}\cdot\{\text{block-diag}\,[X\ I\ I]\}$$

$$=\begin{bmatrix} \begin{pmatrix} XA_i^T+A_iX \\ -B_iM_j-M_j^TB_i^T \\ +XA_j^T+A_jX \\ -B_jM_i-M_i^TB_j^T \end{pmatrix} & XC_i^T & -M_j^T & XC_j^T & -M_i^T \\ C_iX & -W^{-1} & 0 & 0 & 0 \\ -M_j & 0 & -R^{-1} & 0 & 0 \\ C_jX & 0 & 0 & -W^{-1} & 0 \\ -M_i & 0 & 0 & 0 & -R^{-1} \end{bmatrix}$$

$$-2\cdot\text{block-diag}\,(XQ_0X\ 0\ 0\ 0\ 0)$$

$$=\hat{V}_{ij}-2Y_4.$$

Then, the quadratic performance function satisfies

$$J < x^T(0)X^{-1}x(0) < \lambda. \qquad \text{Q.E.D.}$$

Theorem 25 shows that by minimizing λ, we obtain the feedback gains which minimize the upper bound of J. To solve this design problem, the initial values $x(0)$ are assumed known. If not so, Theorem 25 is not directly applicable. In this case, however, if all the vertex points $x_k(0)$ of a polyhedron containing the unknown initial values $x(0)$ are known, that is,

$$x(0) = \sum_{k=1}^{l} \rho_k x_k(0),$$

$$\rho_k \geq 0,$$

$$\sum_{k=1}^{l} \rho_k = 1,$$

$$x_k(0) \in R^n.$$

Theorem 25 can be modified as follows to handle this case.

THEOREM 26 *The feedback gains to minimize the upper bound of the performance function can be obtained by solving the following LMIs. From the solution of the LMIs, we obtain*

$$F_i = M_i X^{-1}$$

for all i. Then, the performance function satisfies $J < x^T(0) P x(0) < \lambda$:

$$\underset{X, M_1, \ldots, M_r, Y_0}{\text{minimize}} \quad \lambda$$

subject to

$$X > 0, \quad Y_0 \geq 0,$$

$$\begin{bmatrix} \lambda & x_k^T(0) \\ x_k(0) & X \end{bmatrix} > 0, \quad k = 1, 2, \ldots, l,$$

$$\hat{U}_{ii} + (s-1)Y_3 < 0,$$

$$\hat{V}_{ij} - 2Y_4 < 0, \quad i < j \text{ s.t. } h_i \cap h_j \neq \phi.$$

Proof. It directly follows from Theorem 25. Q.E.D.

Remark 19 An alternative approach to handle the uncertainty in initial condition is to employ the initial condition independent design [see Chapter 3, equation (3.56)].

An interesting and important theorem is given below.

THEOREM 27 *The following statements are equivalent.*

(1) *There exist a common positive definite X and a common positive semidefinite Y satisfying (3.23) and (3.24).*

(2) *There exist a common positive definite $X' = X/\varepsilon$ and a common positive semidefinite Y_0 satisfying (6.12) and (6.13), where $\exists \varepsilon > 0$.*

Proof. (1) $\Rightarrow$ (2) Assume that (3.23) is satisfied. Since

$$[XC_i^T \ -M_i^T]\begin{bmatrix} W & 0 \\ 0 & R \end{bmatrix}\begin{bmatrix} C_iX \\ -M_i \end{bmatrix} \geq 0,$$

there exists a very small $\varepsilon > 0$ satisfying

$$\varepsilon\bigg(XA_i^T + A_iX - B_iM_i - M_i^TB_i^T + (s-1)Y_0$$

$$+ \varepsilon[XC_i^T \ -M_i^T]\begin{bmatrix} W & 0 \\ 0 & R \end{bmatrix}\begin{bmatrix} C_iX \\ -M_i \end{bmatrix}\bigg) < 0$$

for $i = 1, 2, \ldots, r$. The above condition is equivalent to

$$\begin{bmatrix} \varepsilon\begin{pmatrix} XA_i^T + A_iX \\ -B_iM_i - M_i^TB_i^T \end{pmatrix} & \varepsilon XC_i^T & -\varepsilon M_i^T \\ \varepsilon C_iX & -W^{-1} & 0 \\ -\varepsilon M_i & 0 & -R^{-1} \end{bmatrix}$$

$$+ \varepsilon(s-1)Y_3 < 0, \quad i = 1, 2, \ldots, r.$$

Since $X' = \varepsilon X$, $M_i' = \varepsilon M_i$, and $Y_3' = \varepsilon Y_3$ can be regarded as new X, M_i, and Y_3, respectively, we obtain the condition (6.12).

We can obtain the condition (6.13) from (3.24) as well.

(2) $\Rightarrow$ (1). It is obvious. Q.E.D.

The theorem above says that there exists a common X' satisfying (6.12) and (6.13) for any W and R if conditions (3.23) and (3.24) hold. The optimal fuzzy controller design in Theorem 25 is feasible if the stability conditions (3.23) and (3.24) hold.

A design example for optimal fuzzy control will be discussed in detail in Chapter 7.

APPENDIX TO CHAPTER 6

COROLLARY A.1

$$-C_i^TWC_i - C_j^TWC_j \leq -C_i^TWC_j - C_j^TWC_i,$$

where

$$W > 0.$$

Proof. It is clear.

COROLLARY A.2

$$-\begin{bmatrix} C_i^T & -F_j^T \end{bmatrix} \begin{bmatrix} W & 0 \\ 0 & R \end{bmatrix} \begin{bmatrix} C_i \\ -F_j \end{bmatrix} - \begin{bmatrix} C_j^T & -F_i^T \end{bmatrix} \begin{bmatrix} W & 0 \\ 0 & R \end{bmatrix} \begin{bmatrix} C_j \\ -F_i \end{bmatrix}$$

$$\leq -\begin{bmatrix} C_i^T & -F_j^T \end{bmatrix} \begin{bmatrix} W & 0 \\ 0 & R \end{bmatrix} \begin{bmatrix} C_j \\ -F_i \end{bmatrix} - \begin{bmatrix} C_j^T & -F_i^T \end{bmatrix} \begin{bmatrix} W & 0 \\ 0 & R \end{bmatrix} \begin{bmatrix} C_i \\ -F_j \end{bmatrix},$$

where

$$W > 0 \quad and \quad R > 0.$$

Proof. From Corollary A.1, we have

$$\begin{bmatrix} C_i^T & -F_j^T \end{bmatrix} \begin{bmatrix} W & 0 \\ 0 & R \end{bmatrix} \begin{bmatrix} C_i \\ -F_j \end{bmatrix} - \begin{bmatrix} C_j^T & -F_i^T \end{bmatrix} \begin{bmatrix} W & 0 \\ 0 & R \end{bmatrix} \begin{bmatrix} C_j \\ -F_i \end{bmatrix}$$

$$= -C_i^T W C_i - F_j^T R F_j - C_j^T W C_j - F_i^T R F_i$$

$$\leq -C_i^T W C_j - F_j^T R F_i - C_j^T W C_i - F_i^T R F_j$$

$$= -\begin{bmatrix} C_i^T & -F_j^T \end{bmatrix} \begin{bmatrix} W & 0 \\ 0 & R \end{bmatrix} \begin{bmatrix} C_j \\ -F_i \end{bmatrix} - \begin{bmatrix} C_j^T & -F_i^T \end{bmatrix} \begin{bmatrix} W & 0 \\ 0 & R \end{bmatrix} \begin{bmatrix} C_i \\ -F_j \end{bmatrix}.$$

Q.E.D.

REFERENCES

1. D. E. Kirk, *Optimal Control Theory: An Introduction*, Prentice-Hall, Englewood Cliffs, NJ, 1970.
2. A. J. van der Schaft, "On a State Space Approach to Nonlinear H_∞ Control," *Syst. Control Lett.*, Vol. 16, pp.1–8 (1991).
3. W. M. Lu and J. C. Doyle, "H_∞ Control of Nonlinear Systems: A Convex Characterization," *IEEE Trans. Automatic Control*, Vol. 40, No. 9, pp. 1668–1675 (1995).
4. K. Tanaka, M. Nishimura, and H. O. Wang, "Multi-Objective Fuzzy Control of High Rise/High Speed Elevators Using LMIs," 1998 American Control Conference, 1998, pp. 3450–3454.
5. K. Tanaka, T. Taniguchi, and H. O. Wang, "Model-Based Fuzzy Control of TORA System: Fuzzy Regulator and Fuzzy Observer Design via LMIs That Represent Decay Rate, Disturbance Rejection, Robustness, Optimality," Seventh IEEE International Conference on Fuzzy Systems, Alaska, 1998, pp. 313–318.

6. K. Tanaka, T. Taniguchi, and H. O. Wang, "Fuzzy Control Based on Quadratic Performance Function," 37th IEEE Conference on Decision and Control, Tampa, 1998, pp. 2914–2919.
7. K. Tanaka, T. Taniguchi, and H. O. Wang, "Robust and Optimal Fuzzy Control: A Linear Matrix Inequality Approach," 1999 International Federation of Automatic Control (IFAC) World Congress, Beijing, July 1999, pp. 213–218.
8. J. Li, H. O. Wang, L. Bushnell, K. Tanaka, and Y. Hong, "A Fuzzy Logic Approach to Optimal Control of Nonlinear Systems," *Int. J. Fuzzy Syst.*, Vol. 2, No. 3, pp. 153–163 Sept. (2000).

CHAPTER 7

ROBUST-OPTIMAL FUZZY CONTROL

This chapter discusses the robust-optimal fuzzy control problem [1–3], which combines robust fuzzy control and optimal fuzzy control. The robust-optimal fuzzy control problem is useful for practical control system designs that call for both robustness and optimality. In the last two chapters the robustness and optimality issues have been addressed separately. This chapter presents a unified design procedure to address both issues simultaneously to provide a solution to the robust-optimal fuzzy control problem. A design example is included to illustrate the merits of robust fuzzy control, optimal fuzzy control, and robust-optimal fuzzy control. The well-known nonlinear control benchmark problem, that is, the translational actuator with rotational actuator (TORA) system [4–6], is employed as the design example.

7.1 ROBUST-OPTIMAL FUZZY CONTROL PROBLEM

The robust-optimal fuzzy control design conditions are captured in the following theorem. Naturally these conditions are rendered by combining Theorems 23 (robust fuzzy control) and 25 (optimal fuzzy control).

THEOREM 28 *The PDC controller* (2.23) *that simultaneously considers both the robust fuzzy controller design* (*Theorem* 23) *and the optimal fuzzy controller*

122 ROBUST-OPTIMAL FUZZY CONTROL

design (Theorem 25) can be designed by solving the following LMIs:

$$\underset{\substack{\lambda, \gamma_{ai}^2, \gamma_{bi}^2, X, \\ M_1, \ldots, M_r, Y_0}}{\text{minimize}} \quad \lambda + \sum_{i=1}^{r} \{\alpha_i \gamma_{ai}^2 + \beta_i \gamma_{bi}^2\}$$

subject to

$$X > 0, \quad Y_0 \geq 0,$$

$$\begin{bmatrix} \lambda & x^T(0) \\ x(0) & X \end{bmatrix} > 0, \qquad (7.1)$$

$$\hat{S}_{ii} + (s-1)Y_1 < 0, \quad i = 1, 2, \ldots, r,$$

$$\hat{T}_{ij} - 2Y_2 < 0, \quad i < j \leq r \text{ s.t. } h_i \cap h_j \neq \phi,$$

$$\hat{U}_{ii} + (s-1)Y_3 < 0, \quad i = 1, 2, \ldots, r,$$

$$\hat{V}_{ij} - 2Y_4 < 0, \quad i < j \leq r \text{ s.t. } h_i \cap h_j \neq \phi,$$

where $s > 1$,

$$\hat{S}_{ii} = \begin{bmatrix} \begin{pmatrix} XA_i^T + A_i X \\ -B_i M_i - M_i^T B_i^T \end{pmatrix} & * & * & * & * \\ D_{ai}^T & -I & 0 & 0 & 0 \\ D_{bi}^T & 0 & -I & 0 & 0 \\ E_{ai} X & 0 & 0 & -\gamma_{ai}^2 I & 0 \\ -E_{bi} M_i & 0 & 0 & 0 & -\gamma_{bi}^2 I \end{bmatrix},$$

$$\hat{T}_{ij} = \begin{bmatrix} \begin{pmatrix} XA_i^T + A_i X \\ -B_i M_j - M_j^T B_i^T \\ +XA_j^T + A_j X \\ -B_j M_i - M_i^T B_j^T \end{pmatrix} & D_{ai} & D_{bi} & D_{aj} & D_{bj} & XE_{ai}^T & -M_j^T E_{bi}^T & XE_{aj}^T & -M_i^T E_{bj}^T \\ D_{ai}^T & -I & 0 & 0 & 0 & 0 & 0 & 0 & 0 \\ D_{bi}^T & 0 & -I & 0 & 0 & 0 & 0 & 0 & 0 \\ D_{aj}^T & 0 & 0 & -I & 0 & 0 & 0 & 0 & 0 \\ D_{bj}^T & 0 & 0 & 0 & -I & 0 & 0 & 0 & 0 \\ E_{ai} X & 0 & 0 & 0 & 0 & -\gamma_{ai}^2 I & 0 & 0 & 0 \\ -E_{bi} M_j & 0 & 0 & 0 & 0 & 0 & -\gamma_{bi}^2 I & 0 & 0 \\ E_{aj} X & 0 & 0 & 0 & 0 & 0 & 0 & -\gamma_{aj}^2 I & 0 \\ -E_{bj} M_i & 0 & 0 & 0 & 0 & 0 & 0 & 0 & -\gamma_{bj}^2 I \end{bmatrix},$$

$$Y_1 = \text{block-diag}\begin{pmatrix} Y_0 & 0 & 0 & 0 & 0 \end{pmatrix},$$

$$Y_2 = \text{block-diag}\begin{pmatrix} Y_0 & 0 & 0 & 0 & 0 & 0 & 0 & 0 & 0 \end{pmatrix},$$

$$\hat{U}_{ii} = \begin{bmatrix} \begin{pmatrix} XA_i^T + A_i X \\ -B_i M_i - M_i^T B_i^T \end{pmatrix} & XC_i^T & -M_i^T \\ C_i X & -W^{-1} & 0 \\ -M_i & 0 & -R^{-1} \end{bmatrix},$$

$$\hat{V}_{ij} = \begin{bmatrix} \begin{pmatrix} XA_i^T + A_i X \\ -B_i M_j - M_j^T B_i^T \\ +XA_j^T + A_j X \\ -B_j M_i - M_i^T B_j^T \end{pmatrix} & XC_i^T & -M_j^T & XC_j^T & -M_i^T \\ C_i X & -W^{-1} & 0 & 0 & 0 \\ -M_j & 0 & -R^{-1} & 0 & 0 \\ C_j X & 0 & 0 & -W^{-1} & 0 \\ -M_i & 0 & 0 & 0 & -R^{-1} \end{bmatrix},$$

$$Y_3 = \text{block-diag}\begin{pmatrix} Y_0 & 0 & 0 \end{pmatrix},$$

$$Y_4 = \text{block-diag}\begin{pmatrix} Y_0 & 0 & 0 & 0 & 0 \end{pmatrix},$$

where the asterisk denotes the transposed elements (matrices) for symmetric positions.

Proof. It follows directly from Theorems 23 and 25.

Remark 20 As shown in Chapter 3, the condition (7.1) may be replaced with (3.56) to handle the uncertainty in initial conditions.

When $Q_0 = 0$ (i.e., $Y_0 = XQ_0X$), the relaxed conditions are reduced to the following conditions:

$$\underset{\substack{\lambda,\, \gamma_{ai}^2,\, \gamma_{bi}^2,\, X, \\ M_1, \ldots, M_r}}{\text{minimize}} \quad \lambda + \sum_{i=1}^{r} \left\{ \alpha_i \gamma_{ai}^2 + \beta_i \gamma_{bi}^2 \right\}$$

subject to

$$X > 0,$$

$$\begin{bmatrix} \lambda & x^T(0) \\ x(0) & X \end{bmatrix} > 0,$$

$$\hat{S}_{ii} < 0, \quad i = 1, 2, \ldots, r,$$

$$\hat{T}_{ij} < 0, \quad i < j \le r \text{ s.t. } h_i \cap h_j \ne \phi,$$

$$\hat{U}_{ii} < 0, \quad i = 1, 2, \ldots, r,$$

$$\hat{V}_{ij} < 0, \quad i < j \le r \text{ s.t. } h_i \cap h_j \ne \phi.$$

In the design problem above, the initial conditions $x(0)$ are assumed known. If not so, the theorem is not directly applicable. In this case, if all the vertex points $x_k(0)$ of a polyhedron containing the initial conditions $x(0)$ are known, that is,

$$x(0) = \sum_{k=1}^{l} \rho_k x_k(0),$$

$$\rho_k \ge 0, \quad \sum_{k=1}^{l} \rho_k = 1, \quad x_k(0) \in R^n,$$

Theorem 28 can be modified as follows to handle the uncertain initial conditions.

THEOREM 29 *The PDC controller (2.23) that simultaneously considers both the robust fuzzy controller design (Theorem 23) and the optimal fuzzy control design (Theorem 25) can be designed by solving the following LMIs:*

$$\underset{\lambda,\, \gamma_{ai}^2,\, \gamma_{bi}^2,\, X,\, M_1, \ldots, M_r, Y_0}{\text{minimize}} \quad \lambda + \sum_{i=1}^{r} \{\alpha_i \gamma_{ai}^2 + \beta_i \gamma_{bi}^2\}$$

subject to

$$X > 0 \quad Y_0 \ge 0,$$

$$\begin{bmatrix} \lambda & x_k^T(0) \\ x_k(0) & X \end{bmatrix} > 0, \quad k = 1, 2, \ldots, l,$$

$$\hat{S}_{ii} + (s-1)Y_1 < 0, \quad i = 1, 2, \ldots, r$$

$$\hat{T}_{ij} - 2Y_2 < 0, \quad i < j \le r \text{ s.t. } h_i \cap h_j \ne \phi$$

$$\hat{U}_{ii} + (s-1)Y_3 < 0, \quad i = 1, 2, \ldots, r$$

$$\hat{V}_{ij} - 2Y_4 < 0. \quad i < j \le r \text{ s.t. } h_i \cap h_j \ne \phi$$

Proof. It follows directly from Theorem 28.

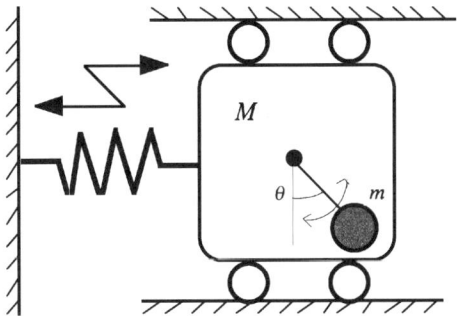

Fig. 7.1 TORA system.

7.2 DESIGN EXAMPLE: TORA

Consider the system shown in Figure 7.1, which represents a translational oscillator with an eccentric rotational proof mass actuator (TORA) [4–6]. The nonlinear coupling between the rotational motion of the actuator and the translational motion of the oscillator provides the mechanism for control.

Let x_1 and x_2 denote the translational position and velocity of the cart with $x_2 = \dot{x}_1$. Let $x_3 = \theta$ and $x_4 = \dot{x}_3$ denote the angular position and velocity of the rotational proof mass. Then the system dynamics can be described by the equation

$$\dot{x} = f(x) + g(x)u + d, \tag{7.2}$$

where u is the torque applied to the eccentric mass, d is the disturbance, and

$$f(x) = \begin{bmatrix} x_2 \\ \dfrac{-x_1 + \varepsilon x_4^2 \sin x_3}{1 - \varepsilon^2 \cos^2 x_3} \\ x_4 \\ \dfrac{\varepsilon \cos x_3 (x_1 - \varepsilon x_4^2 \sin x_3)}{1 - \varepsilon^2 \cos^2 x_3} \end{bmatrix},$$

$$g(x) = \begin{bmatrix} 0 \\ \dfrac{-\varepsilon \cos x_3}{1 - \varepsilon^2 \cos^2 x_3} \\ 0 \\ \dfrac{1}{1 - \varepsilon^2 \cos^2 x_3} \end{bmatrix},$$

$\varepsilon = 0.1$.

126 ROBUST-OPTIMAL FUZZY CONTROL

Consider the case of no disturbance, as in [4–6], introduce new state variables $z_1 = x_1 + \varepsilon \sin x_3$, $z_2 = x_2 + \varepsilon x_4 \cos x_3$, $y_1 = x_3$, $y_2 = x_4$, and employ the feedback transformation

$$v = \frac{1}{1 - \varepsilon^2 \cos^2 y_1} \left[\varepsilon \cos y_1 \left(z_1 - (1 + y_2^2) \varepsilon \sin y_1 \right) + u \right]$$

$$= \alpha(z_1, y_1) + \beta(y_1) u$$

to bring the system into the following form:

$$\dot{z}_1 = z_2, \tag{7.3}$$

$$\dot{z}_2 = -z_1 + \varepsilon \sin y_1, \tag{7.4}$$

$$\dot{y}_1 = y_2, \tag{7.5}$$

$$\dot{y}_2 = v. \tag{7.6}$$

The equilibrium point of system (7.2) can be any point $[0, 0, x_3^0, 0]$, where x_3^0 is an arbitrary constant. Consider $[0, 0, 0, 0]$ as the desired equilibrium point. The linearization around this point has a pair of nonzero imaginary eigenvalues and two zero eigenvalues. Hence the system (7.2) at the origin is an example of a critical nonlinear system. This control problem is interpreted as a regulator problem of $z_1 \to 0$, $z_2 \to 0$, $y_1 \to 0$, and $y_2 \to 0$.

The T-S model of the TORA system can be constructed from (7.3)–(7.6) by using the fuzzy model construction described in Chapter 2:

Rule 1

　　IF $y_1(t)$ is "about $-\pi$ or π rad,"

　　　　THEN

$$\dot{x}(t) = A_1 x(t) + B_1 u(t),$$
$$y(t) = C_1 x(t).$$

Rule 2

　　IF $y_1(t)$ is "about $-\frac{\pi}{2}$ or $\frac{\pi}{2}$ rad,"

　　　　THEN

$$\dot{x}(t) = A_2 x(t) + B_2 u(t),$$
$$y(t) = C_2 x(t).$$

Rule 3

IF $y_1(t)$ is "about 0 rad" and $y_2(t)$ is "about 0,"

THEN
$$\dot{x}(t) = A_3 x(t) + B_3 u(t),$$
$$y(t) = C_3 x(t).$$

Rule 4

IF $y_1(t)$ is "about 0 rad" and $y_2(t)$ is "about $-a$ or a,"

THEN
$$\dot{x}(t) = A_4 x(t) + B_4 u(t),$$
$$y(t) = C_4 x(t),$$

Here, $x^T(t) = [z_1(t), z_2(t), y_1(t), y_2(t)]$,

$$A_1 = \begin{bmatrix} 0 & 1 & 0 & 0 \\ -1 & 0 & \varepsilon\frac{\sin(\alpha\pi)}{\alpha\pi} & 0 \\ 0 & 0 & 0 & 1 \\ \frac{-\varepsilon}{1-\varepsilon^2} & 0 & 0 & 0 \end{bmatrix}, \quad B_1 = \begin{bmatrix} 0 \\ 0 \\ 0 \\ \frac{1}{1-\varepsilon^2} \end{bmatrix},$$

$$A_2 = \begin{bmatrix} 0 & 1 & 0 & 0 \\ -1 & 0 & \varepsilon\frac{2}{\pi} & 0 \\ 0 & 0 & 0 & 1 \\ 0 & 0 & 0 & 0 \end{bmatrix}, \quad B_2 = \begin{bmatrix} 0 \\ 0 \\ 0 \\ 1 \end{bmatrix},$$

$$A_3 = \begin{bmatrix} 0 & 1 & 0 & 0 \\ -1 & 0 & \varepsilon & 0 \\ 0 & 0 & 0 & 1 \\ \frac{\varepsilon}{1-\varepsilon^2} & 0 & \frac{-\varepsilon^2}{1-\varepsilon^2} & 0 \end{bmatrix}, \quad B_3 = \begin{bmatrix} 0 \\ 0 \\ 0 \\ \frac{1}{1-\varepsilon^2} \end{bmatrix},$$

$$A_4 = \begin{bmatrix} 0 & 1 & 0 & 0 \\ -1 & 0 & \varepsilon & 0 \\ 0 & 0 & 0 & 1 \\ \frac{\varepsilon}{1-\varepsilon^2} & 0 & \frac{-\varepsilon^2(1+a^2)}{1-\varepsilon^2} & 0 \end{bmatrix}, \quad B_4 = \begin{bmatrix} 0 \\ 0 \\ 0 \\ \frac{1}{1-\varepsilon^2} \end{bmatrix},$$

$$C_1 = C_2 = C_3 = C_4 = \begin{bmatrix} 1 & 0 & 0 & 0 \\ 0 & 1 & 0 & 0 \\ 0 & 0 & 1 & 0 \\ 0 & 0 & 0 & 1 \end{bmatrix}.$$

In this simulation, $x_4 \in [-a, a]$ ($a = 4$) and $0 < \alpha < 1$ instead of $\alpha = 1$ (e.g., $\alpha = 0.99$) is used to maintain the controllability of the subsystem (A_1, B_1) in Rule 1.

The above fuzzy model is represented as

$$\dot{x}(t) = \sum_{i=1}^{r} h_i(z(t))\{A_i x(t) + B_i u(t)\}, \quad (7.7)$$

$$y(t) = \sum_{i=1}^{r} h_i(z(t)) C_i x(t), \quad (7.8)$$

where $r = 4$ and $z(t) = [y_1(t) \; y_2(t)]$. Here, $h_i(z(t))$ is the weight of the ith rules calculated by the membership values. Figure 7.2 shows the membership functions.

The PDC fuzzy controller is designed as follows:

Control Rule 1

IF $y_1(t)$ is "about $-\pi$ or π rad,"

THEN $u(t) = -F_1 x(t)$.

Control Rule 2

IF $y_1(t)$ is "about $-\frac{\pi}{2}$ or $\frac{\pi}{2}$ rad,"

THEN $u(t) = -F_2 x(t)$.

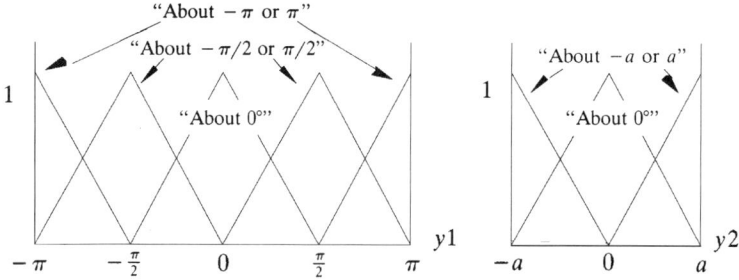

Fig. 7.2 Membership functions.

Control Rule 3

IF $y_1(t)$ is "about 0 rad" and $y_2(t)$ is "about 0,"

THEN $u(t) = -F_3 x(t)$.

Control Rule 4

IF $y_1(t)$ is "about 0 rad" and $y_2(t)$ is "about $-a$ or a,"

THEN $u(t) = -F_4 x(t)$.

Figure 7.3 shows the comparison between a stable fuzzy controller [satisfying (3.23) and (3.24)] and a robust fuzzy controller (satisfying the conditions in Theorem 23) for the TORA system with parameter change $\varepsilon = 0.05$. Figure 7.4 compares the performance of the stable fuzzy controller and an optimal fuzzy controller (satisfying the conditions in Theorem 25) for the nominal TORA system. Figure 7.5 shows the control results of the robust

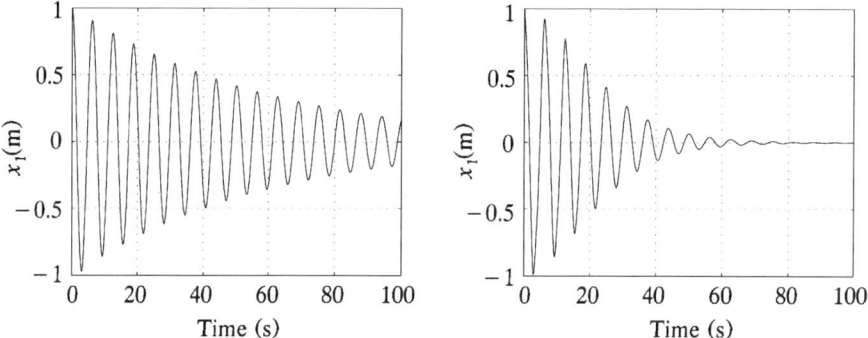

Fig. 7.3 Control results for TORA with parameter change ($\varepsilon = 0.05$).

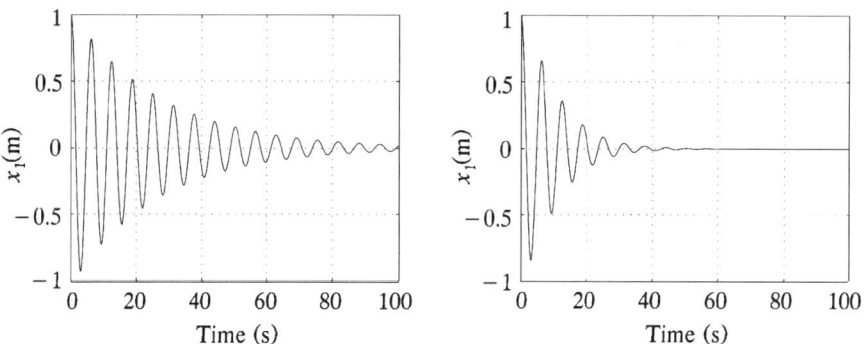

Fig. 7.4 Control results for the nominal TORA.

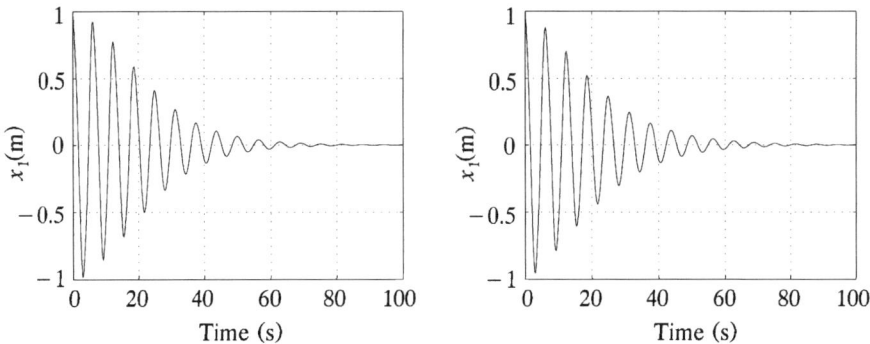

Fig. 7.5 Control results for TORA with parameter change ($\varepsilon = 0.05$).

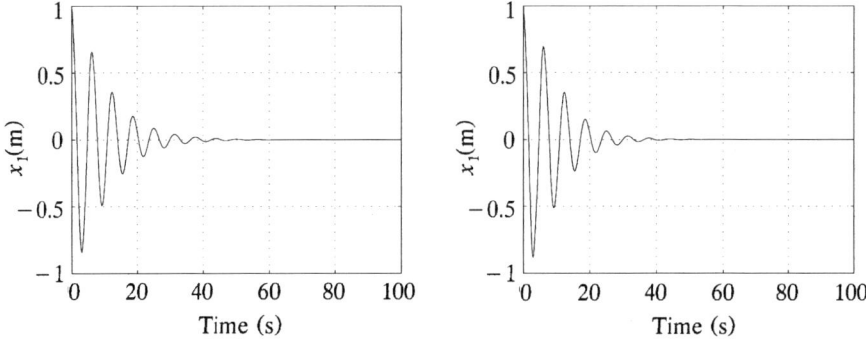

Fig. 7.6 Control results for the nominal TORA.

fuzzy controller and the robust-optimal fuzzy controller (satisfying the conditions in Theorem 28) for the TORA with the parameter change. Figure 7.6 compares the control results of the optimal fuzzy controller and the robust-optimal fuzzy controller for the nominal TORA. In all cases, the fuzzy control designs get the job done but with different performance characteristics. The robust-optimal fuzzy controller is the most versatile in that it addresses both the robustness and the optimality.

REFERENCES

1. K. Tanaka, T. Taniguchi, and H. O. Wang, "Robust and Optimal Fuzzy Control: A Linear Matrix Inequality Approach," 1999 International Federation of Automatic Control (IFAC) World Congress, Beijing, July 1999, pp. 213–218.

2. K. Tanaka, M. Nishimura, and H. O. Wang, "Multi-objective Fuzzy Control of High Rise/High Speed Elevators using LMIs," 1998 American Control Conference, 1998, pp. 3450–3454.
3. K. Tanaka, T. Taniguchi, and H. O. Wang, "Model-Based Fuzzy Control of TORA System: Fuzzy Regulator and Fuzzy Observer Design via LMIs that Represent Decay Rate, Disturbance Rejection, Robustness, Optimality," Seventh IEEE International Conference on Fuzzy Systems, Alaska, 1998, pp. 313–318.
4. R. T. Bupp, D. S. Bernstein, and V. T. Coppola, "A benchmark problem for nonlinear control design: Problem Statement, Experiment Testbed and Passive Nonlinear Compensation," *Proc. 1995 American Control Conference*, Seattle, 1995, pp. 4363–4367.
5. Session on Benchmark Problem for Nonlinear Control Design, *Proc. 1995 American Control Conference*, Seattle, 1995, pp. 4337–4367.
6. M. Jankovic, D. Fontaine, and P. Kokotovic, "TORA Example: Cascade and Passivity Control Design," *Proc. 1995 American Control Conference*, Seattle, 1995, pp. 4347–4351.

CHAPTER 8

TRAJECTORY CONTROL OF A VEHICLE WITH MULTIPLE TRAILERS

This chapter contains an in-depth application study of the fuzzy control methodologies introduced in this book. The system under study is a vehicle with multiple trailers. The control objective is to back the vehicle into a straight-line configuration without forward motion. This is often referred as the problem of backing up control of a truck-trailer. A truck with a single trailer is often used as a testbed to study different control strategies. In this chapter, we consider the more challenging problem of backing up control of a vehicle with *multiple* trailers. Both simulation and experimental results [1-4] are presented. The results demonstrate that the designed fuzzy controller can effectively achieve the backing-up control of the vehicle with multiple trailers while avoiding the saturation of the actuator and "jack-knife" phenomenon. Moreover, the controller guarantees the stability and performance even in the presence of disturbance.

As mentioned above, the backing-up control of "trailer-truck," that is, a vehicle with a trailer, has been used as a testbed for a variety of control design methods [1-11]. In particular, in order to successfully back up the trailer-truck, the so-called jack-knife phenomenon needs to be avoided throughout the operation. In the field of automatic control, a number of control methodologies including nonlinear control, fuzzy control, neural control, and hybrid neural-fuzzy control [5-8] have been applied to this testbed problem. Most of these are simulation-based studies; the important issue of the stability of the control systems was often left out. In our work, stabilizing fuzzy control was applied to the case of a truck with one trailer case in [9] and experimental demonstrations were reported in [1, 10].

134 TRAJECTORY CONTROL OF A VEHICLE WITH MULTIPLE TRAILERS

This chapter mainly deals with the triple-trailer case [3, 4]. The triple-trailer case, that is, backing-up control of a vehicle with triple trailers, is much more challenging than that of the one-trailer case. To the best of our knowledge, experimental results of the triple-trailer case had not been reported in the literature prior to our work. Part of the difficulties associated with multiple-trailer cases, the triple-trailer case included, lie in the exponentially increasing number of jack-knife configurations as the number of trailers increases. In the one-trailer case, only two jack-knife configurations exist. For the triple-trailer case, the number of jack-knife configurations increases to eight. Moreover, we need to address a number of practical constraints, for example, saturation of the steering angle and disturbance rejection, for such difficult control objects. In the control design for the vehicle with triple trailers, we utilize the LMI conditions described in Chapter 3 to explicitly handle the saturation of the steering angle and the jack-knife phenomenon. Both simulation and experimental results demonstrate that the fuzzy controller effectively achieves the backing-up control of the vehicle with triple trailers while avoiding the saturation of the actuator and jack-knife phenomenon. Moreover, the feedback controller guarantees the stability and performance even in the presence of disturbance.

8.1 FUZZY MODELING OF A VEHICLE WITH TRIPLE TRAILERS

Figure 8.1 shows the vehicle model with triple trailers and its coordinate system. We use the following control-oriented model to design a fuzzy controller:

$$x_0(t+1) = x_0(t) + \frac{v \cdot \Delta t}{l} \tan(u(t)), \qquad (8.1)$$

$$x_1(t) = x_0(t) - x_2(t), \qquad (8.2)$$

$$x_2(t+1) = x_2(t) + \frac{v \cdot \Delta t}{L} \sin(x_1(t)), \qquad (8.3)$$

$$x_3(t) = x_2(t) - x_4(t), \qquad (8.4)$$

$$x_4(t+1) = x_4(t) + \frac{v \cdot \Delta t}{L} \sin(x_3(t)), \qquad (8.5)$$

$$x_5(t) = x_4(t) - x_6(t), \qquad (8.6)$$

$$x_6(t+1) = x_6(t) + \frac{v \cdot \Delta t}{L} \sin(x_5(t)), \qquad (8.7)$$

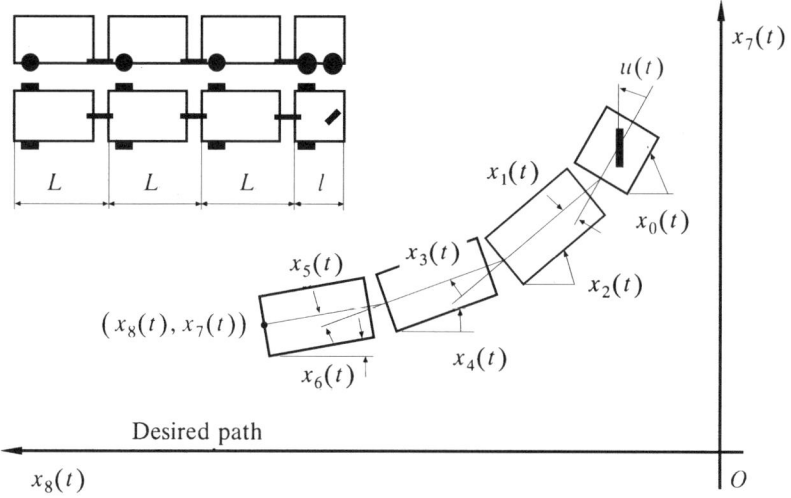

Fig. 8.1 Vehicle model with triple trailers.

$$x_7(t+1) = x_7(t) + v \cdot \Delta t \cos(x_5(t)) \sin\left(\frac{x_6(t+1) + x_6(t)}{2}\right), \quad (8.8)$$

$$x_8(t+1) = x_8(t) + v \cdot \Delta t \cos(x_5(t)) \cos\left(\frac{x_6(t+1) + x_6(t)}{2}\right), \quad (8.9)$$

where

$x_0(t)$ = angle of vehicle,
$x_1(t)$ = angle difference between vehicle and first trailer,
$x_2(t)$ = angle of first trailer,
$x_3(t)$ = angle difference between first trailer and second trailer,
$x_4(t)$ = angle of second trailer,
$x_5(t)$ = angle difference between second trailer and third trailer,
$x_6(t)$ = angle of third trailer,
$x_7(t)$ = vertical position of rear end of third trailer,
$x_8(t)$ = horizontal position of rear end of third trailer,
$u(t)$ = steering angle.

The model presented above is a discretized model with several simplifications. It is not intended to be a model to study the detailed dynamics of the trailer-truck system. Because of the simplicity, its main usage is for control design. This is the same idea as the so-called control-oriented modeling in which some reduced-order type of models are sought instead of the full-fledged dynamic models. The trailer-truck model herein has proven to be effective in designing controllers for the experimental setup which is discussed later in this chapter.

In the simulation and experimental studies the following parameter values are used:

$$l = 0.087 \text{ m}, \quad L = 0.130 \text{ m}, \quad v = -0.10 \text{ m/sec.}, \quad \Delta t = 0.5 \text{ sec.},$$

where l is the length of the vehicle, L is the length of the trailer, Δt is the sampling time, and v is the constant speed of the backward movement. For $x_1(t)$, $x_3(t)$, and $x_5(t)$, 90° and $-90°$ correspond to eight "jack-knife" positions.

The control objective is to back the vehicle into the straight line ($x_7 = 0$) without any forward movement, that is,

$$x_1(t) \to 0, \quad x_3(t) \to 0, \quad x_5(t) \to 0, \quad x_6(t) \to 0, \quad x_7(t) \to 0.$$

To employ the model-based fuzzy control design methodology described in this book, we start with the construction of a Takagi-Sugeno fuzzy model to represent the nonlinear equations (8.1)–(8.8). To facilitate the control design, with the assumption that the values of $u(t)$, $x_1(t)$, $x_3(t)$, and $x_5(t)$ are small, we further simplify the model to be of the following form:

$$x_0(t+1) = x_0(t) + \frac{v \cdot \Delta t}{l} u(t), \tag{8.10}$$

$$x_1(t+1) = \left(1 - \frac{v \cdot \Delta t}{L}\right) x_1(t) + \frac{v \cdot \Delta t}{l} u(t), \tag{8.11}$$

$$x_2(t+1) = x_2(t) + \frac{v \cdot \Delta t}{L} x_1(t), \tag{8.12}$$

$$x_3(t+1) = \left(1 - \frac{v \cdot \Delta t}{L}\right) x_3(t) + \frac{v \cdot \Delta t}{L} x_1(t), \tag{8.13}$$

$$x_4(t+1) = x_4(t) + \frac{v \cdot \Delta t}{L} x_3(t), \tag{8.14}$$

$$x_5(t+1) = \left(1 - \frac{v \cdot \Delta t}{L}\right) x_5(t) + \frac{v \cdot \Delta t}{L} x_3(t), \quad (8.15)$$

$$x_6(t+1) = x_6(t) + \frac{v \cdot \Delta t}{L} x_5(t), \quad (8.16)$$

$$x_7(t+1) = x_7(t) + v \cdot \Delta t \cdot \sin\left(x_6(t) + \frac{v \cdot \Delta t}{2L} x_5(t)\right). \quad (8.17)$$

In this simplified model, the only nonlinear term is in (8.17),

$$v \cdot \Delta t \cdot \sin\left(x_6(t) + \frac{v \cdot \Delta t}{2L} x_5(t)\right). \quad (8.18)$$

This term can be represented by the following Takagi-Sugeno fuzzy model:

$$v \cdot \Delta t \cdot \sin\left(x_6(t) + \frac{v \cdot \Delta t}{2L} x_5(t)\right)$$

$$= w_1(p(t)) \cdot v \cdot \Delta t \cdot \left(x_6(t) + \frac{v \cdot \Delta t}{2L} x_5(t)\right)$$

$$+ w_2(p(t)) \cdot v \cdot \Delta t \cdot g \cdot \left(x_6(t) + \frac{v \cdot \Delta t}{2L} x_5(t)\right), \quad (8.19)$$

where

$$p(t) = x_6(t) + \frac{v \cdot \Delta t}{2L} x_5(t),$$

$$g = 10^{-2}/\pi,$$

$$w_1(p(t)) = \begin{cases} \dfrac{\sin(p(t)) - g \cdot p(t)}{p(t) \cdot (1 - g)}, & p(t) \neq 0, \\ 1, & p(t) = 0, \end{cases} \quad (8.20)$$

$$w_2(p(t)) = \begin{cases} \dfrac{p(t) - \sin(p(t))}{p(t) \cdot (1 - g)}, & p(t) \neq 0, \\ 0, & p(t) = 0. \end{cases} \quad (8.21)$$

From (8.20) and (8.21), it can be seen that $w_1(p(t)) = 1$ and $w_2(p(t)) = 0$ when $p(t)$ is about 0 rad. Similarly, $w_1(p(t)) = 0$ and $w_2(p(t)) = 1$ when $p(t)$ is about π or $-\pi$ rad.

When $w_1(p(t)) = 1$ and $w_2(p(t)) = 0$, that is, $p(t)$ is about 0 rad, substituting (8.19) into (8.17), we have

$$x_7(t+1) = x_7(t) + v \cdot \Delta t \cdot x_6(t) + \frac{(v \cdot \Delta t)^2}{2L} \cdot x_5(t).$$

As a result the simplified nonlinear model can be represented by

$$\begin{bmatrix} x_1(t+1) \\ x_3(t+1) \\ x_5(t+1) \\ x_6(t+1) \\ x_7(t+1) \end{bmatrix} = \begin{bmatrix} 1 - \frac{v \cdot \Delta t}{L} & 0 & 0 & 0 & 0 \\ \frac{v \cdot \Delta t}{L} & 1 - \frac{v \cdot \Delta t}{L} & 0 & 0 & 0 \\ 0 & \frac{v \cdot \Delta t}{L} & 1 - \frac{v \cdot \Delta t}{L} & 0 & 0 \\ 0 & 0 & \frac{v \cdot \Delta t}{L} & 1 & 0 \\ 0 & 0 & \frac{(v \cdot \Delta t)^2}{2L} & v \cdot \Delta t & 1 \end{bmatrix}$$

$$\times \begin{bmatrix} x_1(t) \\ x_3(t) \\ x_5(t) \\ x_6(t) \\ x_7(t) \end{bmatrix} + \begin{bmatrix} \frac{v \cdot \Delta t}{l} \\ 0 \\ 0 \\ 0 \\ 0 \end{bmatrix} u(t). \quad (8.22)$$

When $w_1(p(t)) = 0$ and $w_2(p(t)) = 1$, that is, $p(t)$ is about π or $-\pi$ rad, (8.17) is represented as

$$x_7(t+1) = x_7(t) + g \cdot v \cdot \Delta t \cdot x_6(t) + \frac{g \cdot (v \cdot \Delta t)^2}{2L} \cdot x_5(t).$$

The resulting simplified nonlinear model can be represented by

$$
\begin{bmatrix} x_1(t+1) \\ x_3(t+1) \\ x_5(t+1) \\ x_6(t+1) \\ x_7(t+1) \end{bmatrix} = \begin{bmatrix} 1 - \frac{v \cdot \Delta t}{L} & 0 & 0 & 0 & 0 \\ \frac{v \cdot \Delta t}{L} & 1 - \frac{v \cdot \Delta t}{L} & 0 & 0 & 0 \\ 0 & \frac{v \cdot \Delta t}{L} & 1 - \frac{v \cdot \Delta t}{L} & 0 & 0 \\ 0 & 0 & \frac{v \cdot \Delta t}{L} & 1 & 0 \\ 0 & 0 & \frac{g \cdot (v \cdot \Delta t)^2}{2L} & g \cdot v \cdot \Delta t & 1 \end{bmatrix} \times \begin{bmatrix} x_1(t) \\ x_3(t) \\ x_5(t) \\ x_6(t) \\ x_7(t) \end{bmatrix} + \begin{bmatrix} \frac{v \cdot \Delta t}{l} \\ 0 \\ 0 \\ 0 \\ 0 \end{bmatrix} u(t). \tag{8.23}
$$

In this representation, if $g = 0$, system (8.23) becomes uncontrollable. To alleviate the problem, we select $g = 10^{-2}/\pi$. With this choice of g, the nonlinear term of (8.18) is exactly represented by the expression of (8.19) under the condition

$$-179.4270° < p(t) < 179.4270°.$$

To this end, in application to the vehicle with triple trailers, we arrive at the following Takagi-Sugeno fuzzy model:

Rule 1

IF $p(t)$ is "about 0 rad,"

THEN $x(t+1) = A_1 x(t) + B_1 u(t),$ \hfill (8.24)

Rule 2

IF $p(t)$ is "about π rad or $-\pi$ rad,"

THEN $x(t+1) = A_2 x(t) + B_2 u(t),$

Here,

$$p(t) = x_6(t) + \frac{v \cdot \Delta t}{2L} x_5(t),$$

$$x(t) = \begin{bmatrix} x_1(t) & x_3(t) & x_5(t) & x_6(t) & x_7(t) \end{bmatrix}^T,$$

$$A_1 = \begin{bmatrix} 1 - \frac{v \cdot \Delta t}{L} & 0 & 0 & 0 & 0 \\ \frac{v \cdot \Delta t}{L} & 1 - \frac{v \cdot \Delta t}{L} & 0 & 0 & 0 \\ 0 & \frac{v \cdot \Delta t}{L} & 1 - \frac{v \cdot \Delta t}{L} & 0 & 0 \\ 0 & 0 & \frac{v \cdot \Delta t}{L} & 1 & 0 \\ 0 & 0 & \frac{(v \cdot \Delta t)^2}{2L} & v \cdot \Delta t & 1 \end{bmatrix},$$

$$B_1 = \begin{bmatrix} \frac{v \cdot \Delta t}{l} \\ 0 \\ 0 \\ 0 \\ 0 \end{bmatrix},$$

$$A_2 = \begin{bmatrix} 1 - \frac{v \cdot \Delta t}{L} & 0 & 0 & 0 & 0 \\ \frac{v \cdot \Delta t}{L} & 1 - \frac{v \cdot \Delta t}{L} & 0 & 0 & 0 \\ 0 & \frac{v \cdot \Delta t}{L} & 1 - \frac{v \cdot \Delta t}{L} & 0 & 0 \\ 0 & 0 & \frac{v \cdot \Delta t}{L} & 1 & 0 \\ 0 & 0 & \frac{g \cdot (v \cdot \Delta t)^2}{2L} & g \cdot v \cdot \Delta t & 1 \end{bmatrix},$$

$$B_2 = \begin{bmatrix} \dfrac{v \cdot \Delta t}{l} \\ 0 \\ 0 \\ 0 \\ 0 \end{bmatrix}.$$

The overall fuzzy model is inferred as

$$x(t+1) = \sum_{i=1}^{2} h_i(p(t))\{A_i x(t) + B_i u(t)\}. \tag{8.25}$$

Figure 8.2 shows the membership functions "about 0 rad" and "about π rad or $-\pi$ rad."

Remark 21 As pointed out in Chapters 2–7, the stability conditions for the case of the common B matrix ($B_1 = \cdots = B_r$) can be simplified. In this

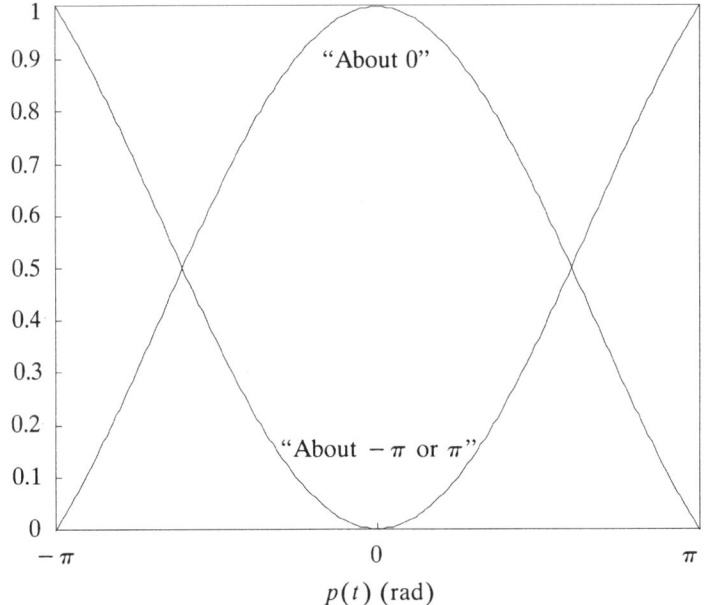

Fig. 8.2 Membership functions.

chapter we employ the general design conditions, that is, not the common **B** matrix case, although the fuzzy model of the vehicle shares common **B** among the rules.

Remark 22 As pointed out in Chapter 2, we construct the fuzzy model for a simplified nonlinear model. The fuzzy model has two rules. If we try to derive a fuzzy model for the original nonlinear system (8.1)–(8.9), 2^6 rules are required to exactly represent the nonlinear dynamics. The rule reduction leads to significant reduction of the effort for the analysis and design of control systems. This approach is useful in practice.

8.1.1 Avoidance of Jack-Knife Utilizing Constraint on Output

Let us recall the LMI constraint on the output (shown in Chapter 3) to avoid the jack-knife phenomenon. The following theorem deals with this aspect of the control design.

THEOREM 30 *Assume that the initial condition $x(0)$ is known. The constraints $\|x_1(t)\| \leq \lambda_1$, $\|x_3(t)\| \leq \lambda_2$, and $\|x_5(t)\| \leq \lambda_3$ are enforced at all times $t \geq 0$ if the LMIs*

$$\begin{bmatrix} 1 & x^T(0) \\ x(0) & X \end{bmatrix} \geq \mathbf{0}, \tag{8.26}$$

$$\begin{bmatrix} X & Xd_1^T \\ d_1 X & \lambda_1^2 I \end{bmatrix} \geq \mathbf{0}, \tag{8.27}$$

$$\begin{bmatrix} X & Xd_2^T \\ d_2 X & \lambda_2^2 I \end{bmatrix} \geq \mathbf{0}, \tag{8.28}$$

$$\begin{bmatrix} X & Xd_3^T \\ d_3 X & \lambda_3^2 I \end{bmatrix} \geq \mathbf{0} \tag{8.29}$$

hold, where $X = P^{-1}$. In the triple-trailer case, we can select $x_1(t)$, $x_3(t)$, and $x_5(t)$ as outputs:

$$x_1(t) = d_1 x(t) = \begin{bmatrix} 1 & 0 & 0 & 0 & 0 \end{bmatrix} \begin{bmatrix} x_1(t) \\ x_3(t) \\ x_5(t) \\ x_6(t) \\ x_7(t) \end{bmatrix},$$

$$x_3(t) = d_2 x(t) = \begin{bmatrix} 0 & 1 & 0 & 0 & 0 \end{bmatrix} \begin{bmatrix} x_1(t) \\ x_3(t) \\ x_5(t) \\ x_6(t) \\ x_7(t) \end{bmatrix},$$

$$x_5(t) = d_3 x(t) = \begin{bmatrix} 0 & 0 & 1 & 0 & 0 \end{bmatrix} \begin{bmatrix} x_1(t) \\ x_3(t) \\ x_5(t) \\ x_6(t) \\ x_7(t) \end{bmatrix}.$$

Proof. The proof of (8.27) is as follows. From $\|x_1(t)\| \le \lambda_1$,

$$x_1^T(t) x_1(t) = x^T(t) d_1^T d_1 x(t) \le \lambda_1^2.$$

Therefore,

$$\frac{1}{\lambda_1^2} x^T(t) d_1^T d_1 x(t) \le 1.$$

In the same way as in the proof of Theorem 12, we have

$$\frac{1}{\lambda_1^2} x^T(t) d_1^T d_1 x(t) \le x^T(t) X^{-1} x(t).$$

The above inequality is

$$x^T(t) \left(\frac{1}{\lambda_1^2} d_1^T d_1 - X^{-1} \right) x(t) \le 0.$$

Therefore, we have

$$X - \frac{1}{\lambda_1^2} X d_1^T d_1 X \ge 0.$$

Inequality (8.27) can then be obtained from the above inequality. We obtain the LMI conditions (8.28) and (8.29) in the same fashion. (Q.E.D.)

As mentioned in Chapter 3, the above LMI design conditions for output constraints depend on the initial states of the system. To alleviate this problem, the initial-state-independent condition given in Theorem 13 may be utilized in the control design.

8.2 SIMULATION RESULTS

In applying the LMI-based fuzzy control design to the backing-up control of a vehicle with triple trailers, we investigate design conditions involving stability, decay rate, constraint on the input and constraints on the output, and disturbance rejection.

The purpose of considering decay rate is to achieve a desired rate of backing up into the straight line. The system settles on to the straight line quicker for a larger decay rate. However, an aggressive decay rate could result in the occurrence of the jack-knife phenomenon and the saturation of the steering angle.

The control input is the steering angle of the vehicle. The objective of the input constraint is to avoid the saturation of the steering angle.

The outputs are the relative angles between the truck and the first trailer, the first trailer and the second trailer, and the second trailer and the third trailer. The purpose of the constraints is to avoid the jack-knife phenomenon.

The following design parameters are used in the simulation:

- The constraint on the input is $\mu = 15°$.
- The constraints on the outputs are $\lambda_i = 90°$ for $i = 1, 2, 3$.

The control input constraint "$\mu = 15°$" is the limitation of the steering angle of the vehicle. The constraint "$\lambda = 90°$" directly means the avoidance of the jack-knife phenomenon. Figure 8.3 shows the simulation results of an easy initial position for the stable fuzzy controller and the decay rate fuzzy controller. Figure 8.4 shows the simulation results of a difficult initial position for the stable fuzzy controller, the decay rate fuzzy controller and

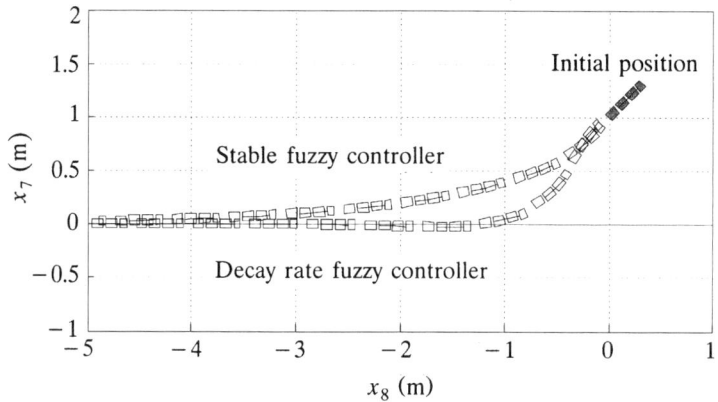

Fig. 8.3 Simulation result 1.

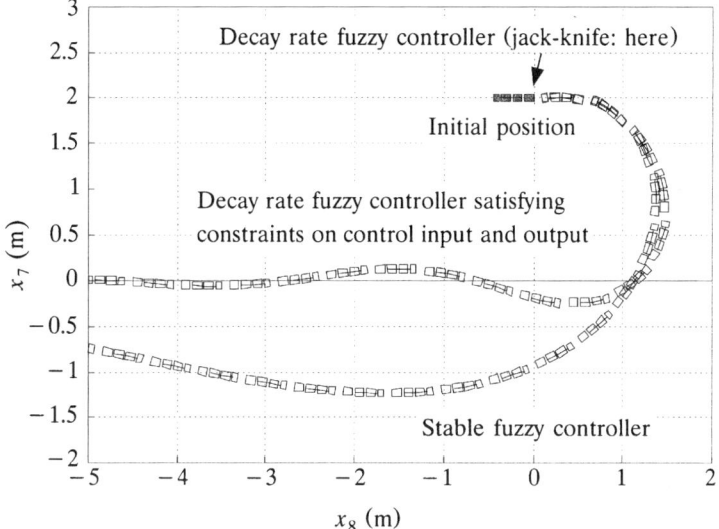

Fig. 8.4 Simulation result 2.

the fuzzy controller satisfying the decay rate and constraint on control input and output. The following important remarks can be made from the simulation results.

Remark 23 When we only invoke the stability conditions in the design, the closed-loop system does not necessarily have the desired performances in terms of decay rate and other specifications. Decay rate condition is included in the design to arrive at a speedy response of the controlled system.

Remark 24 When the vehicle is at an "easy" initial position, the decay rate design is effective, that is, the vehicle approaches the desired straight line quickly. However, if the vehicle starts from a "difficulty" initial position, the following problems occur. The first problem is the occurrence of the saturation of the steering angle. The second problem is the occurrence of the jack-knife phenomenon. In Figure 8.4, the jack-knife phenomenon occurs as soon as the decay rate control starts.

Remark 25 To circumvent these problems, we invoke design conditions involving input constraint (avoiding the steering angle saturation), output constraints (avoiding jack-knife phenomenon), and stability and decay rate. Hence we have a procedure to determine control gains to satisfy the stability and performance of the control system.

146 TRAJECTORY CONTROL OF A VEHICLE WITH MULTIPLE TRAILERS

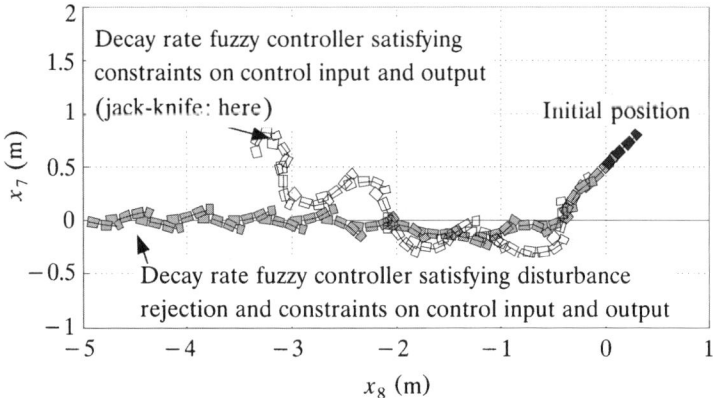

Fig. 8.5 Simulation result 3.

Next, the effect of disturbance rejection is demonstrated. Figure 8.5 shows the control result for the disturbance $v(t) = (8\pi/180)\sin(t)$ rad, where

$$E_i = \begin{bmatrix} 1 & 0 & 0 & 0 & 0 \\ 0 & 1 & 0 & 0 & 0 \\ 0 & 0 & 1 & 0 & 0 \\ 0 & 0 & 0 & 0 & 0 \\ 0 & 0 & 0 & 0 & 0 \end{bmatrix}, \quad C_i = \begin{bmatrix} 1 & 0 & 0 & 0 & 0 \\ 0 & 1 & 0 & 0 & 0 \\ 0 & 0 & 1 & 0 & 0 \\ 0 & 0 & 0 & 0 & 0 \\ 0 & 0 & 0 & 0 & 0 \end{bmatrix}$$

for $i = 1, 2$. This means that $v(t)$'s are added to the angles $x_1(t)$, $x_3(t)$, and $x_5(t)$, where the maximum values of each element in $v(t)$ correspond to $\pm 8°$. The decay rate fuzzy controller could no longer avoid the jack-knife phenomenon. The decay rate fuzzy controller together with disturbance rejection succeeds in the backing-up control though the vehicle oscillates around $x_7(t)$ due to a large disturbance.

Figure 8.6 shows the control result for a larger disturbance $v(t) = (10\pi/180)\sin(t)$ rad, where

$$E_i = \begin{bmatrix} 1 & 0 & 0 & 0 & 0 \\ 0 & 1 & 0 & 0 & 0 \\ 0 & 0 & 1 & 0 & 0 \\ 0 & 0 & 0 & 0 & 0 \\ 0 & 0 & 0 & 0 & 0 \end{bmatrix}, \quad C_i = \begin{bmatrix} 1 & 0 & 0 & 0 & 0 \\ 0 & 1 & 0 & 0 & 0 \\ 0 & 0 & 1 & 0 & 0 \\ 0 & 0 & 0 & 0 & 0 \\ 0 & 0 & 0 & 0 & 0 \end{bmatrix}$$

for $i = 1, 2$. Figure 8.7 shows the magnified area (area A in Figure 8.6) around initial positions. The decay rate fuzzy controller with disturbance rejection performs well even for this large disturbance.

These results demonstrate that the control design is effective for the backing-up control problem.

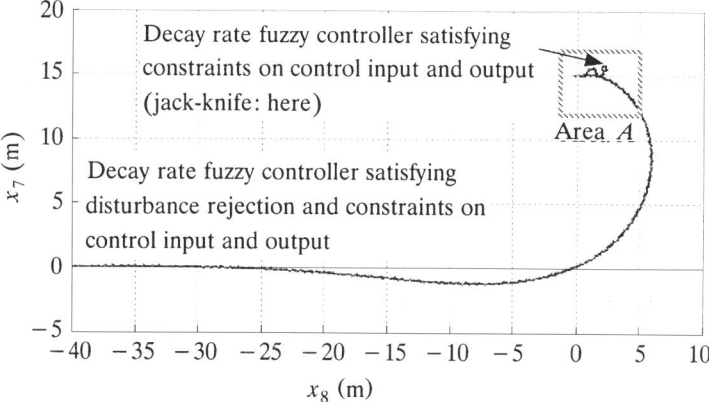

Fig. 8.6 Simulation result 4.

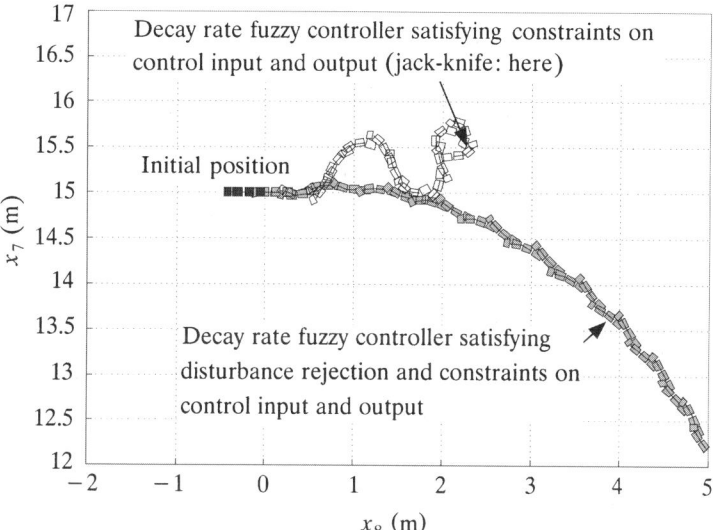

Fig. 8.7 Magnification of Figure 8.6 (area A).

8.3 EXPERIMENTAL STUDY

In this section, we describe the experimental study which is used to validate and evaluate the fuzzy control design methodology presented above. The experimental vehicle with triple trailers is shown in Figure 8.8. The experimental setup is illustrated in Figure 8.9. The forward- and backward-motion control of the vehicle is realized through a DC motor. The steering is done by a stepping motor. The consecutive angle differences $x_1(t), x_3(t), x_5(t)$ are

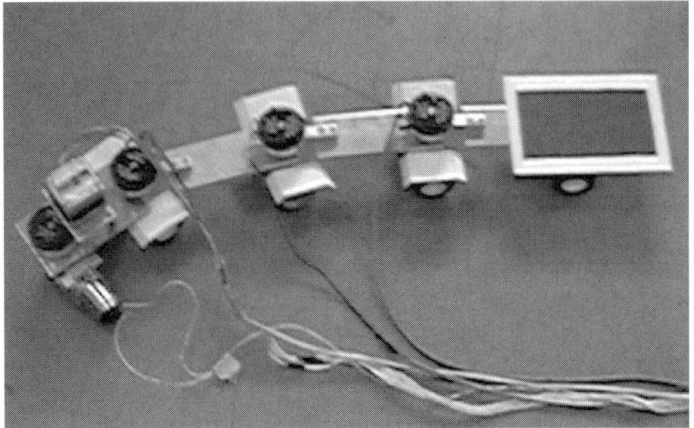

Fig. 8.8 Photograph of articulated vehicle.

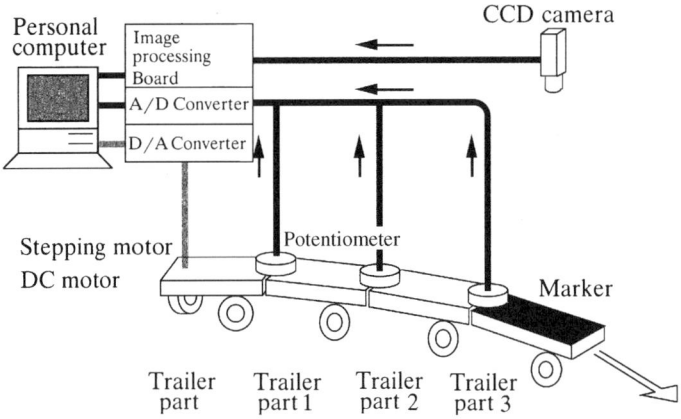

Fig. 8.9 Experimental system.

provided by three potential meters. The third trailer has a marked surface which is tracked by a CCD camera. The variables $x_6(t)$ and $x_7(t)$ are computed successively via the image processing of the CCD camera images. The control input, the steering angle $u(t)$, is determined by the PDC fuzzy controller.

Figures 8.10 and 8.11 show some representative experimental results. It is demonstrated that the backing-up control of the vehicle with triple trailers can be effectively realized by the fuzzy controller.

In the experiments, the CCD camera images are used to compute the angle and position of the third trailer. The image processing speed is slow in the experimental setup. Therefore the vehicle is controlled in a quasi-

EXPERIMENTAL STUDY 149

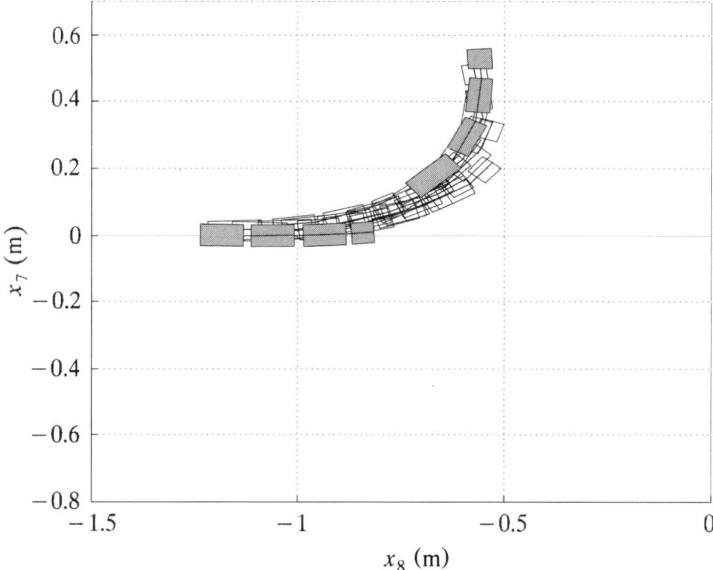

Fig. 8.10 Experimental result.

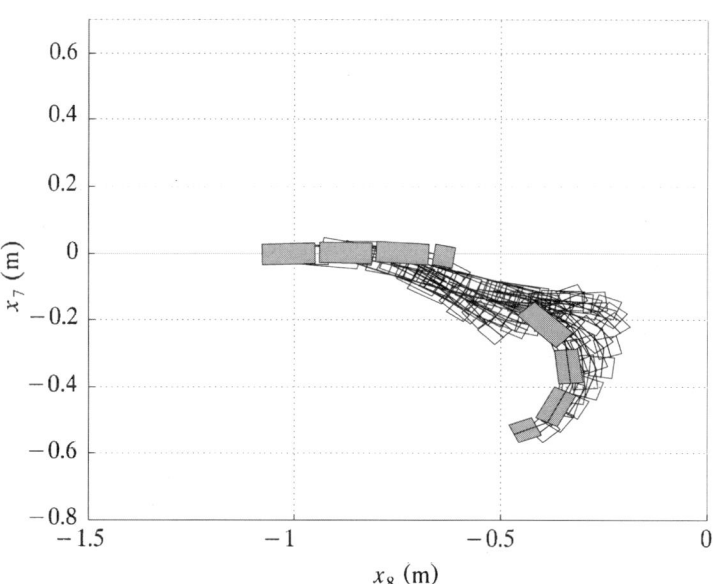

Fig. 8.11 Experimental result.

150 TRAJECTORY CONTROL OF A VEHICLE WITH MULTIPLE TRAILERS

dynamic manner, that is, the vehicle stops momentarily between controls. Also the stepsize of the vehicle movement is kept small. In addition the coverage area of the CCD camera is limited to a small area. As a result, the workspace of the vehicle is also limited so that some configurations cannot be studied within the current setup. A direct benefit of the quasi-dynamic nature of the vehicle motion is that the control-oriented models turn out to be quite suitable and effective in the control design from a practical point of view.

8.4 CONTROL OF TEN-TRAILER CASE

In this section, we present results on the stability analysis and control design for a vehicle with 10 trailers (Figure 8.12). We apply similar design tech-

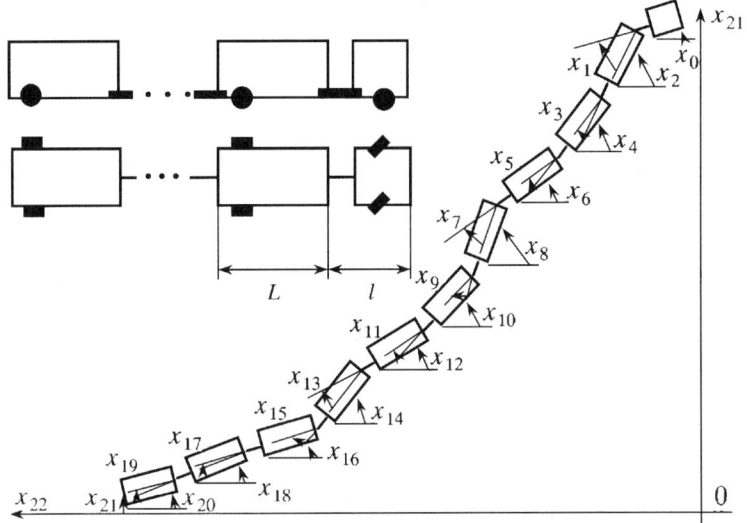

Fig. 8.12 Ten-trailer case.

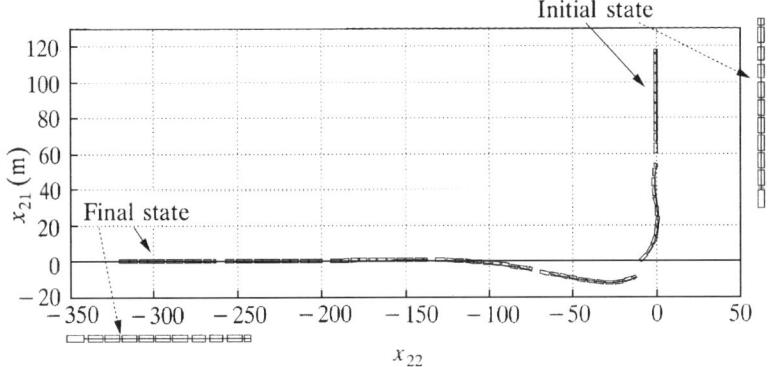

Fig. 8.13 Simulation result 1.

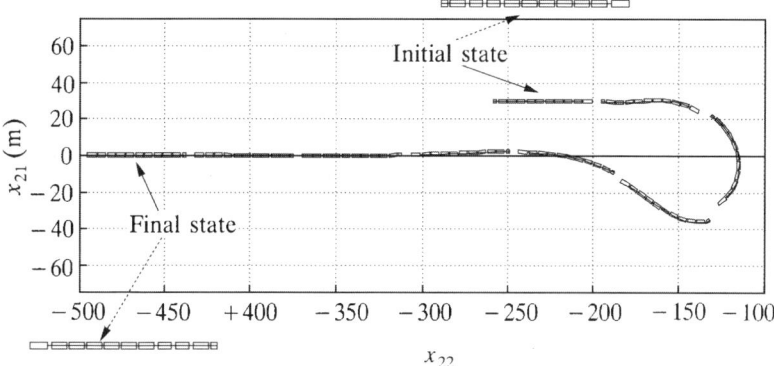

Fig. 8.14 Simulation result 2.

niques as in the triple-trailer case to the 10-trailer case. The backing-up control is very difficult even in theoretical studies. Some simulation results are summarized in Figures 8.13 and 8.14. The simulation results demonstrate the effectiveness of the systematic design techniques [2]. Even for this rather complicated system, the design methodology yields a stabilizing PDC fuzzy controller.

Remark 26 In the 10-trailer case, 2^{13} rules are required to exactly represent the nonlinear dynamics. The stabilizing controller is designed based on a simplified fuzzy model with only two rules [2]. It is demonstrated that the controller performs well for the original nonlinear system. This design example yet again demonstrates the importance of adopting a practical engineering approach to complicated problems.

REFERENCES

1. K. Tanaka and T. Kosaki, "Design of a Stable Fuzzy Controller for an Articulated Vehicle," *IEEE Trans. Syst., Man Cybernet., Part B*, Vol. 27, No. 3, pp. 552–558 (1997).
2. K. Tanaka, T. Kosaki, and H. O. Wang, "Backing Control Problem of a Mobile Robot with Multiple Trailers: Fuzzy Modeling and LMI-Based Design," *IEEE Trans. Syst., Man Cybernet., Part C*, Vol. 28, No. 3, pp. 329–337 (1998).
3. K. Tanaka, T. Taniguchi, and H. O. Wang, "An LMI Approach to Backing Control of a Vehicle with Three Trailers," Eighth International Fuzzy Systems Association World Congress, Vol. 2, Taipei, August 1999, pp. 640–644.
4. K. Tanaka, T. Taniguchi, and H. O. Wang, "Trajectory Control of an Articulated Vehicle with Triple Trailers," 1999 IEEE International Conference on Control Applications, Vol. 2, Hawaii, August, 1999.

5. D. Nguyen and B. Widrow, "The Truck Backer-Upper: An Example of Self-Learning in Neural Networks," *Proc. Int. Joint Conf. Neural Networks* (*IJCNN*-89), Vol. 2, 1989, pp. 357–363.
6. G. S. Kong and B. Kosko, "Adaptive Fuzzy Systems for Backing up a Truck-and-Trailer," *IEEE Trans. Neural Networks*, Vol. 3, No. 2, pp. 211–223 (1992).
7. H. Inoue, K. Kamei, and K. Inoue, "Auto-Generation of Fuzzy Production Rules Using Hyper-Cone Membership Function by Genetic Algorithm," *Proc. Int. Joint Conf. CFSA/IFIS/SOFT'95*, 1995, pp. 53–58.
8. M. Tokunaga and H. Ichihashi, "Backer-Upper Control of a Trailer Truck by Neuro-Fuzzy Optimal Control," *Proc. of 8th Fuzzy System Symposium*, 1992, pp. 49–52, in Japanese.
9. K. Tanaka and M. Sano, "A Robust Stabilization Problem of Fuzzy Controller Systems and Its Applications to Backing Up Control of a Truck-Trailer," *IEEE Trans. Fuzzy Syst.*, Vol. 2, No. 2, pp. 119–134 (1994).
10. K. Tanaka, T. Kosaki, and H. O. Wang, "Fuzzy Control of an Articulated Vehicle and Its Stability Analysis," 13th World Congress International Federation of Automatic Control (IFAC'96), Vol. F, San Francisco, 1996, pp. 115–120.
11. M. Sampei et al., "Arbitrary Path Tracking Control of Articulated Vehicles Using Nonlinear Control Theory," *IEEE Trans. Control Syst. Technol.*, Vol. 3, No. 1, pp. 125–131 (1995).

CHAPTER 9

FUZZY MODELING AND CONTROL OF CHAOTIC SYSTEMS

Chaotic behavior is a seemingly random behavior of a deterministic system that is characterized by sensitive dependence on initial conditions. Chaotic behavior of a physical system can either be desirable or undesirable, depending on the application. It can be beneficial in many circumstances, such as enhanced mixing of chemical reactants. Chaos can, on the other hand, entail large-amplitude motions and oscillations that might lead to system failure. The OGY method [1, 2] for controlling chaos sparked a great number of schemes on controlling chaos in linear and/or nonlinear control frameworks (e.g.. [3]–[9]). In this chapter we explore the interaction between fuzzy control systems and chaos. First, we show that fuzzy modeling techniques can be used to model chaotic dynamical systems, which also implies that fuzzy systems can be chaotic. This is not surprising given the fact that fuzzy systems are essentially nonlinear. On the subject of controlling chaos, this chapter presents a unified approach [10]–[14] using the LMI-based fuzzy control system design.

Up to this point of the book, we have mostly considered the regulation problem in control systems. Regulation is no doubt one of the most important problems in control engineering. For chaotic systems, however, there are a number of interesting nonstandard control problems. In this chapter, we develop a unified approach to address some of these problems, including stabilization, synchronization, and chaotic model following control (CMFC) for chaotic systems. A cancellation technique (CT) is presented as a main result for stabilization. The CT also plays an important role in synchronization and chaotic model following control. Two cases are considered in synchronization. The first one deals with the feasible case of the cancellation problem. The other one addresses the infeasible case of the cancellation problem. Furthermore, the chaotic model following control problem, which is more difficult than the synchronization problem, is discussed using the CT. One of the most important aspects is that the approach described here can be applied not only to stabilization and synchronization but also to the CMFC in the same control framework. That is, it is a unified approach to controlling chaos. In fact, the stabilization and the synchronization discussed here can be regarded as a special case of CMFC. Simulation results show the utility of the unified design approach. This chapter deals with the common B matrix case. Some extended results including the different B matrix case will be given in Chapter 11.

9.1 FUZZY MODELING OF CHAOTIC SYSTEMS

To utilize the LMI-based fuzzy system design techniques, we start with representing chaotic systems using T-S fuzzy models. In this regard, the techniques described in Chapter 2 are employed to construct fuzzy models for chaotic systems. In the following, a number of typical chaotic systems with the control input term added are represented in the T-S modeling framework.

Lorenz's Equation with Input Term

$$\dot{x}_1(t) = -ax_1(t) + ax_2(t) + u(t),$$
$$\dot{x}_2(t) = cx_1(t) - x_2(t) - x_1(t)x_3(t),$$
$$\dot{x}_3(t) = x_1(t)x_2(t) - bx_3(t),$$

where a, b, and c are constants and $u(t)$ is the input term. Assume that $x_1(t) \in [-d \ \ d]$ and $d > 0$. Then, we can have the following fuzzy model which exactly represents the nonlinear equation under $x_1(t) \in [-d \ \ d]$:

Rule 1

  IF $x_1(t)$ is M_1,

   THEN $\dot{x}(t) = A_1 x(t) + Bu(t)$.

Rule 2

  IF $x_1(t)$ is M_2,

   THEN $\dot{x}(t) = A_2 x(t) + Bu(t)$.

Here, $x(t) = [x_1(t) \quad x_2(t) \quad x_3(t)]^T$,

$$A_1 = \begin{bmatrix} -a & a & 0 \\ c & -1 & -d \\ 0 & d & -b \end{bmatrix}, \quad A_2 = \begin{bmatrix} -a & a & 0 \\ c & -1 & d \\ 0 & -d & -b \end{bmatrix}.$$

$$B = \begin{bmatrix} 1 \\ 0 \\ 0 \end{bmatrix}$$

$$M_1(x_1(t)) = \frac{1}{2}\left(1 + \frac{x_1(t)}{d}\right), \quad M_2(x_1(t)) = \frac{1}{2}\left(1 - \frac{x_1(t)}{d}\right).$$

In this chapter, $a = 10$, $b = 8/3$, $c = 28$ and $d = 30$.

Rossler's Equation with Input Term

$$\dot{x}_1(t) = -x_2(t) - x_3(t),$$
$$\dot{x}_2(t) = x_1(t) + ax_2(t),$$
$$\dot{x}_3(t) = bx_1(t) - \{c - x_1(t)\}x_3(t) + u(t),$$

where a, b, and c are constants. Assume that $x_1(t) \in [c - d \quad c + d]$ and $d > 0$. Then, we obtain the following fuzzy model which exactly represents the nonlinear equation under $x_1(t) \in [c - d \quad c + d]$:

Rule 1

   IF $x_1(t)$ is M_1,

   THEN $\dot{x}(t) = A_1 x(t) + Bu(t)$.

Rule 2

   IF $x_1(t)$ is M_2,

   THEN $\dot{x}(t) = A_2 x(t) + Bu(t)$.

Here, $x(t) = [x_1(t) \quad x_2(t) \quad x_3(t)]^T$.

$$A_1 = \begin{bmatrix} 0 & -1 & -1 \\ 1 & a & 0 \\ b & 0 & -d \end{bmatrix}, \quad A_2 = \begin{bmatrix} 0 & -1 & -1 \\ 1 & a & 0 \\ b & 0 & d \end{bmatrix}$$

$$B = \begin{bmatrix} 0 \\ 0 \\ 1 \end{bmatrix}.$$

$$M_1(x_1(t)) = \frac{1}{2}\left(1 + \frac{c - x_1(t)}{d}\right), \quad M_2(x_1(t)) = \frac{1}{2}\left(1 - \frac{c - x_1(t)}{d}\right).$$

In this chapter, $a = 0.34$, $b = 0.4$, and $d = 10$.

Duffing Forced-Oscillation Model

$$\dot{x}_1(t) = x_2(t)$$
$$\dot{x}_2(t) = -x_1^3(t) - 0.1 x_2(t) + 12\cos(t) + u(t)$$

Assume that $x_1(t) \in [-d \ \ d]$ and $d > 0$. Then we can have the following fuzzy model as well:

Rule 1

IF $x_1(t)$ is M_1,

  THEN $\dot{x}(t) = A_1 x(t) + Bu^*(t)$.

Rule 2

IF $x_1(t)$ is M_2,

  THEN $\dot{x}(t) = A_2 x(t) + Bu^*(t)$.

Here, $x(t) = [x_1(t) \ \ x_2(t)]^T$ and $u^*(t) = u(t) + 12\cos(t)$,

$$A_1 = \begin{bmatrix} 0 & 1 \\ 0 & -0.1 \end{bmatrix}, \quad A_2 = \begin{bmatrix} 0 & 1 \\ -d^2 & -0.1 \end{bmatrix},$$

$$B = \begin{bmatrix} 0 \\ 1 \end{bmatrix},$$

$$M_1(x_1(t)) = 1 - \frac{x_1^2(t)}{d^2}, \quad M_2(x_1(t)) = \frac{x_1^2(t)}{d^2}.$$

In this chapter, $d = 50$ in this model.

Henon Mapping Model

$$x_1(t+1) = -x_1^2(t) + 0.3x_2(t) + 1.4 + u(t),$$
$$x_2(t+1) = x_1(t).$$

Assume that $x_1(t) \in [-d \quad d]$ and $d > 0$. The following equivalent fuzzy model can be constructed as well:

Rule 1

IF $x_1(t)$ is M_1,

    THEN $x(t+1) = A_1 x(t) + Bu^*(t).$

Rule 2

IF $x_1(t)$ is M_2,

    THEN $x(t+1) = A_2 x(t) + Bu^*(t).$

Here, $x(t) = [x_1(t) \quad x_2(t)]^T$ and $u^*(t) = u(t) + 1.4$,

$$A_1 = \begin{bmatrix} d & 0.3 \\ 1 & 0 \end{bmatrix}, \quad A_2 = \begin{bmatrix} -d & 0.3 \\ 1 & 0 \end{bmatrix},$$

$$B = \begin{bmatrix} 1 \\ 0 \end{bmatrix},$$

$$M_1(x_1(t)) = \frac{1}{2}\left(1 - \frac{x_1(t)}{d}\right), \quad M_2(x_1(t)) = \frac{1}{2}\left(1 + \frac{x_1(t)}{d}\right).$$

In this chapter, $d = 30$ in this model.

In all cases above, the fuzzy models exactly represent the original systems. As mentioned in Remark 5, the Takagi-Sugeno fuzzy model is a universal approximator for nonlinear dynamical systems. Other chaotic systems can be approximated by the Takagi-Sugeno fuzzy models.

The fuzzy models above have the common B matrix in the consequent parts and $x_1(t)$ in the premise parts. In this chapter, all the fuzzy models are assumed to be the common B matrix case, that is, the fuzzy model (9.1) is considered. The different B matrix case will be discussed in Chapter 11. That is, Chapter 11 deals with the more general setting.

Plant Rule i

IF $z_1(t)$ is M_{i1} and $\cdots$ and $z_p(t)$ is M_{ip},

    THEN $sx(t) = A_i x(t) + Bu(t), \quad i = 1, 2, \ldots, r,$ (9.1)

where $p = 1$ and $z_1(t) = x_1(t)$. Equation (9.1) is represented by the defuzzification form

$$sx(t) = \frac{\sum_{i=1}^{r} w_i(z(t))\{A_i x(t) + Bu(t)\}}{\sum_{i=1}^{r} w_i(z(t))}$$

$$= \sum_{i=1}^{r} h_i(z(t))\{A_i x(t) + Bu(t)\}, \qquad (9.2)$$

where $sx(t)$ denote $\dot{x}(t)$ and $x(t + 1)$ for CFS and DFS, respectively. In the fuzzy models above for chaotic systems, $z(t) = z_1(t) = x_1(t)$.

Remark 27 The fuzzy models above have a single input. We can also consider the multi-input case. For instance, we may consider Lorenz's equation with multi-inputs:

$$\dot{x}_1(t) = -ax_1(t) + ax_2(t) + u_1(t),$$
$$\dot{x}_2(t) = cx_1(t) - x_2(t) - x_1(t)x_3(t) + u_2(t),$$
$$\dot{x}_3(t) = x_1(t)x_2(t) - bx_3(t) + u_3(t).$$

As before, we can derive the following fuzzy model to exactly represent the nonlinear equation under $x_1(t) \in [-d \quad d]$:

Rule 1

  IF $x_1(t)$ is M_1,

  THEN $\dot{x}(t) = A_1 x(t) + Bu(t),$

(9.3)

Rule 2

  IF $x_1(t)$ is M_2,

  THEN $\dot{x}(t) = A_2 x(t) + Bu(t),$

where $u(t) = [u_1(t) \ u_2(t) \ u_3(t)]^T$ and $x(t) = [x_1(t) \ x_2(t) \ x_3(t)]^T$,

$$A_1 = \begin{bmatrix} -a & a & 0 \\ c & -1 & -d \\ 0 & d & -b \end{bmatrix}, \quad A_2 = \begin{bmatrix} -a & a & 0 \\ c & -1 & d \\ 0 & -d & -b \end{bmatrix},$$

$$B = \begin{bmatrix} 1 & 0 & 0 \\ 0 & 1 & 0 \\ 0 & 0 & 1 \end{bmatrix},$$

$$M_1(x_1(t)) = \frac{1}{2}\left(1 + \frac{x_1(t)}{d}\right), \quad M_2(x_1(t)) = \frac{1}{2}\left(1 - \frac{x_1(t)}{d}\right).$$

This fuzzy model with three inputs is used as a design example later in this chapter.

9.2 STABILIZATION

Two techniques for the stabilization of chaotic systems (or nonlinear systems) are presented in this section. We first consider the common **B** stabilization problem followed by a so-called cancellation technique. In particular, the cancellation technique plays an important role in synchronization and chaotic model following control, which are presented in Sections 9.3 and 9.4, respectively.

9.2.1 Stabilization via Parallel Distributed Compensation

Equation (9.4) shows the PDC controller for the fuzzy models given in Section 9.1:

Rule 1

IF $x_1(t)$ is M_1,

THEN $u(t) = -F_1 x(t)$. (9.4)

Rule 2

IF $x_1(t)$ is M_2,

THEN $u(t) = -F_2 x(t)$.

Note that the chaotic systems under consideration in the previous section are represented (coincidentally) by simple T-S fuzzy models with two rules. Therefore the following PDC fuzzy controller also has only two rules:

$$u(t) = -\frac{\sum_{i=1}^{2} w_i(z(t)) F_i x(t)}{\sum_{i=1}^{2} w_i(z(t))} = -\sum_{i=1}^{2} h_i(z(t)) F_i x(t). \quad (9.5)$$

By substituting (9.5) into (9.2), we have

$$sx(t) = \sum_{i=1}^{r} h_i(z(t))(A_i - BF_i) x(t), \quad (9.6)$$

where $r = 2$. We recall stable and decay rate fuzzy controller designs for CFS and DFS cases, where the following conditions are simplified due to the common B matrix case. These design conditions are all given for the general T-S model with r number of rules.

Stable Fuzzy Controller Design: CFS Find $X > 0$ and M_i $(i = 1, \ldots, r)$ satisfying

$$-XA_i^T - A_i X + M_i^T B^T + BM_i > 0,$$

where $X = P^{-1}$ and $M_i = F_i X$.

Stable Fuzzy Controller Design: DFS Find $X > 0$ and M_i $(i = 1, \ldots, r)$ satisfying

$$\begin{bmatrix} X & XA_i^T - M_i^T B^T \\ A_i X - BM_i & X \end{bmatrix} > 0,$$

where $X = P^{-1}$ and $M_i = F_i X$.

Decay Rate Fuzzy Controller Design: CFS

$$\underset{X, M_1, \ldots, M_r}{\text{maximize}} \; \alpha$$

subject to $X > 0$,

$$-XA_i^T - A_i X + M_i^T B^T + BM_i - 2\alpha X > 0,$$

where $\alpha > 0$, $X = P^{-1}$ and $M_i = F_i X$.

Decay Rate Fuzzy Controller Design: DFS

$$\underset{X, M_1, \ldots, M_r}{\text{minimize}} \; \beta$$

subject to $X > 0$,

$$\begin{bmatrix} \beta X & XA_i^T - M_i^T B^T \\ A_i X - BM_i & X \end{bmatrix} > 0,$$

where $X = P^{-1}$ and $M_i = F_i X$. It should be noted that $0 \le \beta < 1$.

Example 10 Let us consider the fuzzy model for Lorenz's equation with the input term. The stable fuzzy controller design for the CFS is feasible. Figure 9.1 shows the control result, where the control input is added at $t > 10$ sec. It can be seen that the designed fuzzy controller stabilizes the chaotic system, that is, $x_1(0) \to 0$, $x_2(0) \to 0$, and $x_3(0) \to 0$.

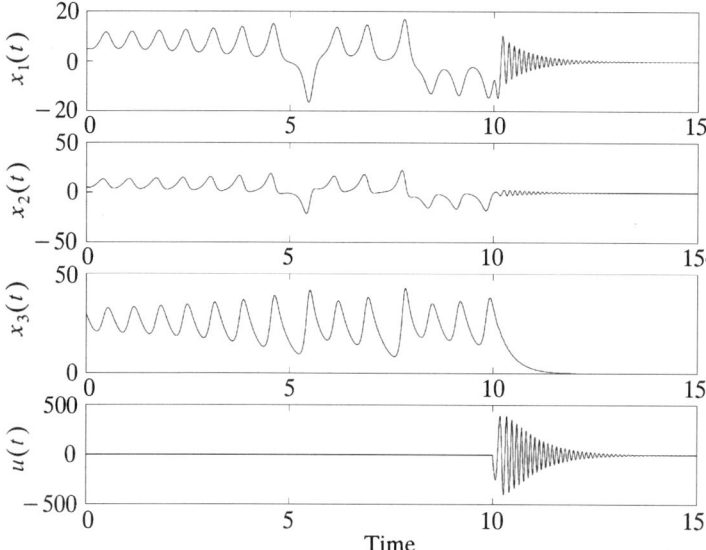

Fig. 9.1 Control result (Example 10).

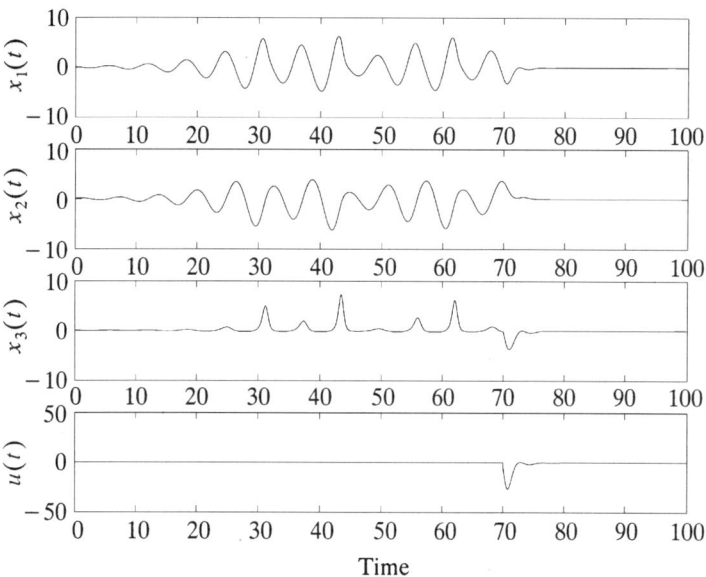

Fig. 9.2 Control result (Example 11).

Example 11 We design a stable fuzzy controller for Rossler's equation with the input as well. The stable fuzzy controller design for the CFS is feasible. Figure 9.2 shows the control result, where the control input is added at $t > 70$ sec. It can be seen that the designed fuzzy controller stabilizes the chaotic system.

162 FUZZY MODELING AND CONTROL OF CHAOTIC SYSTEMS

Example 12 We design a stable fuzzy controller for Duffing forced oscillation with the input. The stable fuzzy controller design for the CFS is feasible. Figure 9.3 shows the control result, where the control input is added at $t > 30$ sec. The designed fuzzy controller stabilizes the chaotic system.

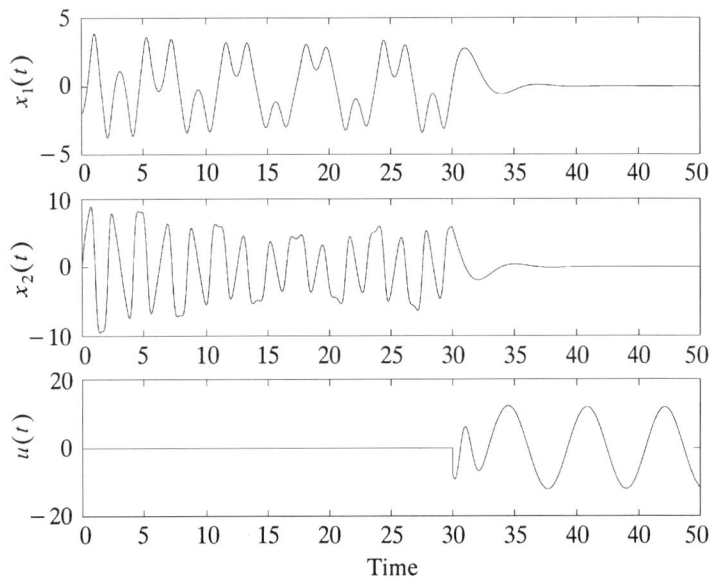

Fig. 9.3 Control result (Example 12).

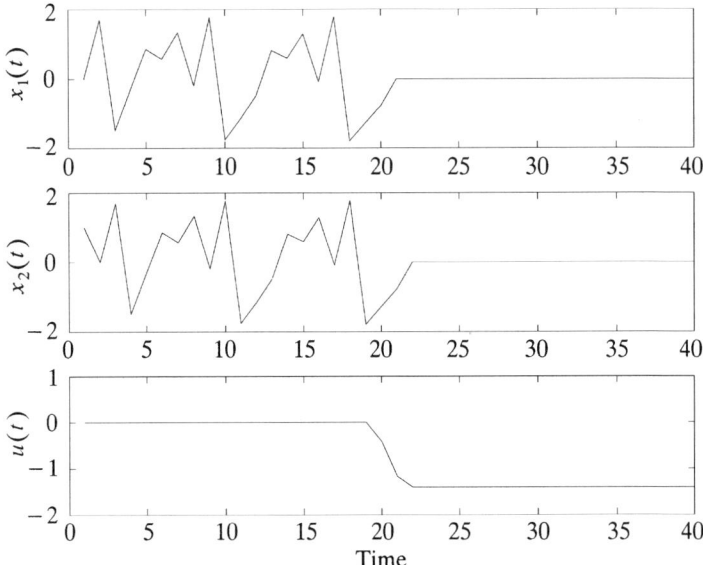

Fig. 9.4 Control result (Example 13).

Example 13 Let us consider the fuzzy model for the Henon mapping model. The stable fuzzy controller design for the DFS is feasible. Figure 9.4 shows the control result, where the control input is added at $t > 20$ sec.

Example 14 Consider the fuzzy model for Lorenz's equation with the input term. The decay rate fuzzy controller design for the CFS is feasible. Figure 9.5 shows the control result, where the control input is added at $t > 10$ sec. Note that the speed of response of the decay rate fuzzy controller is better than that of the stable fuzzy controller in Example 10.

Example 15 Consider the fuzzy model for Lorenz's equation with the input term. The fuzzy controller design satisfying the stability conditions and the constraint on the output for the CFS is feasible, where $\lambda = 9$ and $C = C_1 = C_2 = [1 \ 0 \ 0]$. This means that $x_1(t)$ is selected as the output, that is, $y(t) = x_1(t) = Cx(t)$. Figure 9.6 shows the control result, where the control input is added at $t > 10$ sec. Note that the fuzzy controller satisfies $\max_t \|x_1(t)\| \leq \lambda$, but the control effort is very large.

Example 16 To solve the excessive control effort problem, the constraint on the control input is added to the design of Example 15. The fuzzy controller design satisfying the stability conditions and the constraints on the output and the control input for the CFS is feasible, where $\lambda = 9$, $\mu = 500$, and $C = C_1 = C_2 = [1 \ 0 \ 0]$. Figure 9.7 shows the control result, where the control input is added at $t > 10$ sec. The designed fuzzy controller stabilizes the chaotic system. It should be emphasized that the control input and output satisfy the constraints, that is, $\max_t \|u(t)\|_2 \leq \mu$ and $\max_t \|x_1(t)\|_2 \leq \lambda$.

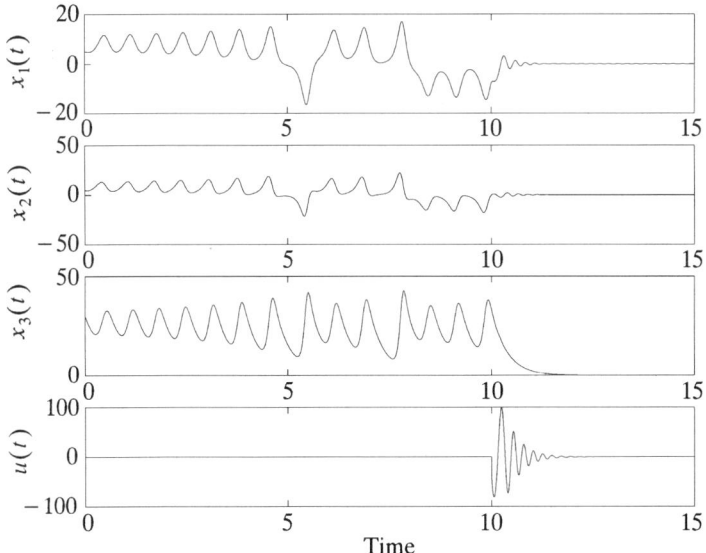

Fig. 9.5 Control result (Example 14).

164 FUZZY MODELING AND CONTROL OF CHAOTIC SYSTEMS

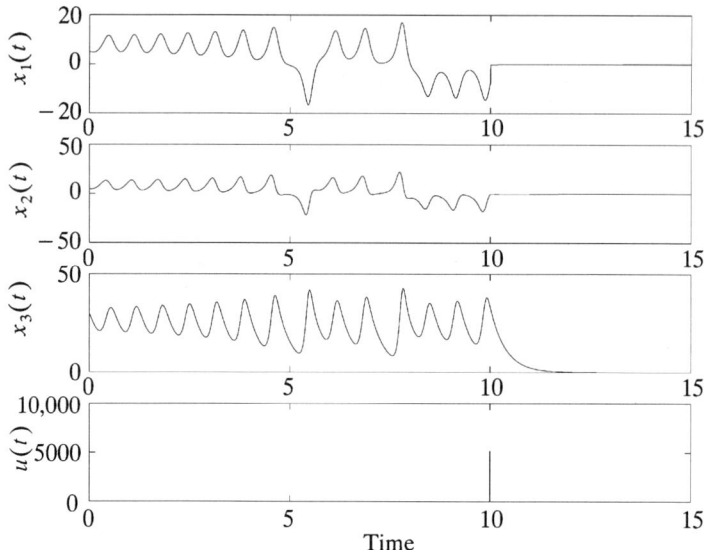

Fig. 9.6 Control result (Example 15).

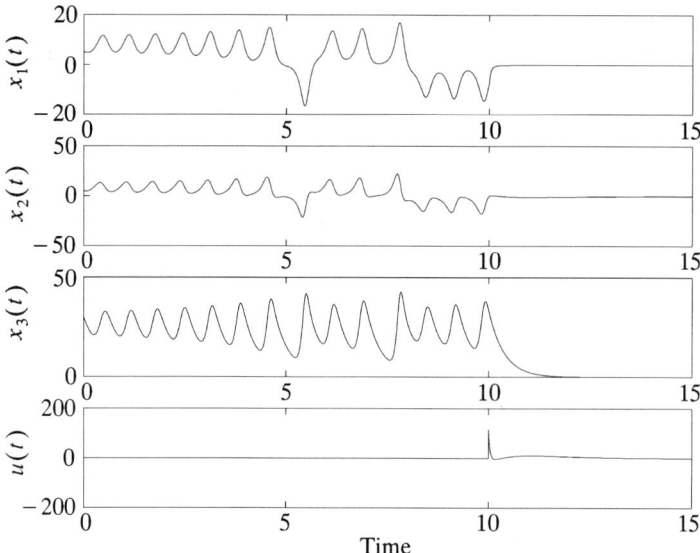

Fig. 9.7 Control result (Example 16).

STABILIZATION   165

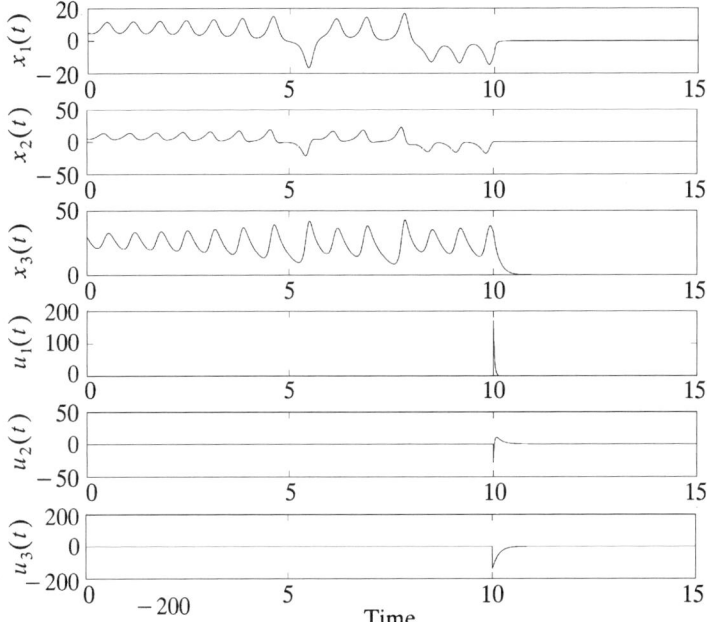

Fig. 9.8 Control result (Example 17).

Example 17 Consider Lorenz's equation with three inputs described in Remark 27. The fuzzy controller design satisfying the stability condition and the constraints on the output and the control input for the CFS is feasible, where $\lambda = 9$, $\mu = 500$, and $C = C_1 = C_2 = [1\ 0\ 0]$. Figure 9.8 shows the control result, where the control input is added at $t > 10$ sec. Note that the control input and output also satisfy the constraints, that is, $\max_t \|u(t)\|_2 \leq \mu$ and $\max_t \|x_1(t)\|_2 \leq \lambda$.

9.2.2 Cancellation Technique

This subsection discusses a cancellation technique (CT). This approach attempts to cancel the nonlinearity of a chaotic system via a PDC controller. If this problem is feasible, the resulting controller can be considered as a solution to the so-called global linearization and the feedback linearization problems. The conditions for realizing the cancellation via the PDC are given in the following theorem.

THEOREM 31 *Chaotic systems represented by the fuzzy system* (9.2) *are exactly linearized via the fuzzy controller* (9.5) *if there exist the feedback gains* F_i

such that

$$\{(A_1 - BF_1) - (A_i - BF_i)\}^T$$
$$\times \{(A_1 - BF_1) - (A_i - BF_i)\} = 0, \quad i = 2, 3, \ldots, r. \quad (9.7)$$

Then, the overall control system is linearized as $sx(t) = Gx(t)$, where $G = A_1 - BF_1 = A_i - BF_i$.

Proof. It is obvious that $G = A_1 - BF_1 = A_i - BF_i$ if the condition (9.7) holds.

The conditions are applicable to both the CFS and the DFS. If B is a nonsingular matrix, the system is exactly linearized using $F_i = B^{-1}(G - A_i)$. However, the assumption that B is a nonsingular matrix is very strict. If B is not a nonsingular matrix, the conditions of Theorem 31 can still be utilized by the following approximation technique. That is, the equality conditions of Theorem 31 are approximate by the following inequality conditions:

$$X\{(A_1 - BF_1) - (A_i - BF_i)\}^T$$
$$\times \{(A_1 - BF_1) - (A_i - BF_i)\}X < \beta S, \quad i = 2, 3, \ldots, r,$$

where X is a positive definite matrix and S is a positive definite matrix such that $S^T S < I$. The conditions (9.7) are likely to be satisfied if the elements in βS are near zero, that is, $\beta S \approx 0$, in the above inequality. Using the Schur complement, we obtain

$$\begin{bmatrix} \beta S & X\{(A_1 - BF_1) - (A_i - BF_i)\}^T \\ \{(A_1 - BF_1) - (A_i - BF_i)\}X & I \end{bmatrix} > 0,$$

$$i = 2, 3, \ldots, r.$$

Define $M_i = F_i X$ so that for $X > 0$ we have $F_i = M_i X^{-1}$. Substituting into the inequalities above yields

$$\begin{bmatrix} \beta S & \{(A_1 X - BM_1) - (A_i X - BM_i)\}^T \\ \{(A_1 X - BM_1) - (A_i X - BM_i)\} & I \end{bmatrix} > 0,$$

$$i = 2, 3, \ldots, r.$$

Note that G is not always a stable matrix even if the condition of Theorem 31 holds.

From the discussion above as well as the stability conditions described in this section, we define the following design problems using the CT:

Stable Fuzzy Controller Design Using the CT: CFS

$$\underset{X, S, M_1, M_2, \ldots, M_r}{\text{minimize}} \beta$$

subject to $X > 0$, $\beta > 0$, $S > 0$,

$$\begin{bmatrix} I & S \\ S & I \end{bmatrix} > 0,$$

$$-A_i X + BM_i - XA_i^T + M_i^T B^T > 0, \quad i = 1, 2, \ldots, r,$$

$$\begin{bmatrix} \beta S & \{(A_1 X - BM_1) - (A_i X - BM_i)\}^T \\ \{(A_1 X - BM_1) - (A_i X - BM_i)\} & I \end{bmatrix} > 0,$$

$$i = 2, 3, \ldots, r.$$

where $X = P^{-1}$ and $M_i = F_i X$.

Stable Fuzzy Controller Design Using the CT: DFS

$$\underset{X, S, M_1, M_2, \ldots, M_r}{\text{minimize}} \beta$$

subject to $X > 0$, $\beta > 0$, $S > 0$,

$$\begin{bmatrix} I & S \\ S & I \end{bmatrix} > 0,$$

$$\begin{bmatrix} X & XA_i - M_i^T B^T \\ A_i X - BM_i & X \end{bmatrix} > 0, \quad i = 1, 2, \ldots, r,$$

$$\begin{bmatrix} \beta S & \{(A_1 X - BM_1) - (A_i X - BM_i)\}^T \\ \{(A_1 X - BM_1) - (A_i X - BM_i)\} & I \end{bmatrix} > 0,$$

$$i = 2, 3, \ldots, r,$$

where $X = P^{-1}$ and $M_i = F_i X$.

Decay Rate Fuzzy Controller Design Using the CT: CFS

$$\underset{X, S, M_1, M_2, \ldots, M_r}{\text{maximize}} \alpha$$

$$\underset{X, S, M_1, M_2, \ldots, M_r}{\text{minimize}} \beta$$

subject to $X > 0$, $\beta > 0$, $\alpha > 0$, $S > 0$,

$$\begin{bmatrix} I & S \\ S & I \end{bmatrix} > 0,$$

$$-A_i X + BM_i - XA_i^T + M_i^T B^T - 2\alpha X > 0, \quad i = 1, 2, \ldots, r,$$

$$\begin{bmatrix} \beta S & \{(A_1 X - BM_1) - (A_i X - BM_i)\}^T \\ \{(A_1 X - BM_1) - (A_i X - BM_i)\} & I \end{bmatrix} > 0,$$

$$i = 2, 3, \ldots, r,$$

where $X = P^{-1}$ and $M_i = F_i X$.

Decay Rate Fuzzy Controller Design Using the CT: DFS

$$\underset{X, S, M_1, M_2, \ldots, M_r}{\text{minimize}} \ \alpha$$

$$\underset{X, S, M_1, M_2, \ldots, M_r}{\text{minimize}} \ \beta$$

subject to $X > 0, \quad \beta > 0, \ 0 \leq \alpha < 1, \ S > 0,$

$$\begin{bmatrix} I & S \\ S & I \end{bmatrix} > 0,$$

$$\begin{bmatrix} \alpha X & XA_i - M_i^T B^T \\ A_i X - BM_i & X \end{bmatrix} > 0, \quad i = 1, 2, \ldots, r,$$

$$\begin{bmatrix} \beta S & \{(A_1 X - BM_1) - (A_i X - BM_i)\}^T \\ \{(A_1 X - BM_1) - (A_i X - BM_i)\} & I \end{bmatrix} > 0,$$

$$i = 2, 3, \ldots, r,$$

where $X = P^{-1}$ and $M_i = F_i X$.

Remark 28 In the LMIs above, if the elements in $\beta \cdot S$ are near zero, that is, $\beta \cdot S \approx 0$, the CT problems are feasible. In this case, $G = A_i - BF_i$ for all i and G is a stable matrix.

Remark 29 The decay rate design problems have two parameters α and β to be maximized or minimized. These problems can be solved as follows: For instance, first minimize β, where $\alpha = 0$. After β is fixed, α can be minimized or maximized. This procedure may be repeated to obtain a tighter solution. Another way is to introduce an idea for mixing α and β as shown in Theorems 28 and 29.

Of course, other LMI conditions, for example, the constraints on control input and output, can be added to the design problem. Thus, by combining a variety of control performances represented by LMIs, we can realize multiobjective control. Chapter 13 will present multiobjective control based on dynamic output feedback.

STABILIZATION 169

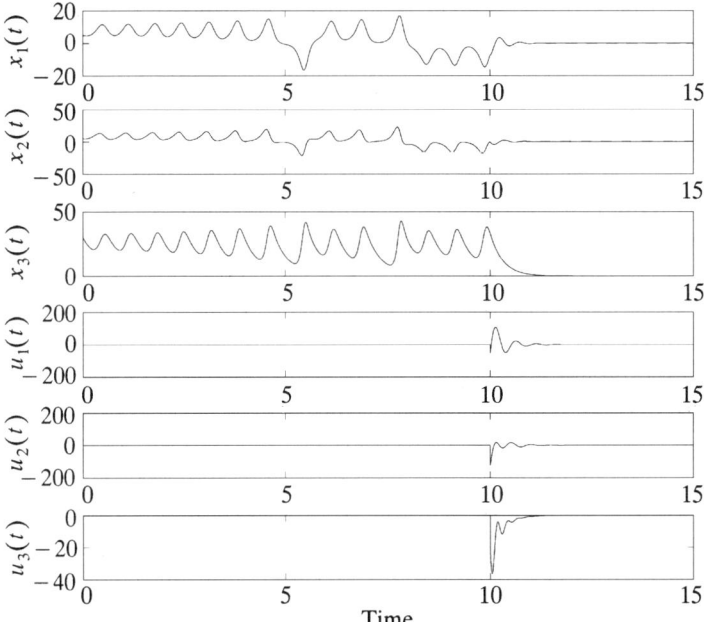

Fig. 9.9 Control result (Example 18).

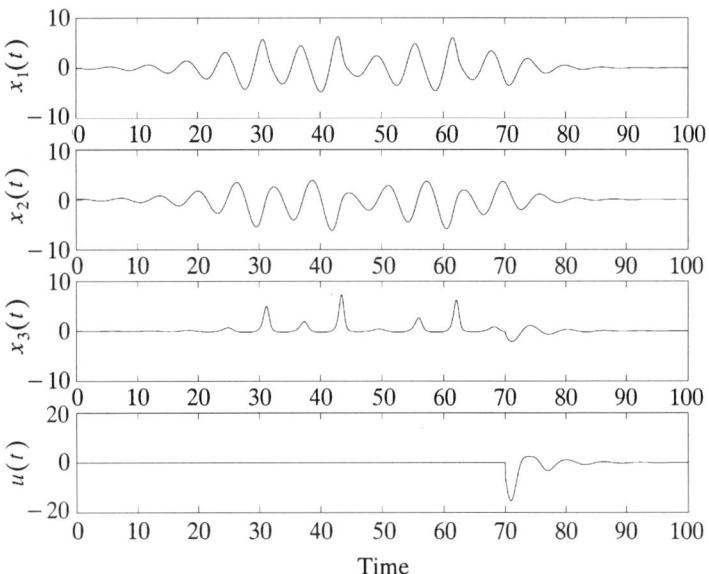

Fig. 9.10 Control result (Example 19).

Example 18 The stable fuzzy controller design to realize the CT for Lorenz's equation with three inputs is feasible. Figure 9.9 shows the control result, where the control input is added at $t > 10$ sec. The designed fuzzy controller linearizes and stabilizes the chaotic system.

Example 19 Let us consider the fuzzy model for Rossler's equation with the input term. The stable fuzzy controller design using the CT is feasible. Figure 9.10 shows the control result, where the control input is added at time > 70 sec. It can be seen that the designed fuzzy controller linearizes and stabilizes the chaotic system.

9.3 SYNCHRONIZATION

In addition to the stabilization of chaotic systems (Section 9.2), chaos synchronization and model following are perhaps more stimulating problems in that chaotic behavior is exploited for potential applications such as secure communications.

In this section, we consider the following synchronization problem: design the control input so that the controlled system achieves asymptotic synchronization with the reference system given that two systems start from different initial conditions. Here the reference system and controlled system are taken to be the same chaotic oscillator except that the controlled system has control input(s) (the controlled system can be viewed as an observer of the reference system). In this section, only the special case of full state feedback based on the CT is considered. Two cases of the cancellation problem are discussed.

Case 1: The cancellation problem is feasible, that is, all the elements in $\beta \cdot S$ are near zero.

Case 2: The cancellation problem is infeasible, that is, all the elements in $\beta \cdot S$ are not near zero.

9.3.1 Case 1

Consider a reference fuzzy model which represents a reference chaotic system.

Reference Rule i

$\quad$ **IF** $z_{R1}(t)$ is M_{i1} and $\cdots$ and $z_{Rp}(t)$ is M_{ip},

$\quad\quad$ **THEN** $sx_R(t) = A_i x_R(t), \quad i = 1, 2, \ldots, r,$ $\quad\quad\quad$ (9.8)

where $z_R(t) = [z_{R1}(t)\ z_{R2}(t)\cdots z_{Rp}(t)]^T$. The defuzzification process is given as

$$sx_R(t) = \sum_{i=1}^{r} h_i(z_R(t))A_i x_R(t). \tag{9.9}$$

Assume that $e(t) = x(t) - x_R(t)$. Then, from (9.2) and (9.9), we have

$$se(t) = \sum_{i=1}^{r} h_i(z(t))A_i x(t) - \sum_{i=1}^{r} h_i(z_R(t))A_i x_R(t) + Bu(t). \tag{9.10}$$

We design two fuzzy subcontrollers to realize the synchronization:

Subcontroller A

Control Rule i

IF $z_1(t)$ is M_{i1} and $\cdots$ and $z_p(t)$ is M_{ip},

THEN $u_A(t) = -F_i x(t), \quad i = 1, 2, \ldots, r.$ \quad (9.11)

Subcontroller B

Control Rule i

IF $z_{R1}(t)$ is M_{i1} and $\cdots$ and $z_{Rp}(t)$ is M_{ip},

THEN $u_B(t) = F_i x_R(t), \quad i = 1, 2, \ldots, r.$ \quad (9.12)

The overall fuzzy controller is constructed by combining the two subcontrollers:

$$u(t) = u_A(t) + u_B(t)$$
$$= -\sum_{i=1}^{r} h_i(z(t))F_i x(t) + \sum_{i=1}^{r} h_i(z_R(t))F_i x_R(t). \tag{9.13}$$

The design is to determine the feedback gains F_i. By substituting (9.13) into (9.10), we obtain

$$se(t) = \sum_{i=1}^{r} h_i(z(t))(A_i - BF_i)x(t)$$
$$- \sum_{i=1}^{r} h_i(z_R(t))(A_i - BF_i)x_R(t). \tag{9.14}$$

172 FUZZY MODELING AND CONTROL OF CHAOTIC SYSTEMS

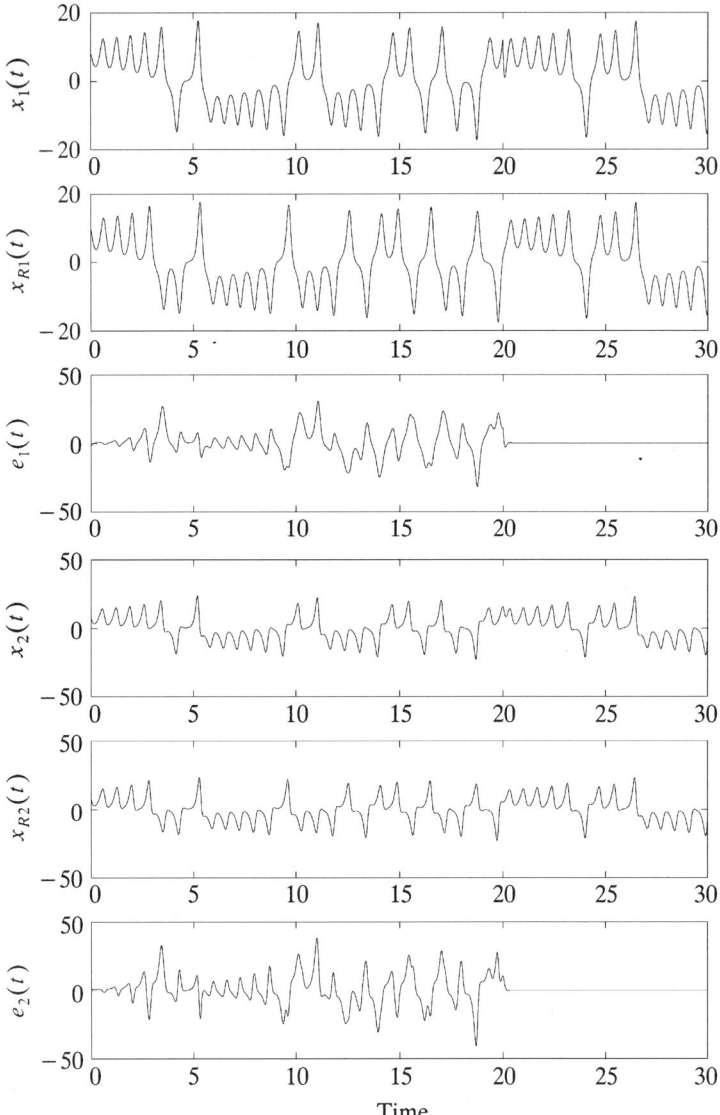

Fig. 9.11 Control result 1 (Example 20).

Applying Theorem 31 to the error system (9.14), we attempt to linearize the error system using the fuzzy control law (9.13). If the conditions of Theorem 31 hold, the linearized error system becomes $se(t) = Ge(t)$, where $G = A_i - BF_i$. As mentioned before, the G is not always a stable matrix even if the conditions of Theorem 31 hold. If we can find feedback gains F_i such that G is a stable matrix, the fuzzy controller linearizes and stabilizes the error system. The linearizable and stable fuzzy controllers with the feedback gains

SYNCHRONIZATION   **173**

F_i can be designed by solving the LMI-based design problems using the approximate CT algorithm described in Section 9.2.

Example 20 The decay rate fuzzy controller design to realize the synchronization for Lorenz's equation with three input terms is feasible. Figures 9.11 and 9.12 show the control result, where the control input is added at $t > 20$

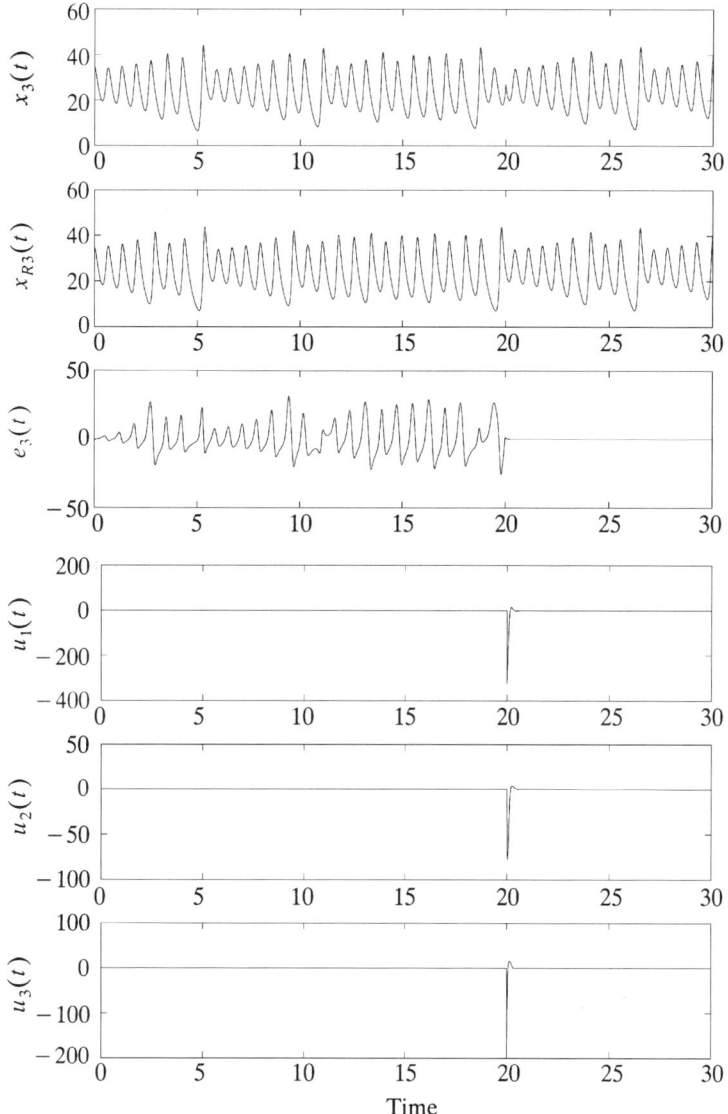

Fig. 9.12 Control result 2 (Example 20).

174 FUZZY MODELING AND CONTROL OF CHAOTIC SYSTEMS

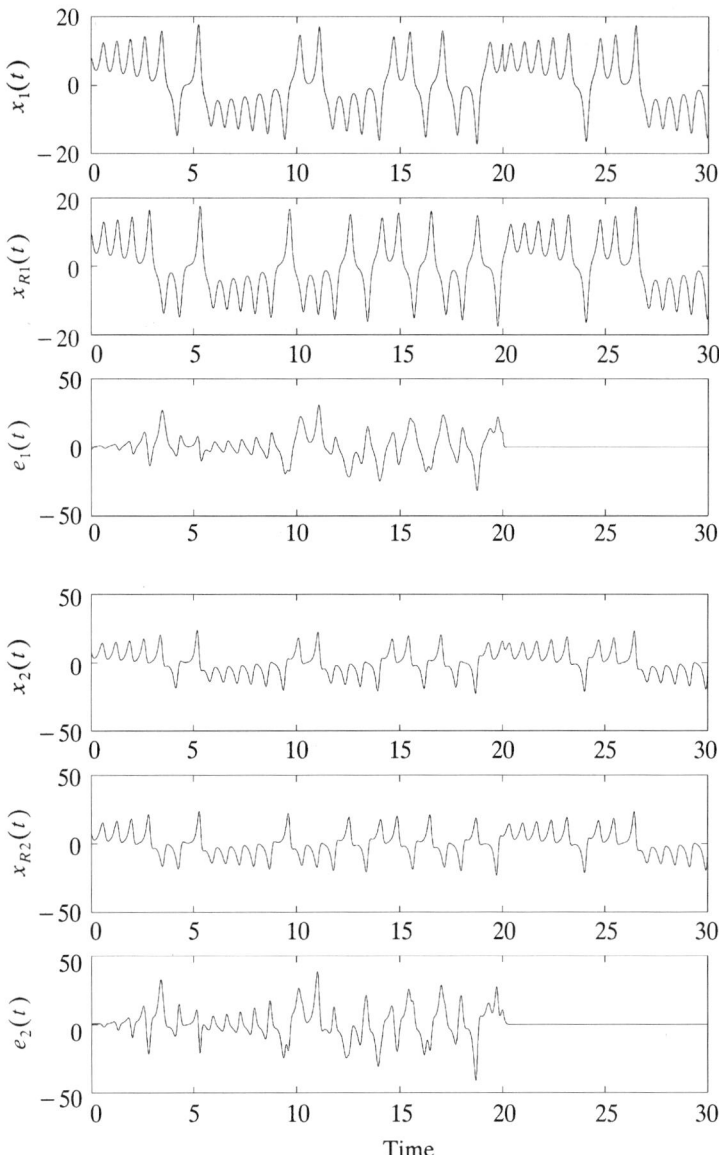

Fig. 9.13 Control result 1 (Example 21).

sec and the initial values of $x(0)$ are slightly different from those of $x_R(0)$. It can be seen that the designed fuzzy controller linearizes and stabilizes the error system, that is, $e_1(t) \to 0$, $e_2(t) \to 0$, and $e_3(t) \to 0$.

Example 21 Consider Lorenz's equation with three inputs. The fuzzy controller design satisfying the stability conditions and the constraints on the output and the control input for the CFS is feasible, where $\lambda = 100$,

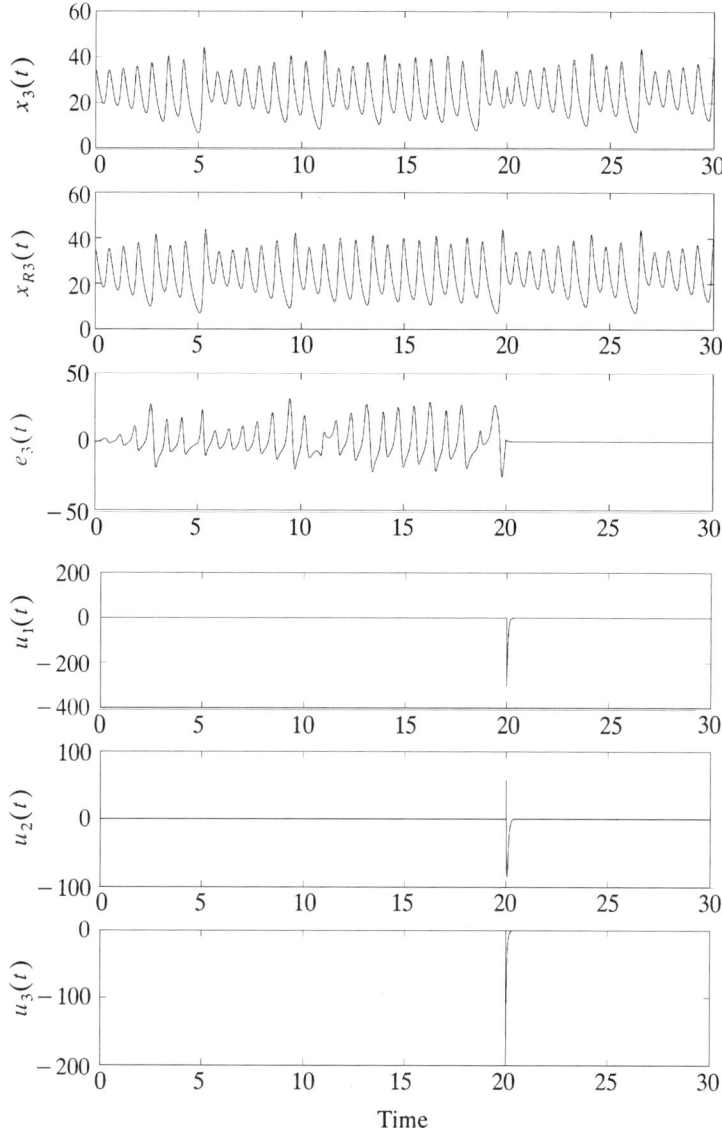

Fig. 9.14 Control result 2 (Example 21).

$\mu = 500$, and $C = C_1 = C_2 = I_3$. This means that $e_1(t)$, $e_2(t)$, and $e_3(t)$ are selected as the outputs, that is, $e(t) = [e_1(t)\ e_2(t)\ e_3(t)] = Cx(t)$. Figures 9.13 and 9.14 show the control result. The designed fuzzy controller linearizes and stabilizes the error system. It should be emphasized that the control input and output satisfy the constraints, that is, $\max_t \|u(t)\|_2 \le \mu$ and $\max_t \|e(t)\|_2 \le \lambda$.

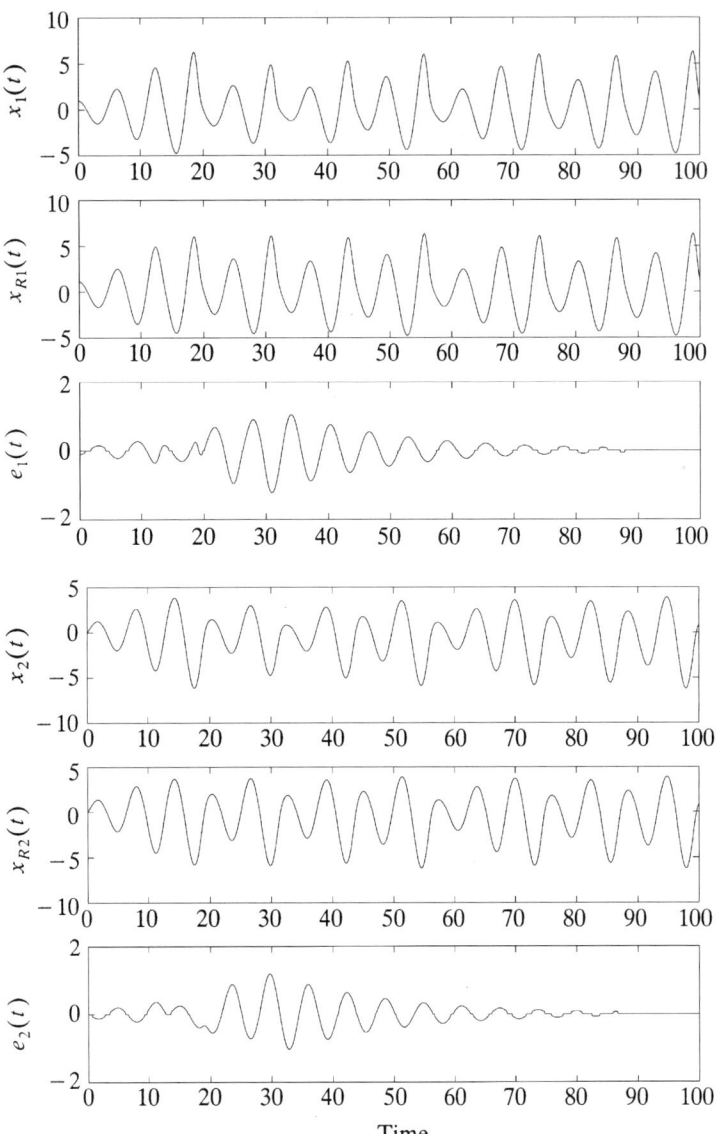

Fig. 9.15 Control result 1 (Example 22).

Example 22 Consider Rossler's equation with the input term. The fuzzy controller design satisfying the stability conditions and the constraints on the output and the control input for the CFS is feasible, where $\lambda = 10$, $\mu = 30$, and $C = C_1 = C_2 = I_3$. Figures 9.15 and 9.16 show the control result, where the control input is added at $t > 30$ sec. It can be seen that the

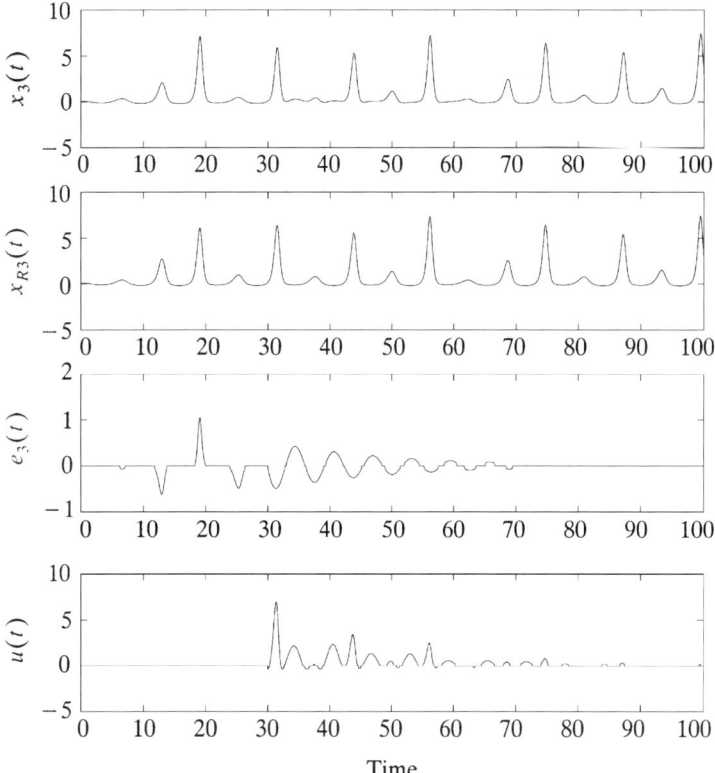

Fig. 9.16 Control result 2 (Example 22).

designed fuzzy controller linearizes and stabilizes the error system. Note that the control input and the output satisfy the constraints, that is, $\max_t \|u(t)\| \leq \mu$ and $\max_t \|e(t)\|_2 \leq \lambda$.

Example 23 Consider Rossler's equation with the input term. The fuzzy controller design satisfying the stability conditions and the constraints on the output and the control input for the CFS is feasible, where $\lambda = 10$, $\mu = 30$, and $C = C_1 = C_2 = I_3$. Figures 9.17 and 9.18 show the control result. It can be seen that the designed fuzzy controller linearizes and stabilizes the error system. It should be emphasized that the control input and the output satisfy the constraints, that is, $\max_t \|u(t)\|_2 \leq \mu$ and $\max_t \|e(t)\|_2 \leq \lambda$. In addition, note that this control result is better than that of Example 22 since the decay rate is considered in the design.

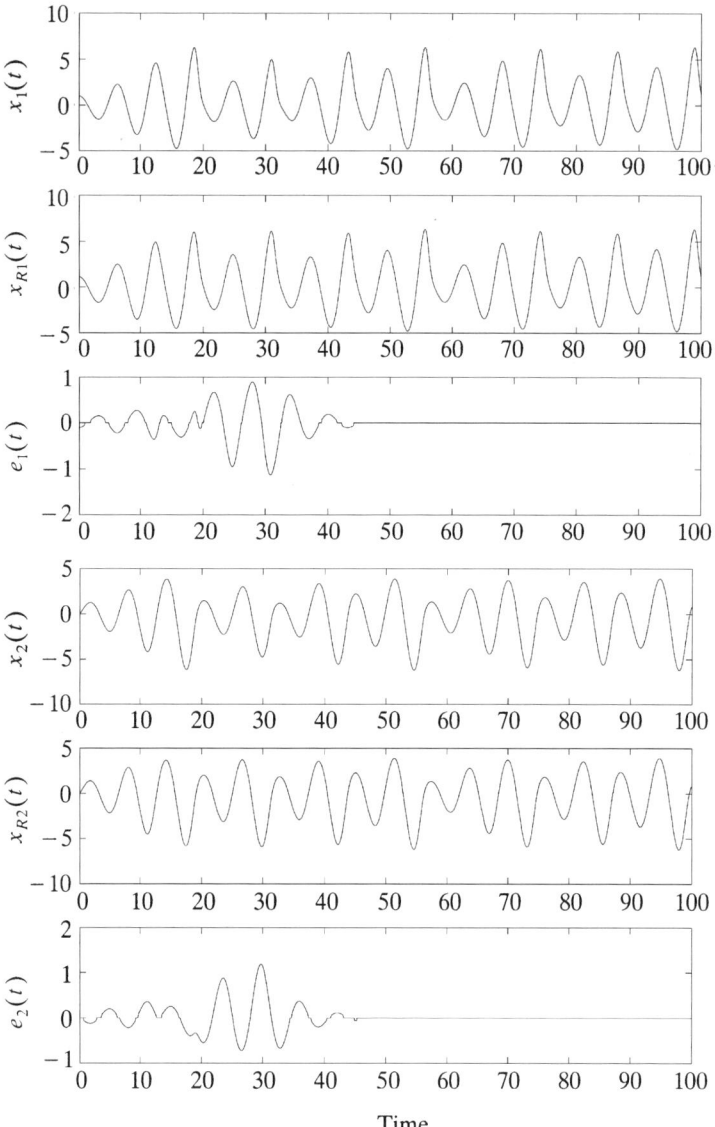

Fig. 9.17 Control result 1 (Example 23).

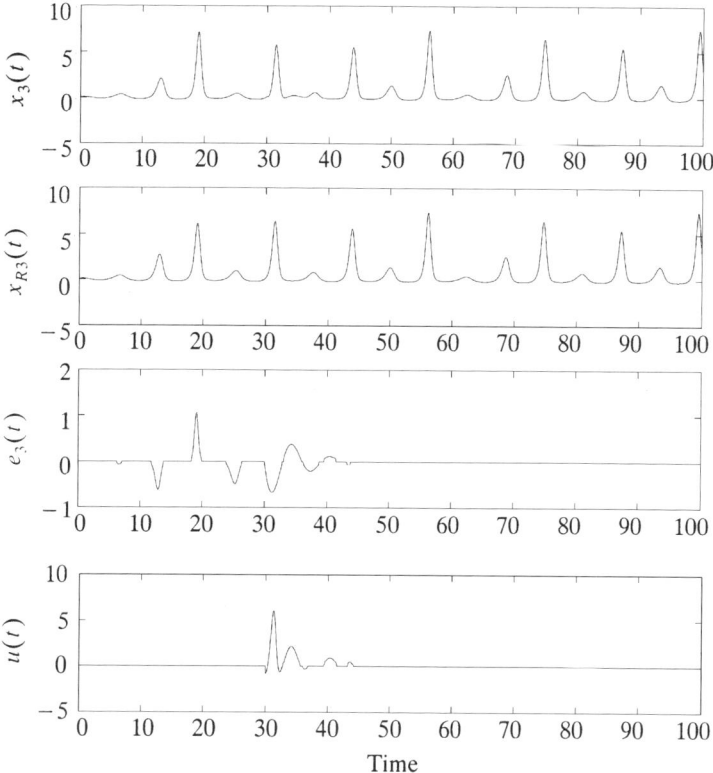

Fig. 9.18 Control result 2 (Example 23).

9.3.2 Case 2

If the cancellation problem is infeasible, that is, all the elements in $\beta \cdot S$ are not near zero, the error system cannot be linearized. Then, we have

$$se(t) = \sum_{i=1}^{r} h_i(z(t))A_i x(t) - \sum_{i=1}^{r} h_i(z_R(t))A_i x_R(t) + Bu(t)$$

$$= \sum_{i=1}^{r} h_i(z(t))A_i e(t)$$

$$+ \sum_{i=1}^{r} \{h_i(z(t)) - h_i(z_R(t))\}A_i x_R(t) + Bu(t). \quad (9.15)$$

Assume that $z(t) = x(t)$ and $z_R(t) = x_R(t)$. Then, the second term is almost zero:

$$\sum_{i=1}^{r} \{h_i(z(t)) - h_i(z_R(t))\}A_i x_R(t) \approx 0$$

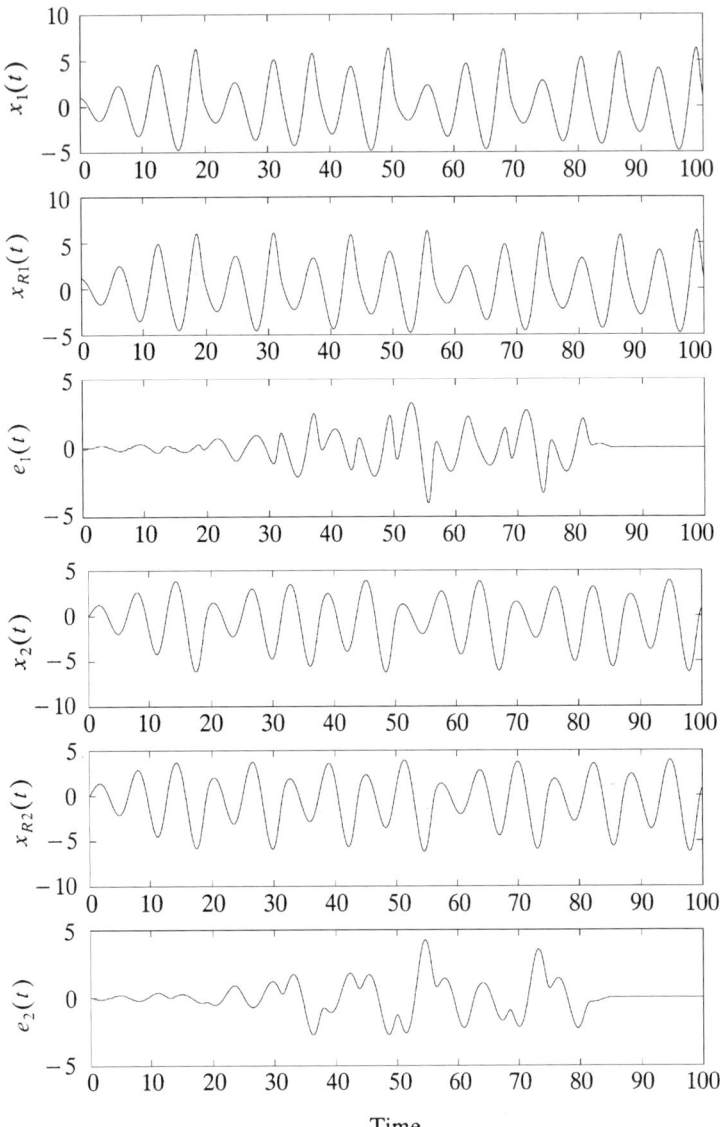

Fig. 9.19 Control result 1 (Example 24).

if $\|e(t)\| \le \delta$, where δ is a small value. As a result, the overall system is approximated as

$$\dot{e}(t) = \sum_{i=1}^{r} h_i(z(t)) A_i e(t) + Bu(t).$$

Consider the following fuzzy feedback law for the error system:

$$u(t) = \begin{cases} -\sum_{i=1}^{r} h_i(z(t)) F_i e(t), & \|e(t)\| \le \delta, \\ 0, & \text{otherwise}. \end{cases}$$

Then, if there exist the feedback gains F_i satisfying the stability conditions described in Chapter 3, the stability of the error system is guaranteed near the equilibrium points, that is, $\|e(t)\| \le \delta$. The feedback gains F_i can be found by solving the design problems in Section 9.2. It should be noted that this approach guarantees only the local stability. This is the same idea as the OGY method [1]. Therefore, the converging time to an equilibrium point is very long in general, but the control effort is small.

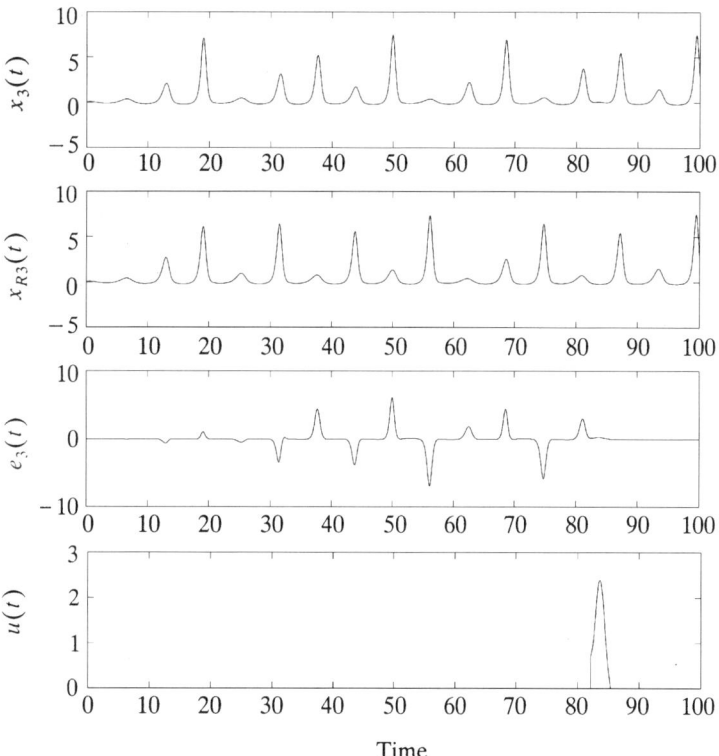

Fig. 9.20 Control result 2 (Example 24).

Example 24 We design a stable fuzzy controller for Rossler's equation with the input using the "case 2" design technique. The design problem is feasible. Figures 9.19 and 9.20 show the control result, where the control starts at $t = 40$ sec. However, the control input is added around 83 seconds and stabilizes the error system and the synchronization is realized.

9.4 CHAOTIC MODEL FOLLOWING CONTROL

Section 9.3 has presented the synchronization of chaotic systems, where A_i matrices of the fuzzy model should be the same as A_i matrices of the fuzzy reference model. This section presents chaotic model following control (CMFC), where A_i matrices of the fuzzy model do not have to be the same as A_i matrices of the fuzzy reference model. Therefore, the CMFC is more difficult than the synchronization. In this section, the controlled objects are assumed to be chaotic systems. However, note that the CMFC can be designed for general nonlinear systems represented by T-S fuzzy models.

Consider a reference fuzzy model which represents a reference chaotic system.

Reference Rule i

IF $z_{R1}(t)$ is N_{i1} and $\cdots$ and $z_{Rp}(t)$ is N_{ip},

THEN $sx_R(t) = D_i x_R(t), \quad i = 1, 2, \ldots, r_R.$ \hfill (9.16)

Assume that $x_R(t) \in R^n$ and $A_i \neq D_i$. The defuzzification process is given as

$$sx_R(t) = \sum_{i=1}^{r_R} v_i(z_R(t)) D_i x_R(t). \quad (9.17)$$

The CMFC can be regarded as nonlinear model following control for the reference fuzzy model (9.17). Assume that $e(t) = x(t) - x_R(t)$. Then, from (9.2) and (9.17), we have

$$se(t) = \sum_{i=1}^{r} h_i(z(t)) A_i x(t)$$

$$- \sum_{i=1}^{r_R} v_i(z_R(t)) D_i x_R(t) + Bu(t). \quad (9.18)$$

Consider two sub-fuzzy controllers to realize the CMFC:

Subcontroller A

Control Rule i

IF $z_1(t)$ is M_{i1} and $\cdots$ and $z_p(t)$ is M_{ip},

THEN $u_A(t) = -F_i x(t),$ $\quad i = 1, 2, \ldots, r.$ (9.19)

Subcontroller B

Control Rule i

IF $z_{R1}(t)$ is N_{i1} and $\cdots$ and $z_{Rp}(t)$ is N_{ip},

THEN $u_B(t) = K_i x_R(t),$ $\quad i = 1, 2, \ldots, r_R.$ (9.20)

The combination of the subcontroller A and the subcontroller B is represented as

$$u(t) = u_A(t) + u_B(t)$$
$$= -\sum_{i=1}^{r} h_i(z(t)) F_i x(t) + \sum_{i=1}^{r_R} v_i(z_R(t)) K_i x_R(t). \quad (9.21)$$

By substituting (9.21) into (9.18), the overall control system is represented as

$$se(t) = \sum_{i=1}^{r} h_i(z(t))(A_i - BF_i)x(t)$$
$$- \sum_{i=1}^{r_R} v_i(z_R(t))(D_i - BK_i)x_R(t). \quad (9.22)$$

THEOREM 32 *The chaotic system represented by the fuzzy system (9.2) is exactly linearized via the fuzzy controller (9.21) if there exist the feedback gains F_i and K_j such that*

$$\{(A_1 - BF_1) - (A_i - BF_i)\}^T$$
$$\times \{(A_1 - BF_1) - (A_i - BF_i)\} = 0, \quad i = 2, 3, \ldots, r, \quad (9.23)$$
$$\{(A_1 - BF_1) - (D_j - BK_j)\}^T$$
$$\times \{(A_1 - BF_1) - (D_j - BK_j)\} = 0, \quad j = 1, 2, \ldots, r_R. \quad (9.24)$$

Then, the overall control system is linearized as $sx(t) = Gx(t)$, where $G = A_1 - BF_1 = A_i - BF_i = D_j - BK_j$.

Proof. It is obvious that $G = A_1 - BF_1 = A_i - BF_i = D_j - BK_j$ if conditions (9.23) and (9.24) hold.

An important remark is in order here.

Remark 30 The CMFC reduces to the synchronization problem when $r = r_R$ and $A_i = D_j$ for $i = 1, \ldots, r$ and $j = 1, \ldots, r_R$. The CMFC reduces to the stabilization problem when $D_i = 0$ and $x_R(0) = 0$ for $i = 1, \ldots, r_R$. Therefore, as mentioned above, the CMFC problem is more general and difficult than the stabilization and synchronization problems. In addition, the controller design described here can be applied not only to stabilization and synchronization but also to the CMFC in the same control framework. Therefore the LMI-based methodology represents a unified approach to the problem of controlling chaos.

If B is a nonsingular matrix, the error system is exactly linearized and stabilized using $F_i = B^{-1}(G - A_i)$ and $K_i = B^{-1}(G - D_i)$. However, the assumption that B is a nonsingular matrix is very strict. On the other hand, if B is not a nonsingular matrix, Theorem 32 can be utilized by the approximation CT technique. The LMI conditions can be derived from Theorem 32 in the same way as described in Section 9.2.

Note that G is not always a stable matrix even if the conditions of Theorem 32 hold. From Theorem 32 and the stability conditions, we define the following design problems:

Stable Fuzzy Controller Design Using the CT: CFS

minimize β
$X, S, M_1, M_2, \ldots, M_r$

subject to $X > 0, \beta > 0, S > 0$,

$$\begin{bmatrix} I & S \\ S & I \end{bmatrix} > 0,$$

$$-A_i X + BM_i - XA_i^T + M_i^T B^T > 0, \quad i = 1, 2, \ldots, r,$$

$$\begin{bmatrix} \beta S & \{(A_1 X - BM_1) - (A_i X - BM_i)\}^T \\ \{(A_1 X - BM_1) - (A_i X - BM_i)\} & I \end{bmatrix} > 0,$$

$$i = 2, 3, \ldots, r,$$

$$\begin{bmatrix} \beta S & \{(A_1 X - BM_1) - (D_j X - BN_j)\}^T \\ \{(A_1 X - BM_1) - (D_j X - BN_j)\} & I \end{bmatrix} > 0,$$

$$j = 1, 2, \ldots, r_R,$$

where $X = P^{-1}$, $M_1 = F_1 X$, $M_i = F_i X$, and $N_j = K_j X$.

Stable Fuzzy Controller Design Using the CT: DFS

$$\underset{X,S,M_1,M_2,\ldots,M_r}{\text{minimize}} \; \beta$$

subject to $X > 0, \beta > 0, S > 0$,

$$\begin{bmatrix} I & S \\ S & I \end{bmatrix} > 0,$$

$$\begin{bmatrix} X & XA_i - M_i^T B^T \\ A_i X - BM_i & X \end{bmatrix} > 0, \quad i = 1, 2, \ldots, r,$$

$$\begin{bmatrix} \beta S & \{(A_1 X - BM_1) - (A_i X - BM_i)\}^T \\ \{(A_1 X - BM_1) - (A_i X - BM_i)\} & I \end{bmatrix} > 0,$$

$$i = 2, 3, \ldots, r,$$

$$\begin{bmatrix} \beta S & \{(A_1 X - BM_1) - (D_j X - BN_j)\}^T \\ \{(A_1 X - BM_1) - (D_j X - BN_j)\} & I \end{bmatrix} > 0,$$

$$j = 1, 2, \ldots, r_R,$$

where $X = P^{-1}$, $M_1 = F_1 X$, $M_i = F_i X$, and $N_j = K_j X$.

Decay Rate Fuzzy Controller Design Using the CT: CFS

$$\underset{X,S,M_1,M_2,\ldots,M_r}{\text{maximize}} \; \alpha$$

$$\underset{X,S,M_1,M_2,\ldots,M_r}{\text{minimize}} \; \beta$$

subject to $X > 0, \beta > 0, \alpha > 0, S > 0$,

$$\begin{bmatrix} I & S \\ S & I \end{bmatrix} > 0,$$

$$-A_i X + BM_i - XA_i^T + M_i^T B^T - 2\alpha X > 0, \quad i = 1, 2, \ldots, r,$$

$$\begin{bmatrix} \beta S & \{(A_1 X - BM_1) - (A_i X - BM_i)\}^T \\ \{(A_1 X - BM_1) - (A_i X - BM_i)\} & I \end{bmatrix} > 0,$$

$$i = 2, 3, \ldots, r,$$

$$\begin{bmatrix} \beta S & \{(A_1 X - BM_1) - (D_j X - BN_j)\}^T \\ \{(A_1 X - BM_1) - (D_j X - BN_j)\} & I \end{bmatrix} > 0,$$

$$j = 1, 2, \ldots, r_R,$$

where $X = P^{-1}$, $M_1 = F_1 X$, $M_i = F_i X$, and $N_j = K_j X$.

Decay Rate Fuzzy Controller Design Using the CT: DFS

$$\underset{X, S, M_1, M_2, \ldots, M_r}{\text{minimize}} \alpha$$

$$\underset{X, S, M_1, M_2, \ldots, M_r}{\text{minimize}} \beta$$

subject to $X > 0$, $\beta > 0$, $0 \leq \alpha < 1$, $S > 0$,

$$\begin{bmatrix} I & S \\ S & I \end{bmatrix} > 0,$$

$$\begin{bmatrix} \alpha X & XA_i - M_i^T B^T \\ A_i X - BM_i & X \end{bmatrix} > 0, \quad i = 1, 2, \ldots, r,$$

$$\begin{bmatrix} \beta S & \{(A_1 X - BM_1) - (A_i X - BM_i)\}^T \\ \{(A_1 X - BM_1) - (A_i X - BM_i)\} & I \end{bmatrix} > 0,$$

$$i = 2, 3, \ldots, r,$$

$$\begin{bmatrix} \beta S & \{(A_1 X - BM_1) - (D_j X - BN_j)\}^T \\ \{(A_1 X - BM_1) - (D_j X - BN_j)\} & I \end{bmatrix} > 0,$$

$$j = 1, 2, \ldots, r_R,$$

where $X = P^{-1}$, $M_1 = F_1 X$, $M_i = F_i X$, and $N_j = K_j X$.

Remark 31 In the LMIs, if all elements in $\beta \cdot S$ are near zero, that is, $\beta \cdot S \approx 0$, the cancellation problems for decay rate fuzzy controller designs are feasible. In this case, $G = A_1 - BF_1 = A_i - BF_i = D_j - BK_j \ \forall_{i,j}$, and G is a stable matrix.

Example 25 Let us consider the fuzzy model for Lorenz's equation with three inputs. The parameters are set as follows:

Rule 1

IF $x_1(t)$ is M_1,

THEN $\dot{x}(t) = A_1 x(t) + Bu(t)$.

Rule 2

IF $x_1(t)$ is M_2,

THEN $\dot{x}(t) = A_2 x(t) + Bu(t)$.

Here, $x(t) = [x_1(t) \; x_2(t) \; x_3(t)]^T$,

$$A_1 = \begin{bmatrix} -0.5 \cdot a & 0.5 \cdot a & 0 \\ 2 \cdot c & -1 & -d \\ 0 & d & -0.5 \cdot b \end{bmatrix},$$

$$A_2 = \begin{bmatrix} -0.5 \cdot a & 0.5 \cdot a & 0 \\ 2 \cdot c & -1 & d \\ 0 & -d & -0.5 \cdot b \end{bmatrix},$$

$$B = \begin{bmatrix} 1 & 0 & 0 \\ 0 & 1 & 0 \\ 0 & 0 & 1 \end{bmatrix},$$

$$M_1(x_1(t)) = \frac{1}{2}\left(1 + \frac{x_1(t)}{d}\right), \quad M_2(x_1(t)) = \frac{1}{2}\left(1 - \frac{x_1(t)}{d}\right).$$

Consider the following reference fuzzy model:

Reference Rule 1

  IF $x_{1R}(t)$ is N_1,

  THEN $\dot{x}_R(t) = D_1 x_R(t)$.

Reference Rule 2

  IF $x_{1R}(t)$ is N_2,

  THEN $\dot{x}_R(t) = D_2 x_R(t)$.

Here, $x_R(t) = [x_{R1}(t) \; x_{R2}(t) \; x_{R3}(t)]^T$,

$$D_1 = \begin{bmatrix} -a & a & 0 \\ c & -1 & -d \\ 0 & d & -b \end{bmatrix}, \quad D_2 = \begin{bmatrix} -a & a & 0 \\ c & -1 & d \\ 0 & -d & -b \end{bmatrix},$$

$$N_1(x_{R1}(t)) = \frac{1}{2}\left(1 + \frac{x_{R1}(t)}{d}\right), \quad N_2(x_{R1}(t)) = \frac{1}{2}\left(1 - \frac{x_{R1}(t)}{d}\right),$$

where $x_{R1}(t) \in [-d \; d]$. The stable fuzzy controller design using the CT is feasible. Figures 9.21 and 9.22 show the control result, where the control input is added at $t > 10$ sec. It can be seen that the designed fuzzy controller realizes chaotic model following control, that is, $e_1(t) \to 0$, $e_2(t) \to 0$, and $e_3(t) \to 0$.

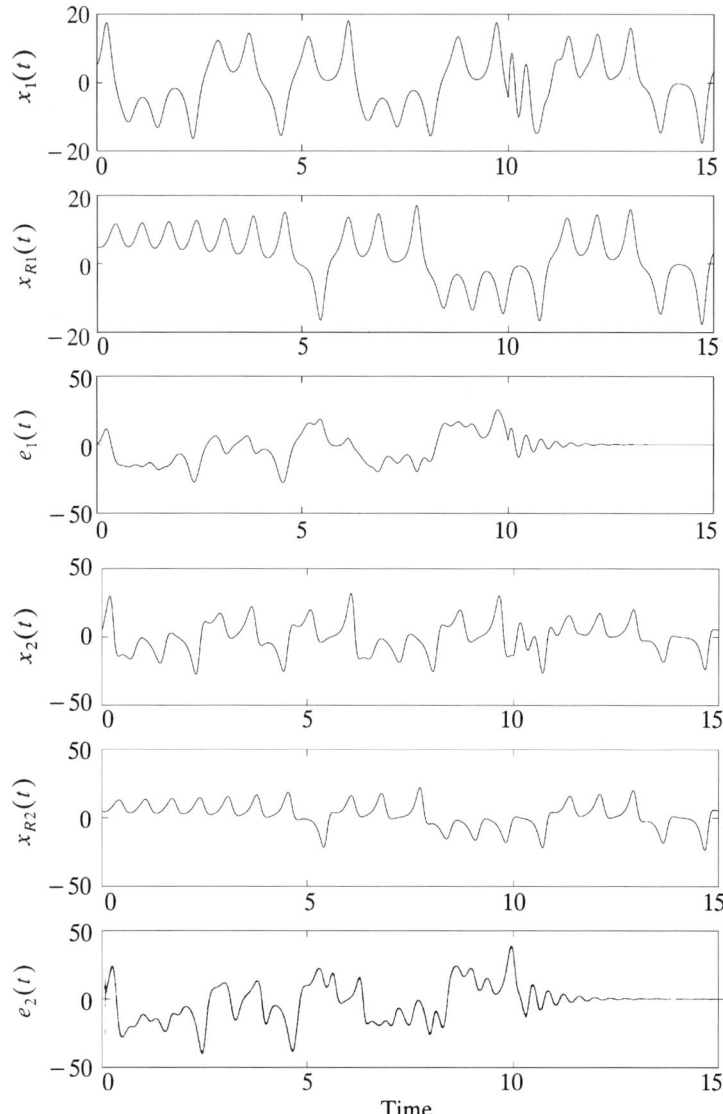

Fig. 9.21 Control result 1 (Example 25).

Example 26 Let us consider the fuzzy model for Rossler's equation with the input term. The parameters are set as follows:

Rule 1

    IF $x_1(t)$ is M_1,

        THEN $\dot{x}(t) = A_1 x(t) + Bu(t)$.

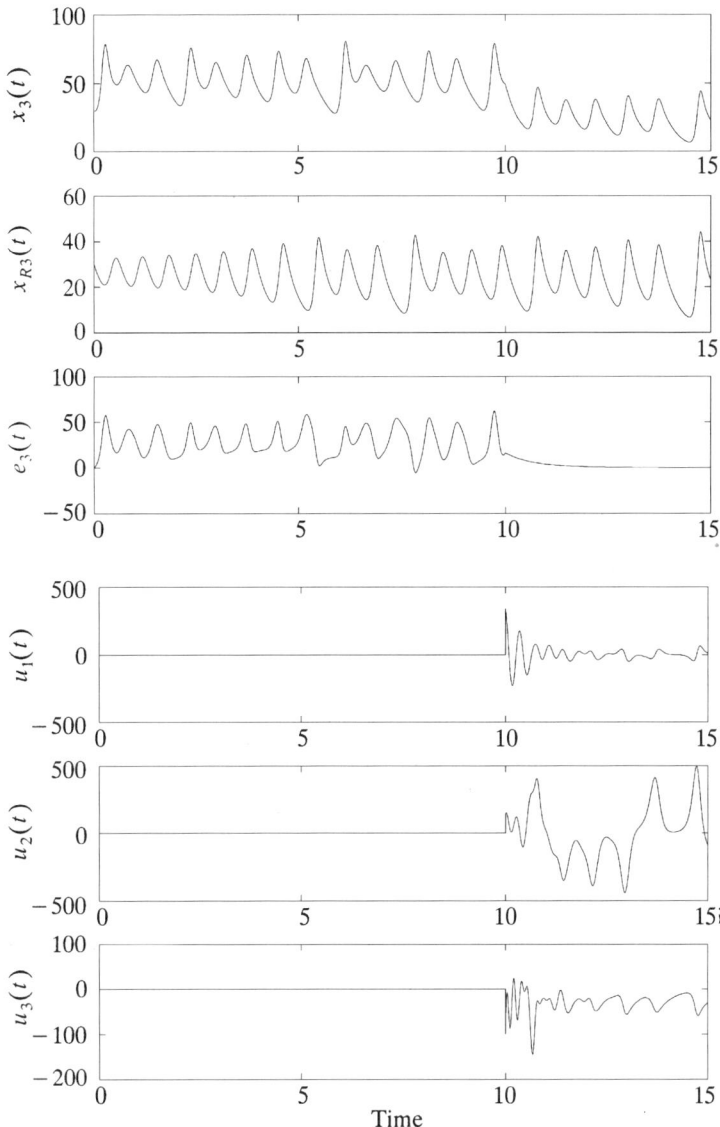

Fig. 9.22 Control result 2 (Example 25).

Rule 2

    ***IF** $x_1(t)$ is M_2,*

        ***THEN** $\dot{x}(t) = A_2 x(t) + Bu(t)$.*

Here, $x(t) = [x_1(t) \ x_2(t) \ x_3(t)]^T$,

$$A_1 = \begin{bmatrix} 0 & -1 & -1 \\ 1 & a & 0 \\ 0.5 \cdot b & 0 & -d \end{bmatrix}, \quad A_2 = \begin{bmatrix} 0 & -1 & -1 \\ 1 & a & 0 \\ 0.5 \cdot b & 0 & d \end{bmatrix},$$

$$B = \begin{bmatrix} 0 \\ 0 \\ 1 \end{bmatrix},$$

$$M_1(x_1(t)) = \frac{1}{2}\left(1 + \frac{2 \cdot c - x_1(t)}{d}\right),$$

$$M_2(x_1(t)) = \frac{1}{2}\left(1 - \frac{2 \cdot c - x_1(t)}{d}\right).$$

Consider the following reference fuzzy model:

Reference Rule 1

IF $x_{1R}(t)$ is N_1,

THEN $\dot{x}_R(t) = D_1 x_R(t)$.

Reference Rule 2

IF $x_{1R}(t)$ is N_2,

THEN $\dot{x}_R(t) = D_2 x_R(t)$.

Here, $x_R(t) = [x_{R1}(t) \ x_{R2}(t) \ x_{R3}(t)]^T$,

$$D_1 = \begin{bmatrix} 0 & -1 & -1 \\ 1 & a & 0 \\ b & 0 & -d \end{bmatrix}, \quad D_2 = \begin{bmatrix} 0 & -1 & -1 \\ 1 & a & 0 \\ b & 0 & d \end{bmatrix},$$

$$N_1(x_{R1}(t)) = \frac{1}{2}\left(1 + \frac{c - x_{R1}(t)}{d}\right),$$

$$N_2(x_{R1}(t)) = \frac{1}{2}\left(1 - \frac{c - x_{R1}(t)}{d}\right),$$

where $x_{R1}(t) \in [c - d \ \ c + d]$. The stable fuzzy controller design using the CT is feasible. Figures 9.23 and 9.24 show the control result, where the control input is added at $t > 30$ sec. The designed fuzzy controller realizes chaotic model following control.

CHAOTIC MODEL FOLLOWING CONTROL

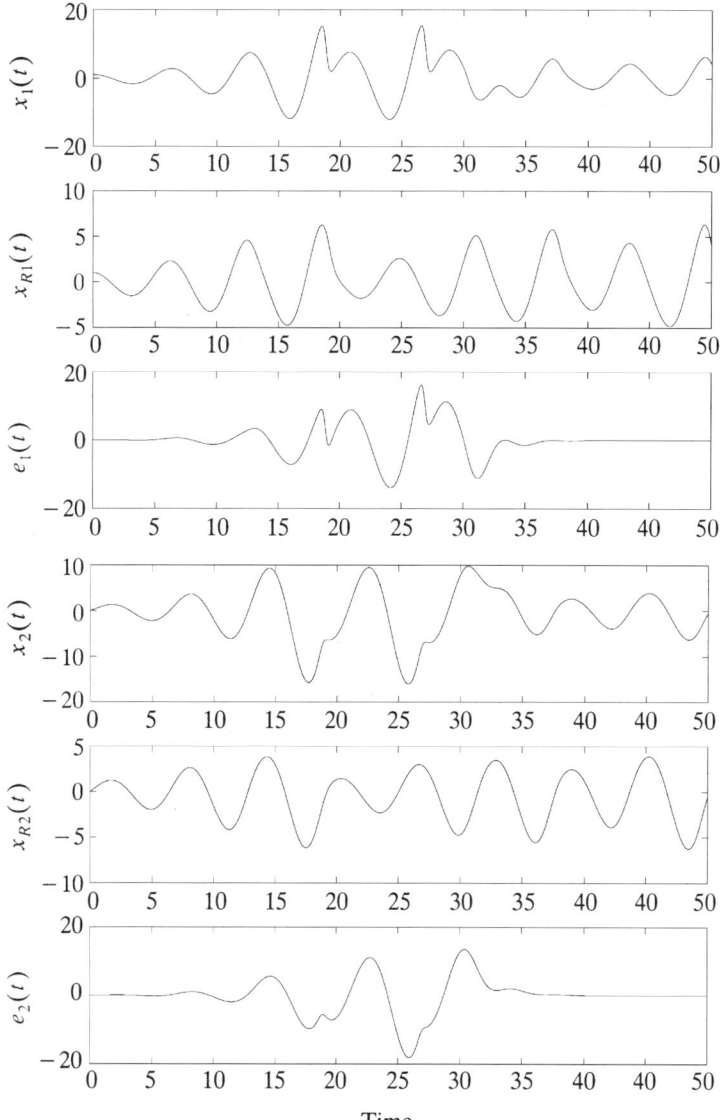

Fig. 9.23 Control result 1 (Example 26).

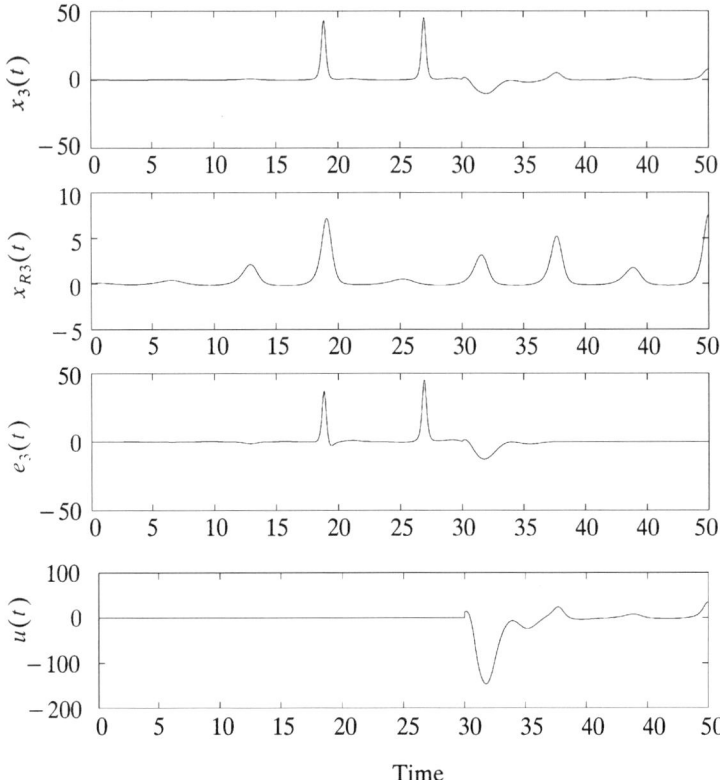

Fig. 9.24 Control result 2 (Example 26).

REFERENCES

1. E. Ott, C. Grebogi, and J. A. Yorke, "Controlling Chaos," *Phys. Rev. Lett.*, Vol. 64, pp. 1196–1199 (1990).
2. T. Shinbort, C. Grebogi, E. Ott, and J. A. Yorke, "Using Small Perturbations to Control Chaos," *Nature*, Vol. 363, pp. 411–417 (1993).
3. G. Chen and X. Dong, "From Chaos to Order—Perspectives and Methodologies in Controlling Chaotic Nonlinear Dynamical Systems," *Int. J. Bifurcation Chaos*, Vol. 3, No. 6, pp. 1363–1409 (1993).
4. E. H. Abed, H. O. Wang, and A. Tesi, "Control of Bifurcations and Chaos," in *The Control Handbook*, W. S. Levine, Editor, CRC Press & IEEE Press, Boca Raton, FL, 1995, pp. 951–966.
5. F. J. Romeiras, C. Grebogi, E. Ott, and W. P. Dayawansa, "Controlling Chaotic Dynamical Systems," *Physica*, Vol. D58, pp. 165–192 (1992).
6. E. Ott, *Chaos in Dynamical Systems*, Cambridge, 1993.
7. H. O. Wang and E. H. Abed, "Bifurcation Control of a Chaotic System," *Automatica*, Vol. 31, No. 9, pp. 1213–1226 (1995).

8. L. O. Chua, M. Komuro, and T. Matsumoto, "The Double Scroll Family: 1 and 2," *IEEE Trans. Circuits Syst*, Vol. 33, pp. 1072–1118 (1996).
9. K. Pyragas, "Continuous Control of Chaos by Self-Controlling Feedback," *Phys. Lett. A*, Vol. 170, pp. 421–428 (1992).
10. H. O. Wang, K. Tanaka, and T. Ikeda, "Fuzzy Modeling and Control of Chaotic Systems," 1996 IEEE International Symposium on Circuits and Systems, Vol. 3, Atlanta, 1996, pp. 209–212.
11. H. O. Wang and K. Tanaka, "An LMI-Based Stable Fuzzy Control of Nonlinear Systems and Its Applications to Control of Chaos," 5th IEEE International Conf. on Fuzzy Systems, Vol. 2, New Orleans, 1996, pp. 1433–1438.
12. K. Tanaka, T. Ikeda, and H. O. Wang, "Controlling Chaos via an LMI-Based Fuzzy Control System Design," 36th IEEE Conference on Decision and Control, Vol. 2, San Diego, 1997, pp. 1488–1493.
13. K. Tanaka, T. Ikeda, and H. O. Wang, "Fuzzy Control of Chaotic Systems Using LMIs: Regulation, Synchronization and Chaos Model Following," Seventh International IEEE Conference on Fuzzy Systems, Alaska, 1998, pp. 434–439.
14. K. Tanaka, T. Ikeda, and H. O. Wang, "A Unified Approach to Controlling Chaos via an LMI-Based Fuzzy Control System Design," *IEEE Trans. Circuits Syst.*, Vol. 45, No. 10, pp. 1021–1040 (1998).

CHAPTER 10

FUZZY DESCRIPTOR SYSTEMS AND CONTROL

This chapter deals with a fuzzy descriptor system defined by extending the original Takagi-Sugeno fuzzy model. A number of stability conditions for the fuzzy descriptor system are derived and represented in terms of LMIs. A motivating example for using the fuzzy descriptor system instead of the original Takagi-Sugeno fuzzy model is presented. An LMI-based design approach is employed to find stabilizing feedback gains and a common Lyapunov function.

The descriptor system, which differs from a state-space representation, has generated a great deal of interest in control systems design. The descriptor system describes a wider class of systems including physical models and nondynamic constraints [1]. It is well known that the descriptor system is much tighter than the state-space model for representing real independent parametric perturbations. There exist a large number of papers on the stability analysis of the T-S fuzzy systems based on the state-space representation. In contrast, the definition of a fuzzy descriptor system and its stability analysis have not been discussed until recently [2]. In [2] we introduced the fuzzy descriptor systems and analyzed the stability of such systems. This chapter presents both the basic framework of [2, 3] as well as some new developments on this topic.

As mentioned in Chapter 1, $h_i \cap v_k \neq \emptyset$ denotes all the pairs (i, k) excepting $h_i(z(t))v_k(z(t)) = 0$ for all $z(t)$; $h_i \cap h_j \cap v_k \neq \emptyset$ denotes all the pairs (i, j, k) excepting $h_i(z(t))h_j(z(t))v_k(z(t)) = 0$ for all $z(t)$; and $i < j$ s.t. $h_i \cap h_j \cap v_k \neq \emptyset$ denotes all $i < j$ excepting $h_i(z(t))h_j(z(t))v_k(z(t)) = 0$, $\forall z(t)$.

10.1 FUZZY DESCRIPTOR SYSTEM

In [4, 5], a fuzzy descriptor system is defined by extending the T-S fuzzy model (2.3) and (2.4). The ordinary Takagi-Sugeno fuzzy model is a special case of the fuzzy descriptor system. We derive stability conditions for the fuzzy descriptor system, where the E matrix in the fuzzy descriptor system is assumed to be not always nonsingular. The fuzzy descriptor system is defined as

$$\sum_{k=1}^{r^e} v_k(z(t)) E_k \dot{x}(t) = \sum_{i=1}^{r} h_i(z(t))(A_i x(t) + B_i u(t)),$$

$$y(t) = \sum_{i=1}^{r} h_i(z(t)) C_i x(t), \quad (10.1)$$

where
$$x(t) \in R^n, \quad y(t) \in R^q, \quad u(t) \in R^m,$$

$$h_i(z(t)) \geq 0, \quad \sum_{i=1}^{r} h_i(z(t)) = 1,$$

$$v_k(z(t)) \geq 0, \quad \sum_{k=1}^{r^e} v_k(z(t)) = 1.$$

Here $x \in R^n$ is the descriptor vector, $u \in R^m$ is the input vector, $y \in R^q$ is the output vector, $E_k \in R^{n \times n}$, $A_i \in R^{n \times n}$, $B_i \in R^{n \times m}$, and $C_i \in R^{q \times n}$. The known premise variables $z_1(t) \sim z_p(t)$ may be functions of the states, external disturbances, and/or time.

Remark 32 A fuzzy descriptor system was first defined in [2]. In [2], a special case, that is, $h_i(z(t)) = v_k(z(t))$ and $r = r^e$, was presented. In [4, 5], the fuzzy descriptor system was generalized as shown in (10.1).

By defining $x^*(t) = [x^T(t) \; \dot{x}^T(t)]^T$, the fuzzy descriptor system (10.1) can be rewritten as

$$E^* \dot{x}^*(t) = \sum_{i=1}^{r} \sum_{k=1}^{r^e} h_i(z(t)) v_k(z(t)) (A_{ik}^* x^*(t) + B_i^* u(t)),$$

$$y(t) = \sum_{i=1}^{r} h_i(z(t)) C_i^* x^*(t), \quad (10.2)$$

where

$$E^* = \begin{bmatrix} I & 0 \\ 0 & 0 \end{bmatrix}, \quad A_{ik}^* = \begin{bmatrix} 0 & I \\ A_i & -E_k \end{bmatrix},$$

$$B_i^* = \begin{bmatrix} 0 \\ B_i \end{bmatrix}, \quad C_i^* = \begin{bmatrix} C_i & 0 \end{bmatrix}.$$

In the following the stability for the fuzzy descriptor system (10.2) is considered.

10.2 STABILITY CONDITIONS

The open-loop systems of (10.2) is defined as follows:

$$E^* \dot{x}^*(t) = \sum_{i=1}^{r} \sum_{k=1}^{r^e} h_i(z(t)) v_k(z(t)) A^*_{ik} x^*(t). \tag{10.3}$$

The fuzzy descriptor system (10.3) is quadratically stable if

$$\frac{dV(x^*(t))}{dt} \leq -\alpha \|x^*(t)\|_2,$$

where

$$V(x^*(t)) = x^{*T}(t) E^{*T} X x^*(t),$$

and the following conditions are satisfied:

1. $\det(sE^* - \sum_{i=1}^{r} \sum_{k=1}^{r^e} h_i(z(t)) v_k(z(t)) A^*_{ik}) \neq 0$ and the open-loop system is impulse free.
2. There exist a common matrix X and $\alpha > 0$ such that

$$X \in \mathbf{R}^{2n \times 2n}, \qquad E^{*T} X = X^T E^* \geq \mathbf{0}, \qquad \det X \neq 0.$$

Theorem 33 gives a sufficient condition for ensuring the stability of (10.3).

THEOREM 33 *The fuzzy descriptor system* (10.3) *is quadratically stable if there exists a common matrix X such that*

$$E^{*T} X = X^T E^* \geq \mathbf{0}, \tag{10.4}$$

$$A^{*T}_{ik} X + X^T A^*_{ik} < \mathbf{0}, \qquad h_i \cap v_k \neq \emptyset. \tag{10.5}$$

Proof. Consider a candidate of the quadratic function

$$V(x^*(t)) = x^{*T}(t) E^{*T} X x^*(t).$$

Then,

$$\dot{V}(x^*(t)) = \sum_{i=1}^{r} \sum_{k=1}^{r^e} h_i(z(t)) v_k(z(t)) x^{*T}(t) \left(A^{*T}_{ik} X + X^T A^*_{ik} \right) x^*(t).$$

Therefore, we have the following stability conditions:

$$A^{*T}_{ik} X + X^T A^*_{ik} < \mathbf{0}, \qquad h_i \cap v_k \neq \emptyset. \qquad \text{(Q.E.D.)}$$

Remark 33 As mentioned before, $h_i \cap v_k \neq \emptyset$ denotes "all the pairs (i, k) excepting $h_i(z(t))v_k(z(t)) = 0$ for all $z(t)$." In other words, we can ignore the condition (10.5) for the pairs (i, k) such that $h_i(z(t))v_k(z(t)) = 0$ for all $z(t)$.

Remark 34 In Theorem 33, X is not required to be positive definite.

Corollary 5 is needed to discuss the stability of closed-loop systems.

COROLLARY 5 *The conditions (10.6) and (10.7) imply (10.4) and (10.5), where S_1 is a positive definite matrix:*

$$S_1 = S_1^T > 0, \tag{10.6}$$

$$\begin{bmatrix} A_i^T S_3 + S_3^T A_i & * \\ S_1 + S_1 A_i - E_k^T S_3 & -E_k^T S_1 - S_1 E_k \end{bmatrix} < 0, \quad h_i \cap v_k \neq \emptyset, \tag{10.7}$$

where the asterisk denotes the transposed elements (matrices) for symmetric positions. For example, in (10.7), it represents $(S_1 + S_1 A_i - E_k^T S_3)^T$.

Proof. Define X as

$$X = \begin{bmatrix} S_1 & 0 \\ S_3 & S_1 \end{bmatrix}.$$

Then, (10.6) is obtained from (10.4) as follows:

$$E^{*T}X = \begin{bmatrix} I & 0 \\ 0 & 0 \end{bmatrix} \begin{bmatrix} S_1 & 0 \\ S_3 & S_1 \end{bmatrix} = \begin{bmatrix} S_1 & 0 \\ 0 & 0 \end{bmatrix} \geq 0,$$

$$X^T E^* = \begin{bmatrix} S_1^T & S_3^T \\ 0 & S_1^T \end{bmatrix} \begin{bmatrix} I & 0 \\ 0 & 0 \end{bmatrix} = \begin{bmatrix} S_1^T & 0 \\ 0 & 0 \end{bmatrix} \geq 0.$$

Equation (10.7) is obtained as follows:

$$A_{ik}^{*T} X + X^T A_{ik}^*$$

$$= \begin{bmatrix} 0 & A_i^T \\ I & -E_k^T \end{bmatrix} \begin{bmatrix} S_1 & 0 \\ S_3 & S_1 \end{bmatrix} + \begin{bmatrix} S_1^T & S_3^T \\ 0 & S_1^T \end{bmatrix} \begin{bmatrix} 0 & I \\ A_i & -E_k \end{bmatrix}$$

$$= \begin{bmatrix} A_i^T S_3 + S_3^T A_i & S_1 + A_i^T S_1 - S_3^T E_k \\ S_1 + S_1 A_i - E_k^T S_3 & -E_k^T S_1 - S_1 E_k \end{bmatrix} < 0. \quad \text{(Q.E.D.)}$$

Remark 35 It is stated in Remark 34 that X is not required to be positive definite. However, in Corollary 5, X is assumed to be invertible since $X = \begin{bmatrix} S_1 & 0 \\ S_3 & S_1 \end{bmatrix}$, where $S_1 > 0$.

Next, we consider stability conditions for closed-loop systems. We propose a modified PDC (10.8) to stabilize the fuzzy descriptor system (10.2):

$$u(t) = -\sum_{i=1}^{r}\sum_{k=1}^{r^e} h_i(z(t))v_k(z(t))F_{ik}^* x^*(t), \qquad (10.8)$$

where $F_{ik}^* = \begin{bmatrix} F_{ik} & 0 \end{bmatrix}$. The fuzzy controller design problem is to determine the local feedback gains F_{ik}.

By substituting (10.8) into (10.2), the fuzzy control system is represented as

$$E^* \dot{x}^*(t) = \sum_{i=1}^{r}\sum_{j=1}^{r}\sum_{k=1}^{r^e} h_i(z(t))h_j(z(t))v_k(z(t))(A_{ik}^* - B_i^* F_{jk}^*)x^*(t). \qquad (10.9)$$

Theorem 34 gives a sufficient condition for ensuring the stability of (10.9).

THEOREM 34 *The fuzzy descriptor system* (10.2) *can be stabilized via the PDC fuzzy controller* (10.8) *if there exist* Z_1, Z_3, *and* M_{ik} *such that*

$$Z_1^T = Z_1 > 0, \qquad (10.10)$$

$$\begin{bmatrix} -Z_3 - Z_3^T & * \\ \begin{pmatrix} Z_1 + A_i Z_1 \\ -B_i M_{ik} + E_k Z_3 \end{pmatrix} & -Z_1 E_k^T - E_k Z_1 \end{bmatrix} < 0,$$

$$h_i \cap v_k \neq \emptyset, \qquad (10.11)$$

$$\begin{bmatrix} -2Z_3 - 2Z_3^T & * \\ \begin{pmatrix} 2Z_1 + A_i Z_1 \\ -B_i M_{jk} + A_j Z_1 \\ -B_j M_{ik} + 2E_k Z_3 \end{pmatrix} & -2Z_1 E_k^T - 2E_k Z_1 \end{bmatrix} \leq 0,$$

$$i < j \leq r \text{ s.t. } h_i \cap h_j \cap v_k \neq \emptyset, \qquad (10.12)$$

where the asterisk denotes the transposed elements (matrices) for symmetric positions.

200 FUZZY DESCRIPTOR SYSTEMS AND CONTROL

Proof. Consider a candidate of a quadratic function

$$V(x^*(t)) = x^{*T}(t) E^{*T} X x^*(t),$$

where

$$X = \begin{bmatrix} S_1 & 0 \\ S_3 & S_1 \end{bmatrix}.$$

Then,

$$\dot{V}(x^*(t)) = \sum_{i=1}^{r} \sum_{j=1}^{r} \sum_{k=1}^{r_e} h_i(z(t)) h_j(z(t)) v_k(z(t)) x^{*T}(t)$$

$$\times \left\{ (A_{ik}^* - B_i^* F_{jk}^*)^T X + X^T (A_{ik}^* - B_i^* F_{jk}^*) \right\} x^*(t)$$

$$= \sum_{i=1}^{r} \sum_{k=1}^{r_e} h_i^2(z(t)) v_k(z(t)) x^{*T}(t)$$

$$\times \left\{ (A_{ik}^* - B_i^* F_{ik}^*)^T X + X^T (A_{ik}^* - B_i^* F_{ik}^*) \right\} x^*(t)$$

$$+ 2 \sum_{i=1}^{r} \sum_{i<j} \sum_{k=1}^{r_e} h_i(z(t)) h_j(z(t)) v_k(z(t)) x^{*T}(t)$$

$$\times \left\{ \left(\frac{A_{ik}^* - B_i^* F_{jk}^* + A_{jk}^* - B_j^* F_{ik}^*}{2} \right)^T X \right.$$

$$\left. + X^T \left(\frac{A_{ik}^* - B_i^* F_{jk}^* + A_{jk}^* - B_j^* F_{ik}^*}{2} \right) \right\} x^*(t).$$

Therefore, the stability conditions are derived as follows:

$$E^{*T} X = X^T E^* \geq 0, \tag{10.13}$$

$$G_{iik}^T X + X^T G_{iik} < 0, \quad h_i \cap v_k \neq \emptyset, \tag{10.14}$$

$$\left(\frac{G_{ijk} + G_{jik}}{2} \right)^T X + X^T \left(\frac{G_{ijk} + G_{jik}}{2} \right) \leq 0,$$

$$i < j \leq r \text{ s.t. } h_i \cap h_j \cap v_k \neq \emptyset, \tag{10.15}$$

where

$$G_{ijk} = A_{ik}^* - B_i^* F_{jk}^* = \begin{bmatrix} 0 & I \\ A_i - B_i F_{jk} & -E_k \end{bmatrix},$$

$$F_{ik}^* = \begin{bmatrix} F_{ik} & 0 \end{bmatrix}.$$

Equation (10.13) can be rewritten as

$$X^{-T}E^{*T} = E^*X^{-1} \geq 0.$$

The above inequality is

$$\begin{bmatrix} S_1 & 0 \\ S_3 & S_1 \end{bmatrix}^{-T} \begin{bmatrix} I & 0 \\ 0 & 0 \end{bmatrix} = \begin{bmatrix} I & 0 \\ 0 & 0 \end{bmatrix} \begin{bmatrix} S_1 & 0 \\ S_3 & S_1 \end{bmatrix}^{-1} \geq 0.$$

Therefore, we obtain

$$\begin{bmatrix} Z_1^T & -Z_3^T \\ 0 & Z_1^T \end{bmatrix} \begin{bmatrix} I & 0 \\ 0 & 0 \end{bmatrix}$$

$$= \begin{bmatrix} I & 0 \\ 0 & 0 \end{bmatrix} \begin{bmatrix} Z_1 & 0 \\ -Z_3 & Z_1 \end{bmatrix} = \begin{bmatrix} Z_1 & 0 \\ 0 & 0 \end{bmatrix} \geq 0,$$

where

$$Z_1 = S_1^{-1} \quad \text{and} \quad Z_3 = S_1^{-1} S_3 S_1^{-1}.$$

Note that the following relation holds:

$$\begin{bmatrix} S_1 & 0 \\ S_3 & S_1 \end{bmatrix} \begin{bmatrix} Z_1 & 0 \\ -Z_3 & Z_1 \end{bmatrix} = \begin{bmatrix} I & 0 \\ 0 & I \end{bmatrix}.$$

Equation (10.14) can be rewritten as

$$X^{-T}G_{iik}^T XX^{-1} + X^{-T}X^T G_{iik} X^{-1}$$

$$= \begin{bmatrix} Z_1^T & -Z_3^T \\ 0 & Z_1^T \end{bmatrix} \begin{bmatrix} 0 & A_i^T - F_{ik}^T B_i^T \\ I & -E_k^T \end{bmatrix}$$

$$+ \begin{bmatrix} 0 & I \\ A_i - B_i F_{ik} & -E_k \end{bmatrix} \begin{bmatrix} Z_1 & 0 \\ -Z_3 & Z_1 \end{bmatrix}$$

$$= \begin{bmatrix} -Z_3 - Z_3^T & * \\ \left(\begin{array}{c} Z_1 + A_i Z_1 \\ -B_i M_{ik} + E_k Z_3 \end{array}\right) & -Z_1 E_k^T - E_k Z_1 \end{bmatrix} < 0.$$

Equation (10.12) is also derived in the same way as condition (10.11).
(Q.E.D.)

The fuzzy controller design problem is to determine F_{ik} ($i = 1, 2, \ldots, r$; $k = 1, 2, \ldots, r^e$) satisfying the conditions of Theorem 34. The feedback gains are obtained as

$$F_{ik} = M_{ik} Z_1^{-1}$$

from the solution Z_1 and M_{ik} of the above LMIs. The matrix $X = \begin{bmatrix} S_1 & 0 \\ S_3 & S_1 \end{bmatrix}$ is obtained as $S_1 = Z_1^{-1}$ and $S_3 = Z_1^{-1} Z_3 Z_1^{-1}$.

Next, we derive stability conditions for (10.9) in the case of $h_i(z(t)) = v_k(z(t))$ and $r = r^e$. In this case, the fuzzy descriptor system (10.2) can be rewritten as

$$E^* \dot{x}^*(t) = \sum_{i=1}^{r} h_i(z(t))(A_{ii}^* x^*(t) + B_i^* u(t)), \qquad (10.16)$$

where

$$E^* = \begin{bmatrix} I & 0 \\ 0 & 0 \end{bmatrix}, \quad A_{ii}^* = \begin{bmatrix} 0 & I \\ A_i & -E_i \end{bmatrix},$$

$$B_i^* = \begin{bmatrix} 0 \\ B_i \end{bmatrix}.$$

In this case, the PDC controller (10.17) instead of (10.8) is used:

$$u(t) = -\sum_{i=1}^{r} h_i(z(t)) F_{ii}^* x^*(t), \qquad (10.17)$$

where $F_{ii}^* = [F_i \quad 0]$. In this case, Theorem 34 can be simplified as follows.

THEOREM 35 *Assume that $h_i(z(t)) = v_k(z(t))$ and $r = r^e$. Then, the fuzzy descriptor system (10.16) can be stabilized via the PDC fuzzy controller (10.17) if there exist Z_1, Z_3, and M_i such that*

$$Z_1^T = Z_1 > 0, \qquad (10.18)$$

$$\begin{bmatrix} -Z_3 - Z_3^T & * \\ \begin{pmatrix} Z_1 + A_i Z_1 \\ -B_i M_i + E_i Z_3 \end{pmatrix} & -Z_1 E_i^T - E_i Z_1 \end{bmatrix} < 0,$$

$$i = 1, 2, \ldots, r, \qquad (10.19)$$

$$\begin{bmatrix} -2Z_3 - 2Z_3^T & * \\ \begin{pmatrix} 2Z_1 + A_iZ_1 \\ -B_iM_j + A_jZ_1 \\ -B_jM_i + 2E_iZ_3 \end{pmatrix} & -2Z_1E_i^T - 2E_iZ_1 \end{bmatrix} < 0,$$

$$i < j \le r \text{ s.t. } h_i \cap h_j \ne \emptyset. \quad (10.20)$$

The feedback gains F_i are obtained as $F_i = M_i Z_1^{-1}$.

Proof. Consider a candidate of quadratic function

$$V(x^*(t)) = x^{*T}(t) E^{*T} X x^*(t).$$

Then,

$$\dot{V}(x^*(t)) = \sum_{i=1}^r \sum_{j=1}^r h_i(z(t)) h_j(z(t)) x^{*T}(t)$$

$$\times \left[(A_{ii}^* - B_i^* F_{jj}^*)^T X + X^T (A_{ii}^* - B_i^* F_{jj}^*) \right] x^*(t)$$

$$= \sum_{i=1}^r h_i^2(z(t)) x^{*T}(t)$$

$$\times \left[(A_{ii}^* - B_i^* F_{ii}^*)^T X + X^T (A_{ii}^* - B_i^* F_{ii}^*) \right] x^*(t)$$

$$+ 2 \sum_{i=1}^r \sum_{i<j} h_i(z(t)) h_j(z(t)) x^{*T}(t)$$

$$\times \left[\left(\frac{A_{ii}^* - B_i^* F_{jj}^* + A_{jj}^* - B_j^* F_{ii}^*}{2} \right)^T X \right.$$

$$\left. + X^T \left(\frac{A_{ii}^* - B_i^* F_{jj}^* + A_{jj}^* - B_j^* F_{ii}^*}{2} \right) \right] x^*(t) < 0.$$

Therefore, we have the following stability conditions:

$$E^{*T} X = X^T E^* \ge 0,$$

$$G_{ii}^T X + X^T G_{ii} < 0, \quad i = 1, 2, \ldots, r,$$

$$\left(\frac{G_{ij} + G_{ji}}{2} \right)^T X + X^T \left(\frac{G_{ij} + G_{ji}}{2} \right) \le 0, \quad i < j \le r \text{ s.t. } h_i \cap h_j \ne \emptyset,$$

where
$$G_{ij} = A^*_{ii} - B^*_i F^*_{jj},$$

$$A^*_{ii} = \begin{bmatrix} 0 & I \\ A_i & -E_i \end{bmatrix}, \quad B^*_i = \begin{bmatrix} 0 \\ B_i \end{bmatrix},$$

$$F^*_{ii} = \begin{bmatrix} F_i & 0 \end{bmatrix}.$$

We can obtain the conditions (10.18)–(10.20) in the same way as in Theorem 34. (Q.E.D.)

Now consider the common B matrix case, that is, $B_1 = B_2 = \cdots = B_r$ in (10.2). The stability analysis for the common B matrix case is simpler and easier in comparison with that of the general case. Keep this in mind because we will refer to this when discussing the motivation behind the introduction of the fuzzy descriptor system.

In the common B matrix case, the stability conditions of Theorems 34 and 35 can be simplified as Theorems 36 and 37, respectively. Theorem 37 gives stability conditions for the case of $h_i(z(t)) = v_k(z(t))$ and $r = r^e$.

THEOREM 36 *The fuzzy descriptor system (10.2) with the common B matrix, that is, $B_1 = B_2 = \cdots = B_r = B$, can be stabilized via the PDC fuzzy controller (10.8) if there exist Z_1, Z_3, and M_{ik} such that*

$$Z_1^T = Z_1 > 0, \tag{10.21}$$

$$\begin{bmatrix} -Z_3 - Z_3^T & * \\ \begin{pmatrix} Z_1 + A_i Z_1 \\ -BM_{ik} + E_k Z_3 \end{pmatrix} & -Z_1 E_k^T - E_k Z_1 \end{bmatrix} < 0, \quad h_i \cap v_k \neq \emptyset. \tag{10.22}$$

The feedback gains F_{ik} are obtained as $F_{ik} = M_{ik} Z_1^{-1}$.

Proof. Consider a candidate of quadratic function
$$V(x^*(t)) = x^{*T}(t) E^{*T} X x^*(t).$$
Then,
$$\dot{V}(x^*(t)) = \sum_{i=1}^{r} \sum_{k=1}^{r^e} h_i(z(t)) v_k(z(t)) x^{*T}(t)$$
$$\times \left[(A^*_{ik} - B^* F^*_{ik})^T X + X^T (A^*_{ik} - B^* F^*_{ik}) \right] x^*(t) < 0,$$

where
$$B^* = \begin{bmatrix} 0 \\ B \end{bmatrix}.$$

Therefore, the fuzzy control system is stable if

$$E^{*T}X = X^T E^* \geq 0,$$

$$(A_{ik}^* - B^* F_{ik}^*)^T X + X^T (A_{ik}^* - B^* F_{ik}^*) < 0, \qquad h_i \cap v_k \neq \emptyset.$$

In the same way as in the proof of Theorem 34, we obtain the LMI condition (10.21) and (10.22). (Q.E.D.)

Next, we discuss the stability of the fuzzy descriptor system with the common B matrix in the case of $h_i(z(t)) = v_k(z(t))$ and $r = r^e$. By utilizing the property of $h_i(z(t)) = v_k(z(t))$, Theorem 36 can be simplified as follows.

THEOREM 37 *Assume that $h_i(z(t)) = v_k(z(t))$ and $r = r^e$. The fuzzy descriptor system (10.2) with the common B matrix can be stabilized via the PDC fuzzy controller (10.17) if there exist Z_1, Z_3, and M_i such that*

$$Z_1^T = Z_1 > 0, \tag{10.23}$$

$$\begin{bmatrix} -Z_3 - Z_3^T & * \\ \begin{pmatrix} Z_1 + A_i Z_1 \\ -BM_i + E_i Z_3 \end{pmatrix} & -Z_1 E_i^T - E_i Z_1 \end{bmatrix} < 0,$$

$$i = 1, 2, \ldots, r. \tag{10.24}$$

The feedback gains F_i are obtained as $F_i = M_i Z_1^{-1}$.

Proof. Consider a candidate of a quadratic function

$$V(x^*(t)) = x^{*T}(t) E^{*T} X x^*(t).$$

Then,

$$\dot{V}(x^*(t)) = \sum_{i=1}^{r} h_i(z(t)) x^{*T}(t)$$

$$\times \left[(A_{ii}^* - B^* F_{ii}^*)^T X + X^T (A_{ii}^* - B^* F_{ii}^*) \right] x^*(t) < 0.$$

Therefore, the fuzzy control system is stable if

$$E^{*T}X = X^T E^* \geq 0,$$

$$(A_{ii}^* - B^* F_{ii}^*)^T X + X^T (A_{ii}^* - B^* F_{ii}^*) < 0, \qquad i = 1, 2, \ldots, r.$$

In the same way as in the proof of Theorem 35, we obtain the LMI conditions (10.23) and (10.24). (Q.E.D.)

10.3 RELAXED STABILITY CONDITIONS

This section derives relaxed stability conditions by utilizing properties of membership functions. Theorem 38 is a relaxed stability condition for Theorem 34.

THEOREM 38 *Assume that the number of rules that fire for all t is less than or equal to s, where $1 < s \le r$. The fuzzy descriptor system (10.2) can be stabilized via the PDC fuzzy controller (10.8) if there exist a common matrix Z_1, Z_3, Y_1, Y_2, and Y_3 such that*

$$Z_1^T = Z_1 > 0,$$

$$Y = \begin{bmatrix} Y_1 & Y_3^T \\ Y_3 & Y_2 \end{bmatrix} \ge 0,$$

$$\begin{bmatrix} -Z_3 - Z_3^T + (s-1)Y_1 & * \\ \begin{pmatrix} Z_1 + A_i Z_1 - B_i M_{ik} \\ + E_k Z_3 + (s-1)Y_3 \end{pmatrix} & -Z_1 E_k^T - E_k Z_1 + (s-1)Y_2 \end{bmatrix} < 0,$$

$$h_i \cap v_k \ne \emptyset, \quad (10.25)$$

$$\begin{bmatrix} -2Z_3 - 2Z_3^T - 2Y_1 & * \\ \begin{pmatrix} 2Z_1 + A_i Z_1 - B_i M_{jk} \\ + A_j Z_1 - B_j M_{ik} \\ + 2 E_k Z_3 - 2Y_3 \end{pmatrix} & -2Z_1 E_k^T - 2 E_k Z_1 - 2Y_2 \end{bmatrix} < 0,$$

$$i < j \le r \text{ s.t. } h_i \cap h_j \cap v_k \ne \emptyset. \quad (10.26)$$

The feedback gains are obtained as $F_{ik} = M_{ik} Z_1^{-1}$.

Proof. Consider a candidate of quadratic function

$$V(x^*(t)) = x^{*T}(t) E^{*T} X x^*(t).$$

Now assume that

$$\left(\frac{G_{ijk} + G_{jik}}{2}\right)^T X + X^T \left(\frac{G_{ijk} + G_{jik}}{2}\right) - U \le 0,$$

$$i < j \le r \text{ s.t. } h_i \cap h_j \cap v_k \ne \emptyset,$$

where

$$G_{ijk} = \begin{bmatrix} 0 & I \\ A_i - B_i F_{ik} & -E_k \end{bmatrix},$$

$$U = \begin{bmatrix} Q_1 & Q_3^T \\ Q_3 & Q_2 \end{bmatrix} \ge 0.$$

From the above assumption, we have

$$\dot{V}(x^*(t)) = \sum_{i=1}^{r} \sum_{k=1}^{r^e} h_i^2(z(t)) v_k(z(t)) x^{*T}(t) (G_{iik}^T X + X^T G_{iik}) x^*(t)$$

$$+ 2 \sum_{i=1}^{r} \sum_{i<j} \sum_{k=1}^{r^e} h_i(z(t)) h_j(z(t)) v_k(z(t)) x^{*T}(t)$$

$$\times \left\{ \left(\frac{G_{ijk} + G_{jik}}{2} \right)^T X + X^T \left(\frac{G_{ijk} + G_{jik}}{2} \right) \right\} x^{*T}(t)$$

$$\leq \sum_{i=1}^{r} \sum_{k=1}^{r^e} h_i^2(z(t)) v_k(z(t)) x^{*T}(t) (G_{iik}^T X + X^T G_{iik}) x^*(t)$$

$$+ 2 \sum_{i=1}^{r} \sum_{i<j} \sum_{k=1}^{r^e} h_i(z(t)) h_j(z(t)) v_k(z(t)) x^{*T}(t) U x^*(t)$$

$$\leq \sum_{i=1}^{r} \sum_{k=1}^{r^e} h_i^2(z(t)) v_k(z(t)) x^{*T}(t)$$

$$\times (G_{iik}^T X + X^T G_{iik} + (s-1)U) x^*(t)$$

since

$$\sum_{i=1}^{r} h_i^2(z(t)) - \frac{1}{s-1} \sum_{i=1}^{r} \sum_{i<j} 2 h_i(z(t)) h_j(z(t)) \geq 0, \quad i < s \leq r.$$

Therefore, the closed-loop system is stable if

$$E^{*T} X = X^T E^* \geq 0,$$

$$G_{iik}^T X + X^T G_{iik} + (s-1)U < 0, \quad h_i \cap v_k \neq \emptyset, \quad (10.27)$$

$$\left(\frac{G_{ijk} + G_{jik}}{2} \right)^T X + X^T \left(\frac{G_{ijk} + G_{jik}}{2} \right) - U \leq 0,$$

$$i < j \leq r \text{ s.t. } h_i \cap h_j \cap v_k \neq \emptyset. \quad (10.28)$$

In the same way as in the proof of Theorem 34, we obtain the LMI conditions of Theorem 38. Therefore, only derivation of condition (10.25) is given below. Equation (10.27) can be rewritten as well:

$$X^{-T}G_{iik}^T XX^{-1} + X^{-T}X^T G_{iik} X^{-1} + (s-1)X^{-T}UX^{-1}$$

$$= \begin{bmatrix} Z_1^T & -Z_3^T \\ 0 & Z_1^T \end{bmatrix} \begin{bmatrix} 0 & A_i^T - F_{ik}^T B_i^T \\ I & -E_k^T \end{bmatrix}$$

$$+ \begin{bmatrix} 0 & I \\ A_i - B_i F_{ik} & -E_k \end{bmatrix} \begin{bmatrix} Z_1 & 0 \\ -Z_3 & Z_1 \end{bmatrix}$$

$$+ (s-1)\begin{bmatrix} Z_1^T & -Z_3^T \\ 0 & Z_1^T \end{bmatrix} \begin{bmatrix} Q_1 & Q_3^T \\ Q_3 & Q_2 \end{bmatrix} \begin{bmatrix} Z_1 & 0 \\ -Z_3 & Z_1 \end{bmatrix}$$

$$= \begin{bmatrix} -Z_3 - Z_3^T + (s-1)Y_1 & * \\ \begin{pmatrix} Z_1 + A_i Z_1 - B_i M_{ik} \\ + E_k Z_3 + (s-1)Y_3 \end{pmatrix} & \begin{pmatrix} -Z_1 E_k^T - E_k Z_1 \\ +(s-1)Y_2 \end{pmatrix} \end{bmatrix} < 0,$$

where

$$Y_1 = Z_1 Q_1 Z_1 - Z_3^T Q_3 Z_1 - Z_1 Q_3^T Z_3 + Z_3^T Q_2 Z_3,$$
$$Y_2 = Z_1 Q_2 Z_1,$$
$$Y_3 = Z_1 Q_3 Z_1 - Z_1 Q_2 Z_3. \tag{Q.E.D.}$$

Theorem 38 is reduced to Theorem 34 when $Y_1 = Y_2 = Y_3 = 0$. This means that Theorem 38 gives more relaxed conditions.

Next, we derive stability conditions for Theorem 38 in the case of $h_i(z(t)) = v_k(z(t))$ and $r = r^e$.

THEOREM 39 *Assume that the number of rules that fire for all t is less than or equal to s, where $1 < s \leq r$. Moreover, assume that $h_i(z(t)) = v_k(z(t))$ and $r = r^e$. Then, the fuzzy descriptor system (10.16) can be stabilized via the PDC fuzzy controller (10.17) if there exist Z_1, Z_3, Y_1, Y_2, and Y_3 such that*

$$Z_1^T = Z_1 > 0,$$

$$Y = \begin{bmatrix} Y_1 & Y_3^T \\ Y_3 & Y_2 \end{bmatrix} \geq 0,$$

$$\begin{bmatrix} -Z_3 - Z_3^T + (s-1)Y_1 & * \\ \begin{pmatrix} Z_1 + A_i Z_1 - B_i M_i \\ + E_i Z_3 + (s-1)Y_3 \end{pmatrix} & -Z_1 E_i^T - E_i Z_1 + (s-1)Y_2 \end{bmatrix} < 0,$$

$$i = 1, 2, \ldots, r,$$

RELAXED STABILITY CONDITIONS

$$\begin{bmatrix} -2Z_3 - 2Z_3^T - 2Y_1 & * \\ \begin{pmatrix} 2Z_1 + A_i Z_1 - B_i M_j \\ + A_j Z_1 - B_j M_i \\ + 2E_i Z_3 - 2Y_3 \end{pmatrix} & -2Z_1 E_i^T - 2E_i Z_1 - 2Y_2 \end{bmatrix} < 0,$$

$$i < j \le r \text{ s.t. } h_i \cap h_j \ne \emptyset.$$

The feedback gains are obtained as $F_i = M_i Z_1^{-1}$.

Proof. Consider a candidate of a quadratic function

$$V(x^*(t)) = x^{*T}(t) E^{*T} X x^*(t).$$

Now assume that

$$\left(\frac{G_{ij} + G_{ji}}{2} \right)^T X + X^T \left(\frac{G_{ij} + G_{ji}}{2} \right) - U \le 0, \quad i < j \le r \text{ s.t. } h_i \cap h_j \ne \emptyset,$$

where

$$G_{ij} = \begin{bmatrix} 0 & I \\ A_i - B_i F_j & -E_i \end{bmatrix},$$

$$U = \begin{bmatrix} Q_1 & Q_3^T \\ Q_3 & Q_2 \end{bmatrix} \ge 0.$$

From the above assumption, we have

$$\dot{V}(x^*(t)) = \sum_{i=1}^{r} h_i^2(z(t)) x^{*T}(t) (G_{ii}^T X + X^T G_{ii}) x^*(t)$$

$$+ 2 \sum_{i=1}^{r} \sum_{i<j} h_i(z(t)) h_j(z(t)) x^{*T}(t)$$

$$\times \left\{ \left(\frac{G_{ij} + G_{ji}}{2} \right)^T X + X^T \left(\frac{G_{ij} + G_{ji}}{2} \right) \right\} x^{*T}(t)$$

$$\le \sum_{i=1}^{r} h_i^2(z(t)) x^{*T}(t) (G_{ii}^T X + X^T G_{ii}) x^*(t)$$

$$+ 2 \sum_{i=1}^{r} \sum_{i<j} h_i(z(t)) h_j(z(t)) x^{*T}(t) U x^*(t)$$

$$\le \sum_{i=1}^{r} h_i^2(z(t)) x^{*T}(t) (G_{ii}^T X + X^T G_{ii} + (s-1) U) x^*(t)$$

since

$$\sum_{i=1}^{r} h_i^2(z(t)) - \frac{1}{s-1} \sum_{i=1}^{r} \sum_{i<j} 2h_i(z(t))h_j(z(t)) \geq 0, \quad i < s \leq r.$$

Therefore, the closed-loop system is stable if

$$E^{*T}X = X^T E^* \geq 0,$$

$$G_{ii}^T X + X^T G_{ii} + (s-1)U < 0, \quad i = 1, 2, \ldots, r, \quad (10.29)$$

$$\left(\frac{G_{ij} + G_{ji}}{2}\right)^T X + X^T \left(\frac{G_{ij} + G_{ji}}{2}\right) - U \leq 0,$$

$$i < j \leq r \text{ s.t. } h_i \cap h_j \neq \emptyset. \quad (10.30)$$

In the same way as in the proof of Theorem 38, we obtain the LMI conditions of Theorem 39. (Q.E.D.)

Theorem 39 is reduced to Theorem 35 when $Y_1 = Y_2 = Y_3 = 0$. This means that Theorem 39 gives more relaxed conditions.

Consider the common B matrix case, that is, $B_1 = B_2 = \cdots = B_r$. It should be emphasized that stability conditions for the common B matrix case become very easy.

THEOREM 40 *The fuzzy descriptor system* (10.2) *with the common B matrix, that is,* $B_1 = B_2 = \cdots = B_r = B$, *can be stabilized via the PDC fuzzy controller* (10.8) *if there exist* Z_1, Z_3, *and* M_{ik} *such that*

$$Z_1^T = Z_1 > 0,$$

$$\begin{bmatrix} -Z_3 - Z_3^T & * \\ \begin{pmatrix} Z_1 + A_i Z_1 \\ -BM_{ik} + E_k Z_3 \end{pmatrix} & -Z_1 E_k^T - E_k Z_1 \end{bmatrix} < 0, \quad h_i \cap h_j \neq \emptyset.$$

The feedback gains F_{ik} *are obtained as* $F_{ik} = M_{ik} Z_1^{-1}$.

Proof. Theorem 40 is derived in the same way as in the proof of Theorem 36. (Q.E.D.)

Next, we discuss the stability of the fuzzy descriptor system with the common B matrix in the case of $h_i(z(t)) = v_k(z(t))$ and $r = r^e$. By utilizing the property of $h_i(z(t)) = v_k(z(t))$, Theorem 40 can be simplified as follows.

THEOREM 41 *Assume that $h_i(z(t)) = v_k(z(t))$ and $r = r^e$. The fuzzy descriptor system (10.2) with the common B matrix can be stabilized via the PDC fuzzy controller (10.17) if there exist Z_1, Z_3, and M_i such that*

$$Z_1^T = Z_1 > 0,$$

$$\begin{bmatrix} -Z_3 - Z_3^T & * \\ \begin{pmatrix} Z_1 + A_i Z_1 \\ -BM_i + E_i Z_3 \end{pmatrix} & -Z_1 E_i^T - E_i Z_1 \end{bmatrix} < 0, \quad i = 1, 2, \ldots, r.$$

The feedback gains F_i are obtained as $F_i = M_i Z_1^{-1}$.

Proof. Theorem 41 is derived in the same way as in the proof of Theorem 37.

Note that Theorems 40 and 41 are the same as Theorems 36 and 37, respectively.

10.4 WHY FUZZY DESCRIPTOR SYSTEMS?

We present a motivating example of the need of the fuzzy descriptor system instead of the ordinary fuzzy model. Consider a simple nonlinear system,

$$(1 + a\cos\theta(t))\ddot{\theta}(t) = -b\dot{\theta}^3(t) + c\theta(t) + du(t), \quad (10.31)$$

where $a < 1$ and assume the range of $\dot{\theta}(t)$ as $|\dot{\theta}(t)| < \phi$.

First, we replace the nonlinear dynamics (10.31) with the ordinary Takagi-Sugeno fuzzy model. From (10.31), we have

$$\ddot{\theta}(t) = -\frac{b}{1 + a\cos\theta(t)}\dot{\theta}^3(t)$$

$$+ \frac{c}{1 + a\cos\theta(t)}\theta(t) + \frac{d}{1 + a\cos\theta(t)}u(t). \quad (10.32)$$

Equation (10.32) can be exactly represented by the following fuzzy model:

$$\dot{x}(t) = \sum_{i=1}^{4} m_i(x_1(t), x_2(t))(A_i x(t) + B_i u(t)), \quad (10.33)$$

where $x(t) = [x_1(t) \ x_2(t)]^T = [\theta(t) \ \dot{\theta}(t)]^T$,

$$A_1 = \begin{bmatrix} 0 & 1 \\ c/(1+a) & -b \cdot \phi^2/(1+a) \end{bmatrix},$$

$$A_2 = \begin{bmatrix} 0 & 1 \\ c/(1-a) & -b \cdot \phi^2/(1-a) \end{bmatrix},$$

$$A_3 = \begin{bmatrix} 0 & 1 \\ c/(1+a) & 0 \end{bmatrix}, \quad A_4 = \begin{bmatrix} 0 & 1 \\ c/(1-a) & 0 \end{bmatrix},$$

$$B_1 = \begin{bmatrix} 0 \\ d/(1+a) \end{bmatrix}, \quad B_2 = \begin{bmatrix} 0 \\ d/(1-a) \end{bmatrix},$$

$$B_3 = \begin{bmatrix} 0 \\ d/(1+a) \end{bmatrix}, \quad B_4 = \begin{bmatrix} 0 \\ d/(1-a) \end{bmatrix},$$

$$m_1(x_1(t), x_2(t)) = \frac{x_2^2(t)(1+a)(1+\cos x_1(t))}{2\phi^2(a + \cos x_1(t))},$$

$$m_2(x_1(t), x_2(t)) = \frac{x_2^2(t)(1-a)(1-\cos x_1(t))}{2\phi^2(a + \cos x_1(t))},$$

$$m_3(x_1(t), x_2(t)) = \frac{(\phi^2 - x_2^2(t))(a+1)(1+\cos x_1(t))}{2\phi^2(a + \cos x_1(t))},$$

$$m_4(x_1(t), x_2(t)) = \frac{(\phi^2 - x_2^2(t))(1-a)(1-\cos x_1(t))}{2\phi^2(a + \cos x_1(t))}.$$

The PDC fuzzy controller is constructed from the fuzzy model (10.33):

$$u(t) = -\sum_{i=1}^{4} m_i(x_1(t), x_2(t)) F_i x(t). \tag{10.34}$$

By substituting (10.34) into (10.33), the fuzzy control system is represented as

$$\dot{x}(t) = \sum_{i=1}^{4} \sum_{j=1}^{4} m_i(x_1(t), x_2(t)) m_j(x_1(t), x_2(t)) \{A_i - B_i F_j\} x(t). \tag{10.35}$$

The stability conditions for (10.35) were given in Chapter 3 as follows.

Assume that the number of rules that fire for all t is less than or equal to s, where $1 < s \leq r$. The fuzzy system (10.33) can be stabilized via the PDC fuzzy controller (10.34) if there exist X, Y, and M_i such that

$$X > 0, \quad Y \geq 0,$$

$$-XA_i^T - A_iX + M_i^T B_i^T + B_i M_i - (s-1)Y > 0, \quad i = 1, \ldots, 4,$$

$$2Y - XA_i^T - XA_j^T - A_iX - A_jX$$

$$+ M_i^T B_j^T + M_j^T B_i^T + B_i M_j + B_j M_i \geq 0,$$

$$i = 1, \ldots, 4, \quad i < j \leq 4,$$

where $M_i = F_i X$. Note that 12 LMI conditions are required to find stable feedback gains F_i.

Next, we replace the nonlinear dynamics (10.31) with the fuzzy descriptor system. Equation (10.31) can be exactly represented by the following fuzzy descriptor system:

$$\sum_{k=1}^{2} v_k(x_2(t)) E_k \dot{x}(t) = \sum_{i=1}^{2} h_i(x_1(t))(A_i x(t) + B_i u(t)), \quad (10.36)$$

where $x(t) = [x_1(t) \quad x_2(t)]^T = [\theta(t) \quad \dot{\theta}(t)]^T$,

$$E_1 = \begin{bmatrix} 1 & 0 \\ 0 & 1+a \end{bmatrix}, \quad E_2 = \begin{bmatrix} 1 & 0 \\ 0 & 1-a \end{bmatrix},$$

$$A_1 = \begin{bmatrix} 0 & 1 \\ c & -b \cdot \phi^2 \end{bmatrix}, \quad A_2 = \begin{bmatrix} 0 & 1 \\ c & 0 \end{bmatrix},$$

$$B_1 = \begin{bmatrix} 0 \\ d \end{bmatrix}, \quad B_2 = \begin{bmatrix} 0 \\ d \end{bmatrix},$$

$$h_1(x_2(t)) = \frac{x_2^2(t)}{2}, \quad h_2(x_2(t)) = 1 - \frac{x_2^2(t)}{2},$$

$$v_1(x_1(t)) = \frac{1 + \cos x_1(t)}{2}, \quad v_2(x_1(t)) = \frac{1 - \cos x_1(t)}{2}.$$

Note that the fuzzy descriptor system has the common B matrix. The simpler stability condition, Theorem 36 or 41, is applicable for designing a stable fuzzy controller for (10.36). In contrast, the ordinary fuzzy system (10.33) has different B matrices. The fuzzy descriptor system (10.36) can be stabilized

214 FUZZY DESCRIPTOR SYSTEMS AND CONTROL

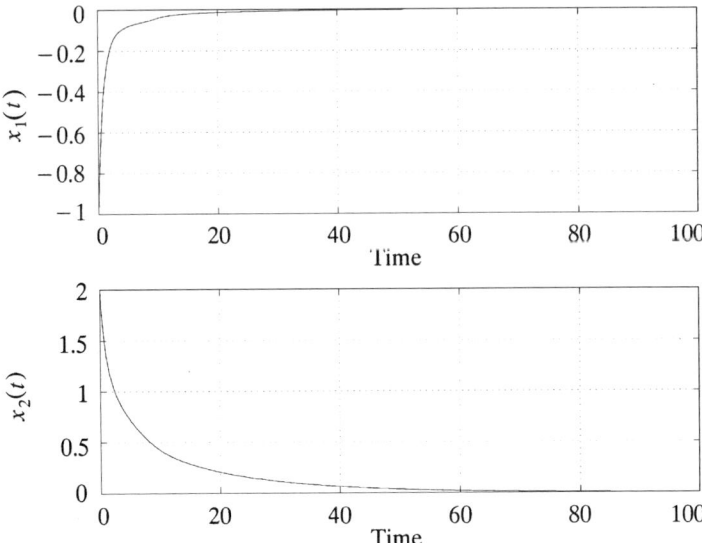

Fig. 10.1 Control result 1.

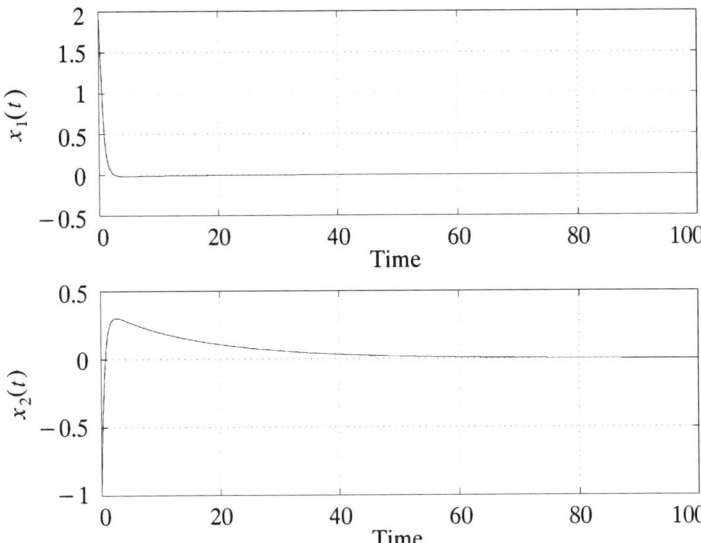

Fig. 10.2 Control result 2.

via the fuzzy controller (10.8) if there exist Z_1, Z_3, and M_{ik} such that

$$Z_1 = Z_1^T > 0,$$

$$\begin{bmatrix} -Z_3 - Z_3^T & * \\ \begin{pmatrix} Z_1 + A_i Z_1 \\ -BM_{ik} + E_k Z_3 \end{pmatrix} & -Z_1 E_k^T - E_k Z_1 \end{bmatrix} < 0, \quad i = 1, 2 \quad k = 1, 2.$$

Note that five LMI conditions are required to find feedback gains F_{ik}.

Therefore the fuzzy descriptor system is suitable for modeling and analysis of complex systems represented in the form (10.31). The form is often observed in nonlinear mechanical systems [6, 7].

Figures 10.1 and 10.2 show the control results for the fuzzy descriptor system. The fuzzy controller is designed using Theorem 41. The designed controller stabilizes the fuzzy descriptor system (10.36), that is, the nonlinear system (10.31).

REFERENCES

1. D. G. Luenberger, "Dynamic Equations in Descriptor Form," *IEEE Trans. Automatic Control*, Vol. AC-22, No. 3, pp. 312–321 (1977).
2. T. Taniguchi, K. Tanaka, K. Yamafuji, and H. O. Wang, "Fuzzy Descriptor Systems: Stability Analysis and Design via LMIs," 1999 American Control Conference, San Diego, June 1999, pp. 1827–1831.
3. T. Taniguchi, K. Tanaka, and H. O. Wang, "Fuzzy Descriptor Systems and Fuzzy Controller Designs," Eighth International Fuzzy Systems Association World Congress, Taipei, Vol. 2, Aug. 1999, pp. 655–659.
4. T. Taniguchi, K. Tanaka, and H. O. Wang, "Universal Trajectory Tracking Control Using Fuzzy Descriptor Systems," 38th IEEE Conference on Decision and Control, Phoenix, Dec. 1999, pp. 4852–4857.
5. T. Taniguchi, K. Tanaka, and H. O. Wang, "Fuzzy Descriptor Systems and Nonlinear Model Following Control," *IEEE Trans. on Fuzzy Syst.*, Vol. 8, No. 4, pp. 442–452 (2000).
6. A. Bedford and W. Fowler, *Statics—Engineering Mechanics*, Addison-Wesley Publishing Company, Reading, MA, 1995.
7. A. Bedford and W. Fowler, *Dynamics—Engineering Mechanics*, Addison-Wesley Publishing Company, Reading, MA, 1995.

CHAPTER 11

NONLINEAR MODEL FOLLOWING CONTROL

In Chapter 9, the model following control for chaotic systems based on the Takagi-Sugeno fuzzy models with the common B matrix is discussed. In this chapter, we present a more general framework [1, 2] to address the nonlinear model following control problem for the fuzzy descriptor systems introduced in Chapter 10. Specifically, these extended results deal with nonlinear model following control for fuzzy descriptor systems with *different* B matrices. A new parallel distributed compensation, the so-called twin parallel distributed compensation (TPDC), is proposed to solve the nonlinear model following control. The TPDC fuzzy controller mirrors the structures of the fuzzy descriptor systems which represent a nonlinear plant and a nonlinear reference model. A design procedure based on the TPDC is presented. As in the usual spirit of this book, all design conditions are rendered in terms of LMIs. The proposed method represents a unified approach to nonlinear model following control. It contains the regulation and servo control problems as special cases. Several design examples are included to show the utility of the nonlinear model following control.

11.1 INTRODUCTION

This chapter presents a unified approach to nonlinear model following control that is much more difficult than the regulation problem. In this chapter, the nonlinear model following control means nonlinear control to reduce the error between the states of a nonlinear system and those of a nonlinear reference model, that is, $\lim_{t \to \infty} x(t) - x_R(t) = 0$, where $x(t)$ and

$x_R(t)$ denote the states of the nonlinear system and those of the nonlinear reference model, respectively. The important feature is that $x_R(t)$ is not necessarily zero or a constant. The nonlinear system and the nonlinear reference model are allowed to be linear, nonlinear, or even chaotic if the nonlinear models are represented in the form of the fuzzy descriptor systems. Thus, to execute the nonlinear model following control, we need the fuzzy descriptor systems for a nonlinear system and a nonlinear reference model. Now the question that needs to be addressed is "Is it possible to approximate any smooth nonlinear systems with the Takagi-Sugeno fuzzy model having no consequent constant terms." The answer is yes in the C^0 or C^1 context. As mentioned in Chapter 2, it was proven in [3] and [4] that any smooth nonlinear systems plus their first-order derived systems can be approximated using the Takagi-Sugeno fuzzy model (having no consequent constant terms) with any desired accuracy (for more details, see Chapter 14). Thus, the nonlinear model following control discussed here is a unified approach containing the regulation and servo control problems as special cases, where "servo control" means control for step inputs of reference signals.

As mentioned in Chapters 1 and 10, $h_i \cap v_k \neq \emptyset$ denotes all the pairs (i, k) excepting $h_i(z(t))v_k(z(t)) = 0$ for all $z(t)$; $h_i \cap h_j \cap v_k \neq \emptyset$ denotes all the pairs (i, j, k) excepting $h_i(z(t))h_j(z(t))v_k(z(t)) = 0$ for all $z(t)$; and $i < j$ s.t $h_i \cap h_j \cap v_k \neq \emptyset$ denotes all $i < j$ excepting $h_i(z(t))h_j(z(t))v_k(z(t)) = 0$, $\forall z(t)$.

11.2 DESIGN CONCEPT

In the nonlinear model following control, we use the fuzzy descriptor system model introduced in Chapter 10 to describe both the plant and the reference system. The plant is represented by the fuzzy descriptor system (10.1). To facilitate the analysis, system (10.1) is rewritten as (10.2). In the following, we develop the fuzzy descriptor system model for the reference system.

11.2.1 Reference Fuzzy Descriptor System

Consider a nonlinear reference model described via a descriptor fuzzy system:

$$\sum_{l=1}^{r_R^e} v_{R_l}(z_R(t)) E_{R_l} \dot{x}_R(t) = \sum_{p=1}^{r_R} h_{R_p}(z_R(t)) D_p x_R(t), \quad (11.1)$$

where $x_R(t) \in \mathbf{R}^{n_R}$ and $\mathbf{D}_p \in \mathbf{R}^{n_R \times n_R}$,

$$v_{R_l}(z_R(t)) \geq 0, \quad \sum_{l}^{r_R^e} v_{R_l}(z_R(t)) = 1,$$

$$h_{R_p}(z_R(t)) \geq 0, \quad \sum_{p=1}^{r_R} h_{R_p}(z_R(t)) = 1.$$

We use $z_R(t)$ to denote the vector containing all the individual elements $z_{R_j}(t)$ ($j = 1, 2, \ldots, p_R$).

The augmented system with the new state $x_R^*(t) = [x_R^T(t) \quad \dot{x}_R^T(t)]^T$ is described as

$$E^* \dot{x}_R^*(t) = \sum_{p=1}^{r_R} \sum_{\ell=1}^{r_R^e} h_{R_p}(z_R(t)) \, v_{R_\ell}(z_R(t)) D_{p\ell}^* x_R^*(t), \qquad (11.2)$$

where

$$E^* = \begin{bmatrix} I & 0 \\ 0 & 0 \end{bmatrix}, \qquad D_{p\ell}^* = \begin{bmatrix} 0 & I \\ D_p & -E_{R\ell} \end{bmatrix}.$$

11.2.2 Twin-Parallel Distributed Compensations

This section introduces the so-called twin parallel distributed compensation (TPDC) to realize nonlinear model following control. The main difference for the ordinary PDC controller presented in Chapter 2 is to add a control term feeding back the signal of $x_R(t)$. It might be reminded that a similar controller structure as TPDC was first employed in Chapter 9 in the nonlinear model following control for chaotic systems.

Specifically, the TPDC fuzzy controller consists of two subcontrollers:

$$u_A(t) = -\sum_{i=1}^{r} \sum_{k=1}^{r^e} h_i(z(t)) v_k(z(t)) F_{ik}^* x^*(t), \quad \text{subcontroller A}$$

$$u_B(t) = \sum_{p=1}^{r_R} \sum_{\ell=1}^{r_R^e} h_{R_p}(z_R(t)) v_{R_\ell}(z_R(t)) K_{p\ell}^* x_R^*(t), \quad \text{subcontroller B}$$

where

$$F_{ik}^* = \begin{bmatrix} F_{ik} & 0 \end{bmatrix}, \qquad K_{p\ell}^* = \begin{bmatrix} K_{p\ell} & 0 \end{bmatrix}.$$

Note that $u_A(t)$ is the same as (10.8). The TPDC controller is obtained as

$$u(t) = u_A(t) + u_B(t)$$

$$= -\sum_{i=1}^{r} \sum_{k=1}^{r^e} h_i(z(t)) v_k(z(t)) F_{ik}^* x^*(t)$$

$$+ \sum_{p=1}^{r_R} \sum_{\ell=1}^{r_R^e} h_{R_p}(z_R(t)) v_{R_\ell}(z_R(t)) K_{p\ell}^* x_R^*(t). \qquad (11.3)$$

220 NONLINEAR MODEL FOLLOWING CONTROL

The error system consisting of (10.2), (11.2), and (11.3) is as follows:

$$E^*\dot{e}(t) = \sum_{i=1}^{r}\sum_{k=1}^{r^e} h_i^2(z(t))v_k(z(t))(A_{ik}^* - B_i^*F_{ik}^*)x^*(t)$$

$$+ 2\sum_{i=1}^{r}\sum_{i<j}\sum_{k=1}^{r^e} h_i(z(t))h_j(z(t))v_k(z(t))$$

$$\times \left(\frac{A_{ik}^* - B_i^*F_{jk}^* + A_{jk}^* - B_j^*F_{ik}^*}{2}\right)x^*(t)$$

$$- \sum_{i=1}^{r}\sum_{p=1}^{r_R}\sum_{\ell=1}^{r_R^e} h_i(z(t))h_{R_p}(z_R(t))v_{R_\ell}(z_R(t))$$

$$\times (D_{p\ell}^* - B_i^*K_{p\ell}^*)x_R^*(t), \qquad (11.4)$$

where $e(t) = x^*(t) - x_R^*(t)$.

THEOREM 42 *If conditions (11.6) hold, the error system becomes*

$$E^*\dot{e}(t) = \{Gx^*(t) - Gx_R^*(t)\} = Ge(t) \qquad (11.5)$$

by the TPDC fuzzy controller (11.3),

$$G = A_{ik}^* - B_i^*F_{ik}^*, \qquad h_i \cap v_k \neq \emptyset,$$

$$= \tfrac{1}{2}(A_{ik}^* - B_i^*F_{jk}^* + A_{jk}^* - B_j^*F_{ik}^*),$$

$$i < j \leq r \text{ s.t. } h_i \cap h_j \cap v_k \neq \emptyset,$$

$$= D_{p\ell}^* - B_i^*K_{p\ell}^*, \qquad h_i \cap h_{R_p} \cap v_{R_\ell} \neq \emptyset. \qquad (11.6)$$

Proof. We naturally arrive at the conditions (11.6) to cancel the nonlinearity of the error system (11.4). (Q.E.D.)

Note that G is not always a stable matrix. The TPDC fuzzy controller (11.3) with the feedback gains F_{ik} and $K_{p\ell}$ should be designed so as to guarantee the condition (11.6) and the stability of the error system (11.4).

THEOREM 43 *The feedback gains F_{ik} and $K_{p\ell}$ can be determined by solving the following eigenvalue problem (EVP):*

minimize β
Y, Z, M_i, N_k

subject to $\beta > 0$,

$$\begin{bmatrix} Z_1 & 0 \\ 0 & Z_1 \end{bmatrix} < I,$$

$$Z_1^T = Z_1 > 0, \quad Y = \begin{bmatrix} Y_1 & Y_3^T \\ Y_3 & Y_2 \end{bmatrix} \geq 0, \qquad (11.7)$$

$$\begin{bmatrix} -Z_3 - Z_3^T + (s-1)Y_1 & * \\ Z_1 + A_i Z_1 - B_i M_{ik} + E_k Z_3 + (s-1)Y_3 & -Z_1 E_k^T - E_k Z_1 + (s-1)Y_2 \end{bmatrix} < 0,$$

$$h_i \cap v_k \neq \emptyset, \qquad (11.8)$$

$$\begin{bmatrix} -2Z_3 - 2Z_3^T - 2Y_1 & * \\ \begin{pmatrix} 2Z_1 + A_i Z_1 - B_i M_{jk} \\ + A_j Z_1 - B_j M_{ik} + 2E_k Z_3 - 2Y_3 \end{pmatrix} & -2Z_1 E_k^T - 2E_k Z_1 - 2Y_2 \end{bmatrix} < 0,$$

$$i < j \leq r \text{ s.t. } h_i \cap h_j \cap v_k \neq \emptyset, \qquad (11.9)$$

$$\begin{bmatrix} \beta I & * & * & * \\ 0 & \beta I & * & * \\ 0 & 0 & \beta I & * \\ A_1 Z_1 - B_1 M_{11} - A_i Z_1 + B_i M_{ik} & -E_1 Z_1 + E_k Z_1 & 0 & I \end{bmatrix} > 0,$$

$$h_{i-\{1\}} \cap v_k \neq \emptyset, \qquad (11.10)$$

$$\begin{bmatrix} \beta I & * & * & * \\ 0 & \beta I & * & * \\ 0 & 0 & I & * \\ \begin{pmatrix} A_i Z_1 - B_i M_{ik} \\ -\frac{1}{2}(A_i Z_1 - B_i M_{jk} + A_j Z_1 - B_j M_{ik}) \end{pmatrix} & 0 & 0 & I \end{bmatrix} > 0,$$

$$i < j \leq r \text{ s.t. } h_i \cap h_j \cap v_k \neq \emptyset, \qquad (11.11)$$

$$\begin{bmatrix} \beta I & * & * & * \\ 0 & \beta I & * & * \\ 0 & 0 & I & * \\ A_i Z_1 - B_i M_{ik} - D_p Z_1 + B_i N_{p\ell} & -E_k Z_1 + E_{R_\ell} Z_1 & 0 & I \end{bmatrix} > 0,$$

$$h_i \cap v_k \cap h_{R_p} \cap v_{R_\ell} \neq \emptyset, \qquad (11.12)$$

where $h_{i-\{1\}} \cap h_k \neq \emptyset$ denotes all the pairs excepting $h_i(z(t))v_k(z(t)) \neq 0$, $\forall z(t)$ for $i = 2, 3, \ldots, r$ and $k = 1, 2, \ldots, r^c$. The feedback gains are obtained as $F_{ik} = M_{ik} Z_1^{-1}$ and $K_{p\ell} = N_{p\ell} Z_1^{-1}$.

Proof. Consider the condition of (11.6). The condition (11.6) to cancel the nonlinearity of the error system is satisfied if (11.13), (11.14), and (11.15) hold for

$$\left\| \beta \cdot (\text{block-diag}[Z_1 \ Z_1])^{-1} (\text{block-diag}[Z_1 \ Z_1])^{-1} \right\|_2 \simeq 0$$

under $\begin{bmatrix} Z_1 & 0 \\ 0 & Z_1 \end{bmatrix} < I.$

$$\beta I - \begin{bmatrix} Z_1 & 0 \\ 0 & Z_1 \end{bmatrix}^T \{A_{11}^* - B_1^* F_{11}^* - (A_{ik}^* - B_i^* M_{ik}^*)\}^T$$

$$\times \{A_{11}^* - B_1^* F_{11}^* - (A_{ik}^* - B_i^* F_{ik}^*)\} \begin{bmatrix} Z_1 & 0 \\ 0 & Z_1 \end{bmatrix} > 0,$$

$$h_{i-\{1\}} \cap v_k \neq \emptyset, \qquad (11.13)$$

$$\beta I - \begin{bmatrix} Z_1 & 0 \\ 0 & Z_1 \end{bmatrix}^T \{A_{ik}^* - B_i^* F_{ik}^* - \tfrac{1}{2}(A_{ik}^* - B_i^* M_{jk}^* + A_{jk}^* - B_j^* M_{ik}^*)\}^T$$

$$\times \{A_{ik}^* - B_i^* F_{ik}^* - \tfrac{1}{2}(A_{ik}^* - B_i^* F_{jk}^* + A_{jk}^* - B_j^* F_{ik}^*)\} \begin{bmatrix} Z_1 & 0 \\ 0 & Z_1 \end{bmatrix} > 0,$$

$$i < j \leq r \quad \text{s.t.} \quad h_i \cap h_j \cap v_k \neq \emptyset, \qquad (11.14)$$

$$\beta I - \begin{bmatrix} Z_1 & 0 \\ 0 & Z_1 \end{bmatrix}^T \{A_{ik}^* - B_i^* F_{ik}^* - (D_{p\ell}^* - B_i^* K_{p\ell}^*)\}^T$$

$$\times \{A_{ik}^* - B_i^* F_{ik}^* - (D_{p\ell}^* - B_i^* K_{p\ell}^*)\} \begin{bmatrix} Z_1 & 0 \\ 0 & Z_1 \end{bmatrix} > 0,$$

$$h_i \cap h_{R_p} \cap v_{R_\ell} \neq \emptyset, \qquad (11.15)$$

where $\beta > 0$. By the Schur complement, the above conditions (11.13)–(11.15) can be converted into (11.10)–(11.12). (Q.E.D.)

From the solutions Z_1, M_{ik}, and $N_{p\ell}$, we obtain the feedback gains as follows: $F_{ik} = M_{ik} Z_1^{-1}$ and $K_{p\ell} = N_{p\ell} Z_1^{-1}$. If the LMI design problem is feasible and

$$\left\| \beta \cdot \left(\text{bloc-diag}[Z_1 \ Z_1]^{-1}\right)\left(\text{block-diag}[Z_1 \ Z_1]^{-1}\right) \right\|_2 \simeq 0,$$

the nonlinear model following control based on the cancellation technique can be realized. Then, the TPDC fuzzy controller with the feasible solutions F_{ik} and $K_{p\ell}$ provides a tractable means to achieve $\lim_{t \to \infty} e(t) = 0$. As shown in Theorem 38, equations (11.7)–(11.9) are stability conditions of the error system.

DESIGN CONCEPT

The nonlinear model following control is reduced to the servo control problem when we select D_p ($p = 1, 2, \ldots, r_R$) such that $x_R(t) = c$, where $c \neq 0$ in general. It is reduced to the regulation problem when we select D_p ($p = 1, 2, \ldots, r_R$) such that $x_R(t) = 0$. In these cases, note that $r_R = 1$. The fact will be seen in design examples.

As mentioned above, this method contains the typical regulation and servo control problems as special cases. However, it realizes not only stabilization but also cancellation of the nonlinearity for the error system. If only stabilization (regulation) is required in controller designs, the feedback gains should be determined only by using the stability conditions (11.7)-(11.9), that is, Theorem 38.

Remark 36 The condition (11.6) to cancel the nonlinearity might often be conservative since it completely requires the cancellation of nonlinearity. A relaxed approach was reported in [5].

11.2.3 The Common B Matrix Case

Consider the common B matrix case, that is, $B_1 = B_2 = \cdots = B_r$. In this case, the cancellation technique of Theorem 43 can be simplified as follows.

THEOREM 44 *The feedback gains F_{ik} and $K_{p\ell}$ can be determined by solving the following EVP:*

$$\underset{Y, Z, M_i, N_k}{\text{minimize}} \beta$$

subject to $\beta > 0$,

$$Z_1^T = Z_1 > 0, \quad \begin{bmatrix} Z_1 & 0 \\ 0 & Z_1 \end{bmatrix} < I, \tag{11.16}$$

$$\begin{bmatrix} -Z_3 - Z_3^T & * \\ Z_1 + A_i Z_1 - BM_{ik} + E_k Z_3 & -Z_1 E_k^T - E_k Z_1 \end{bmatrix} < 0,$$

$$h_i \cap v_k \neq \emptyset, \tag{11.17}$$

$$\begin{bmatrix} \beta I & & * & * & * \\ 0 & \beta I & & * & * \\ 0 & 0 & I & 0 \\ \begin{pmatrix} A_1 Z_1 - BM_{11} \\ -A_i Z_1 + BM_{ik} \end{pmatrix} & -E_1 Z_1 + E_k Z_1 & 0 & I \end{bmatrix} > 0,$$

$$h_{i-\{1\}} \cap v_k \neq \emptyset, \tag{11.18}$$

$$\begin{bmatrix} \beta I & * & * & * \\ 0 & \beta I & * & * \\ 0 & 0 & I & 0 \\ \begin{pmatrix} A_i Z_1 - BM_{ik} \\ -D_p Z_1 + BN_{p\ell} \end{pmatrix} & -E_k Z_1 + E_{R_\ell} Z_1 & 0 & I \end{bmatrix} > 0,$$

$$h_i \cap v_k \cap h_{R_p} \cap v_{R_\ell} \neq \emptyset. \quad (11.19)$$

The feedback gains are obtained as $F_{ik} = M_{ik} Z_1^{-1}$ and $K_{p\ell} = N_{p\ell} Z_1^{-1}$.

Proof. Consider the condition of (11.6). The condition (11.6) to cancel the nonlinearity of the error system is satisfied if (11.20) and (11.21) hold for

$$\left\| \beta \cdot (\text{block-diag}[Z_1 \ Z_1])^{-1} (\text{block-diag}[Z_1 \ Z_1])^{-1} \right\|_2 \simeq 0$$

under $\begin{bmatrix} Z_1 & 0 \\ 0 & Z_1 \end{bmatrix} < I.$

$$\beta I - \begin{bmatrix} Z_1 & 0 \\ 0 & Z_1 \end{bmatrix}^T \{A_{11}^* - B^* F_{11}^* - (A_{ik}^* - B^* M_{ik}^*)\}^T$$

$$\times \{A_{11}^* - B^* F_{11}^* - (A_{ik}^* - B_i^* F_{ik}^*)\} \begin{bmatrix} Z_1 & 0 \\ 0 & Z_1 \end{bmatrix} > 0,$$

$$h_{i-\{1\}} \cap v_k \neq \emptyset, \quad (11.20)$$

$$\beta I - \begin{bmatrix} Z_1 & 0 \\ 0 & Z_1 \end{bmatrix}^T \{A_{ik}^* - B^* F_{ik}^* - (D_{p\ell}^* - B^* K_{p\ell}^*)\}^T$$

$$\times \{A_{ik}^* - B^* F_{ik}^* - (D_{p\ell}^* - B^* K_{p\ell}^*)\} \begin{bmatrix} Z_1 & 0 \\ 0 & Z_1 \end{bmatrix} > 0,$$

$$h_i \cap v_k \neq \emptyset, \quad (11.21)$$

where $\beta > 0$. By the Schur complement, conditions (11.20) and (11.21) can be converted into (11.18) and (11.19). (Q.E.D.)

11.3 DESIGN EXAMPLES

This section gives design examples for the nonlinear model following control. Recall the simple nonlinear system (10.31):

$$(1 + a\cos\theta(t))\ddot{\theta}(t) = -b\dot{\theta}^3(t) + c\theta(t) + du(t),$$

where $a = 0.2$ and assume the range of $\dot{\theta}(t)$ as $|\dot{\theta}(t)| < \phi$. We also recall the fuzzy descriptor system (10.36),

$$\sum_{k=1}^{2} v_k(x_2(t))E_k \dot{x}(t) = \sum_{i=1}^{2} h_i(x_1(t))(A_i x(t) + B_i u(t)),$$

where $x(t) = [x_1(t) \ x_2(t)]^T = [\theta(t) \ \dot{\theta}(t)]^T$,

$$E_1 = \begin{bmatrix} 1 & 0 \\ 0 & 1+a \end{bmatrix}, \quad E_2 = \begin{bmatrix} 1 & 0 \\ 0 & 1-a \end{bmatrix},$$

$$A_1 = \begin{bmatrix} 0 & 1 \\ c & -b \cdot \phi^2 \end{bmatrix}, \quad A_2 = \begin{bmatrix} 0 & 1 \\ c & 0 \end{bmatrix},$$

$$B_1 = \begin{bmatrix} 0 \\ d \end{bmatrix}, \quad B_2 = \begin{bmatrix} 0 \\ d \end{bmatrix},$$

$$h_1(x_2(t)) = \frac{x_2^2(t)}{2}, \quad h_2(x_2(t)) = 1 - \frac{x_2^2(t)}{2},$$

$$v_1(x_1(t)) = \frac{1 + \cos x_1(t)}{2}, \quad v_2(x_1(t)) = \frac{1 - \cos x_1(t)}{2}.$$

We use $a = 0.2$, $b = 1$, $c = -1$, $d = 10$, and $\phi = 4$. Note that the fuzzy descriptor system has the common B matrix.

We consider three cases of reference nonlinear models.

Case 1: Descriptor reference system:

$$(1 + \xi \cos \theta_R(t))\ddot{\theta}_R(t) = -\theta_R(t) + k(1 - \theta_R^2(t))\dot{\theta}_R(t). \quad (11.22)$$

Case 2: Constant output model (servo control problem).
Case 3: Zero output model (regulator control problem).

All the cases of the reference nonlinear models can be represented by the following fuzzy model:

$$\sum_{\ell=1}^{r_R^e} v_{R_\ell}(z_R(t))E_{R_\ell}\dot{x}_R(t) = \sum_{p=1}^{r_R} h_{R_p}(z_R(t))D_p x_R(t),$$

where $x_R(t) = [x_{R_1}(t) \ x_{R_2}(t)]^T = [\theta_{R_1}(t) \ \dot{\theta}_{R_2}(t)]^T$.

In Case 1, $r_R^e = r_R = 2$,

$$E_{R_1} = \begin{bmatrix} 1 & 0 \\ 0 & 1+\xi \end{bmatrix}, \qquad E_{R_2} = \begin{bmatrix} 1 & 0 \\ 0 & 1-\xi \end{bmatrix},$$

$$D_1 = \begin{bmatrix} 0 & 1 \\ -1 & k(1-\psi^2) \end{bmatrix}, \qquad D_2 = \begin{bmatrix} 0 & 1 \\ -1 & k \end{bmatrix},$$

$$h_{R_1}(x_R(t)) = \frac{1}{\psi^2} x_{R_1}^2(t), \qquad h_{R_2}(x_R(t)) = 1 - \frac{1}{\psi^2} x_{R_1}^2(t),$$

$$v_{R_1}(x_{R_1}(t)) = \frac{1 + \cos x_{R_1}(t)}{2}, \qquad v_{R_2}(x_{R_1}(t)) = \frac{1 - \cos x_{R_1}(t)}{2},$$

where it is assumed that $x_{R_2}(t) \in [-\psi \quad \psi]$. We use $k = 1$ and $\psi = 4$. This reference system is reduced to the van del Pol equation when $\xi = 0$ for all t.

Cases 2 and 3 are special cases of nonlinear model following control. By considering the condition of $\ddot{x}_R(t) = \dot{x}_R(t) = 0$, we select E_{R_1} and D_1 as follows, where $r_R^e = r_R = 1$,

$$E_{R_1} = \begin{bmatrix} \zeta_1 & 0 \\ 0 & \zeta_2 \end{bmatrix}, \qquad D_1 = \begin{bmatrix} 0 & 1 \\ 0 & 0 \end{bmatrix}.$$

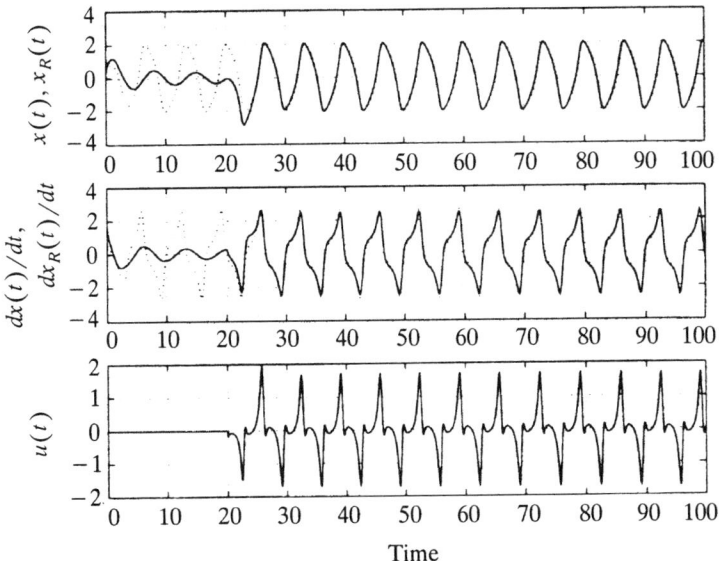

Fig. 11.1 Simulation result 1 (Case 1 for $\xi = 0$).

In the servo control problem (Case 2), $x_R(0) = [0 \quad c]^T$, $c \neq 0$. In this example, $c = 1.5$. In the regulator design problem (Case 3), $x_R(0) = [0 \quad 0]^T$.

Two kinds of ξ are selected: $\xi = 0$ and $\xi = 0.5$ in Case 1. Figures 11.1 and 11.2 show the control results for Case 1 ($\xi = 0$ and $\xi = 0.5$). In Cases 2 and 3, $\zeta_1 = 1$ and $\zeta_2 = 1$. Figure 11.3 shows the control result for Case 2.

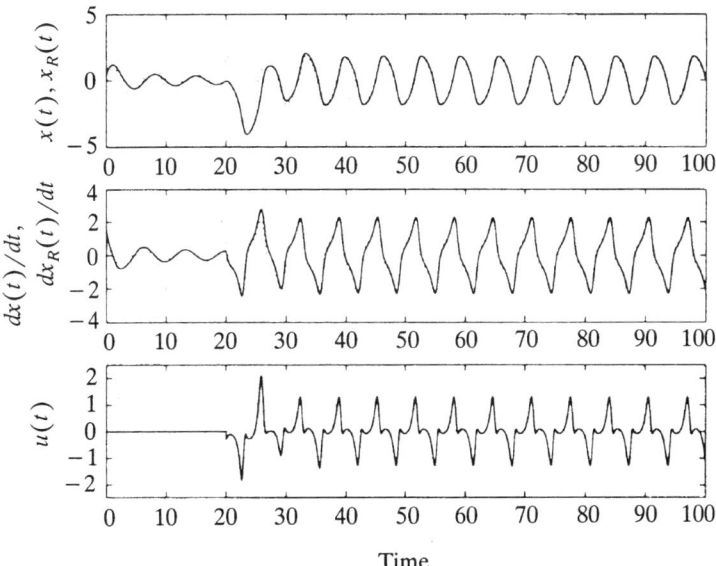

Fig. 11.2 Simulation result 2 (Case 1 for $\xi = 0.5$).

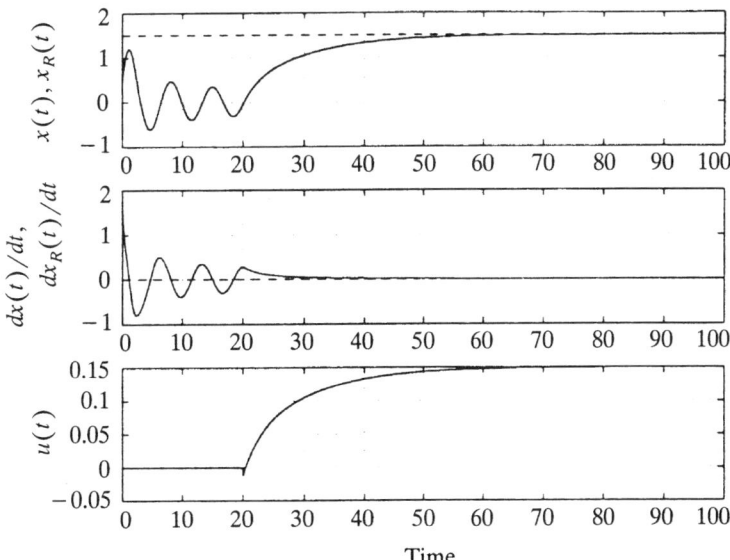

Fig. 11.3 Simulation result 3 (Case 2 for $\zeta_1 = 1$ and $\zeta_2 = 1$).

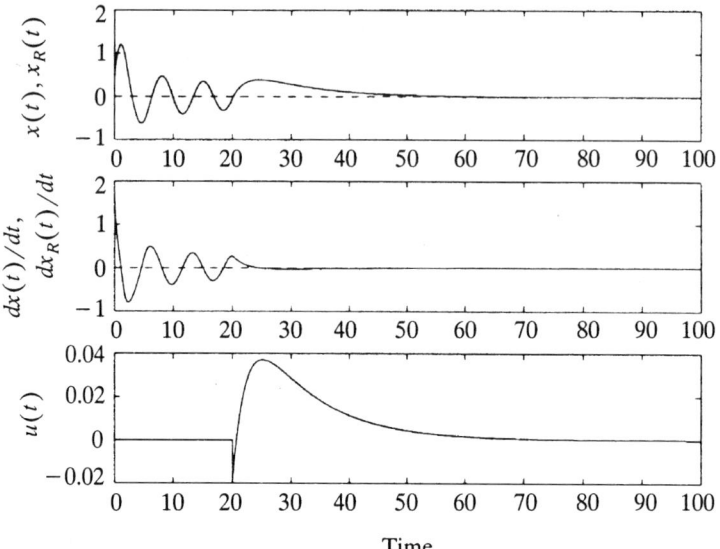

Fig. 11.4 Simulation result 4 (Case 3 for $\zeta_1 = 1$ and $\zeta_2 = 1$).

Figure 11.4 shows the control result for Case 3. In these figures, the dotted and real lines denote $x_R(t)$ and $x(t)$, respectively. The control input $u(t)$ is added after 20 sec in these simulations. It can be seen that the nonlinear model following control is effectively realized even for the complex descriptor reference system (11.22).

REFERENCES

1. T. Taniguchi, K. Tanaka, and H. O. Wang, "Universal Trajectory Tracking Control Using Fuzzy Descriptor Systems," 38th IEEE Conference on Decision and Control, 1999.
2. T. Taniguchi, K. Tanaka, and H. O. Wang, "Fuzzy Descriptor Systems and Nonlinear Model Following Control." *IEEE Trans. on Fuzzy Syst.*, Vol. 8, No. 4, pp. 442–452 (2000).
3. H. O. Wang, D. Niemann, J. Li, and K. Tanaka, "T-S Fuzzy Model with Linear Rule Consequence and PDC Controller: A Universal Framework for Nonlinear Control Systems," 18th International Conference of the North American Fuzzy Information Processing Society (*NAFIPS '99*), 1999, to appear.
4. J. Li, H. O. Wang, D. Niemann, and K. Tanaka, "Using Linear Takagi-Sugeno Fuzzy Systems to Approximate Nonlinear Functions–Applications to Modeling and Control of Nonlinear Systems," *IEEE Trans. Fuzzy Syst.*, submitted.
5. T. Taniguchi, K. Tanaka, K. Yamafuji, and H. O. Wang, "A New PDC for Fuzzy Reference Models," *1999 IEEE International Conference on Fuzzy Systems*, Vol. 2, Seoul, August 1999, pp. 898–903.

CHAPTER 12

NEW STABILITY CONDITIONS AND DYNAMIC FEEDBACK DESIGNS

This chapter presents a unified systematic framework of control synthesis [1–5] for dynamic systems described by the Takagi-Sugeno fuzzy model. In comparison with preceding chapters, this chapter provides two significant extensions. First we provide a new sufficient condition for the existence of a quadratically stabilizing state feedback PDC controller which is more general and relaxed than the existing conditions. Second, we introduce the notion of dynamic parallel distributed compensation (DPDC) and we provide a set of sufficient LMI conditions for the existence of quadratically stabilizing dynamic compensators.

In this chapter, the notation $M > 0$ stands for a positive definite symmetric matrix M; $\mathcal{L}(A, P) = A^T P^T + PA$ is defined as a mapping from $\Re^{n \times n} \times \Re^{n \times n}$ to $\Re^{n \times n}$. The same holds for $\mathcal{L}(A^T, Q^T) = AQ + Q^T A^T$. The term P^{-T} is the same as $(P^{-1})^T$. From this chapter onward, we will use italic symbols such as A and B instead of **A** and **B**. In addition, to lighten the notation, we will use x, y, z, p, and u instead of $x(t)$, $y(t)$, $z(t)$, $p(t)$, and $u(t)$, respectively. Another notable point regarding the notation is that we will use $p(t)$ or p instead of $z(t)$ as premise variables. This is because z is used as performance variables in Chapters 13 and 15 which are based on the setting presented in this chapter. The symbol x' denotes the transposed vector of x. We often drop the p and just write h_i, but it should be kept in mind that the h_i's are functions of the variable p.

The summation process associated with the center of gravity defuzzification in system (2.3) and (2.4) can also be viewed as an interpolation between the vectors $A_i x + B_i u$ based on the value of the parameter p. The parameter p can be given several different interpretations. First, we can assume that the parameter p is a measurable external disturbance signal which does not depend on the state or control input of the system (2.3) and (2.4). Using this interpretation, equations (2.3) and (2.4) describe a time-varying linear system. Second, we can assume that the parameter p is a function of the state, $p = f(x)$. Using this interpretation, equations (2.3) and (2.4) describe a nonlinear system. As a slight modification to this interpretation, we can assume that the parameter p is a function of the measurable outputs of the system, $p = f(y)$. Finally, we can assume that p is an unknown constant value, in which case equations (2.3) and (2.4) describe a linear differential inclusion (LDI). In most cases, we can only derive a benefit from the fuzzy rule base description if we know the values of the parameters, so we will not usually consider this last interpretation. It is also possible to interpret p using a combination of these approaches.

12.1 QUADRATIC STABILIZABILITY USING STATE FEEDBACK PDC

In this section, we consider the special form of parameter-dependent state feedback which mirrors the structure of the T-S model, that is, parallel distributed compensation (PDC) [19, 20]

The PDC controller structure consists of fuzzy rules:

Control Rule i

IF $p_1(t)$ is M_{1i} $\cdots$ and $p_l(t)$ is M_{il},

  THEN $u(t) = K_i x(t)$,

where $i = 1, 2, \ldots, r$. The output of the PDC controller is

$$u = \sum_{i=1}^{r} h_i K_i x. \qquad (12.1)$$

Remark 37 Note that the notation for PDC here is in slightly different form from earlier chapters where the PDC controller is of the following form

$$u = -\sum_{i=1}^{r} h_i F_i x. \qquad (12.2)$$

Let us consider the Lyapunov function candidate $V(x) = x'Px$, where $P > 0$. Taking the time derivative of this function along the flow of the system,

$$\frac{d}{dt}V(x) = \sum_{i=1}^{r}\sum_{j=1}^{r} h_i h_j x'\left(\mathcal{L}(A_i,P) + K_j^T B_i^T P + PB_i K_j\right)x \quad (12.3)$$

$$= \frac{1}{2}\sum_{i=1}^{r}\sum_{j=1}^{r} h_i h_j x'\left(\mathcal{L}(A_i + A_j, P) + K_j^T B_i^T P\right.$$

$$\left. + PB_i K_j + K_i^T B_j^T P + PB_j K_i\right)x. \quad (12.4)$$

If for each $1 \le i \le j \le r$ there exists a symmetric $n \times n$ matrix $T_{ij} = T_{ij}^T$ such that

$$\mathcal{L}(A_i + A_j, P) + K_j^T B_i^T P + PB_i K_j + K_i^T B_j^T P + PB_j K_i < T_{ij}, \quad \forall i, \forall j$$
$$(12.5)$$

and

$$\mathbf{T} = \begin{bmatrix} T_{11} & \cdots & T_{1r} \\ \vdots & \ddots & \vdots \\ T_{1r} & \cdots & T_{rr} \end{bmatrix} < 0, \quad (12.6)$$

then

$$\frac{d}{dt}V(x) < \sum_{i=1}^{r}\sum_{j=1}^{r} h_i h_j x' T_{ij} x$$

$$= [h_1 x' \ \ldots \ h_r x'] \mathbf{T} [h_1 x' \ \ldots \ h_r x']^T$$

$$< 0.$$

In order to express these inequalities as LMI conditions, we need to use a transformation. Define $Q = P^{-1}$, $\hat{T}_{ij} = QT_{ij}Q$, and $M_i = K_i Q$. Pre- and postmultiplying equation (12.5) by Q produces the expression

$$\mathcal{L}(A_i^T + A_j^T, Q) + M_j^T B_i^T + B_i M_j + M_i^T B_j^T + B_j M_i < \hat{T}_{ij},$$

$$i \le j \text{ s.t. } h_i \cap h_j \neq \emptyset. \quad (12.7)$$

We also know that $\mathbf{T} < 0$ if and only if

$$\hat{\mathbf{T}} = \begin{bmatrix} \hat{T}_{11} & \cdots & \hat{T}_{1r} \\ \vdots & \ddots & \vdots \\ \hat{T}_{1r} & \cdots & \hat{T}_{rr} \end{bmatrix} < 0. \quad (12.8)$$

The resulting LMI conditions are summarized in the following theorem:

THEOREM 45 *The T-S model (2.3) is quadratically stabilizable in the large via a state feedback PDC controller (12.1) if there exist $Q > 0$, M_i, $i = 1, 2, \ldots, r$, and $\hat{T}_{ij}$ such that the LMI conditions (12.7) and (12.8) have feasible solutions. The ith gain of the PDC controller is given by*

$$K_i = M_i Q^{-1} \tag{12.9}$$

and the Lyapunov function is given by

$$V = x^T Q^{-1} x. \tag{12.10}$$

Remark 38 The above theorem is a generalization of the stability condition given in [17] and [20]. It is also weaker than the LMI condition given in [25], in which case T_{ij} becomes $t_{ij} I$. The above theorem can be further relaxed if we know the structure of the fuzzy membership function:

- Sometimes there is no overlap between two rules, that is, the product of the h_i and the h_j may be identically zero. In this case, the above theorem can be relaxed by dropping the condition (12.7) corresponding to the i and j in (12.7).
- If only $s < r$ rules can fire at the same time, then the conditions of this theorem can be further relaxed to only require that all the diagonal $s \times s$ principal submatrices of **T** are negative definite.

12.2 DYNAMIC FEEDBACK CONTROLLERS

In this section we introduce the concept of a DPDC, and we derive a set of LMI conditions which can be used to design a stabilizing DPDC.

In order to derive the LMI design conditions, it is useful to begin with a parameter-dependent linear model described by the equations

$$\dot{x}(t) = A(p)x(t) + B(p)u(t),$$
$$y(t) = C(p)x(t), \tag{12.11}$$

where $x(t)$, $y(t)$, and $u(t)$ denote the state, measurement, and input vectors, respectively. The variable $p(t)$ is a vector of measurable parameters. In general, these parameters may be functions of the system states, external disturbances, and time. Note that the T-S model is in this form.

A parameter-dependent dynamic compensator is a parameter-dependent linear system of the form

$$\dot{x}_c(t) = A_c(p)x_c(t) + B_c(p)y(t),$$
$$u(t) = C_c(p)x_c(t) + D_c(p)y(t). \tag{12.12}$$

Defining the augmented system matrix

$$A_{cl}(p) = \begin{bmatrix} A(p) + B(p)D_c(p)C(p) & B(p)C_c(p) \\ B_c(p)C(p) & A_c(p) \end{bmatrix}$$

and the augmented state vector

$$x_{cl}(t) = \begin{bmatrix} x^T(t) & x_c^T(t) \end{bmatrix}^T,$$

the resulting closed-loop dynamic equations are described by the equation

$$\dot{x}_{cl}(t) = A_{cl}(p)x_{cl}(t). \tag{12.13}$$

The system (12.11) is said to be quadratically stabilizable via an s-dimensional parameter-dependent linear compensator if and only if there exists an s-dimensional parameter-dependent controller and a positive definite matrix $P_{cl} > 0$ such that

$$P_{cl}A_{cl}(p) + A_{cl}^T(p)P_{cl} < 0. \tag{12.14}$$

Remark 39 If we fix the value of p, equation (12.14) represents a sufficient condition for the existence of a set of linear, time-invariant controller matrices $A_c(p)$, $B_c(p)$, $C_c(p)$, and $D_c(p)$ which will stabilize the system (2.3) and (2.4) at the fixed value of p. The unknown controller does not enter linearly into equation (12.14), so this equation does not represent an LMI condition. However, the authors of the paper [14] present a transformation procedure which results in a modified set of inequalities which are linear in the unknown data. In what follows, we perform this transformation pointwise with respect to p.

We will first partition the constant matrices P and P^{-1} into components:

$$P_{cl} = \begin{bmatrix} P_{11} & P_{12} \\ P_{12}^T & P_{22} \end{bmatrix}$$

and

$$P_{cl}^{-1} = \begin{bmatrix} Q_{11} & Q_{12} \\ Q_{12}^T & Q_{22} \end{bmatrix},$$

and we will also define the matrices

$$\Pi_1 = \begin{bmatrix} Q_{11} & I \\ Q_{12}^T & 0 \end{bmatrix}$$

and

$$\Pi_2 = P_{cl}\Pi_1 = \begin{bmatrix} I & P_{11} \\ 0 & P_{12}^T \end{bmatrix}.$$

Equation (12.14) will hold if and only if

$$\Pi_1^T P_{cl} A_{cl}(p)\Pi_1 + \Pi_1^T A_{cl}^T(p) P_{cl} \Pi_1 < 0.$$

This equation can also be rewritten as

$$\Pi_2^T A_{cl}(p)\Pi_1 + \Pi_1^T A_{cl}^T(p)\Pi_2 < 0.$$

Writing out the first term on the left-hand side of this equation, we have

$$\begin{bmatrix} I & 0 \\ P_{11} & P_{12} \end{bmatrix} \begin{bmatrix} (A(p) + B(p)D_c(p)C(p)) & B(p)C_c(p) \\ B_c(p)C(p) & A_c(p) \end{bmatrix} \begin{bmatrix} Q_{11} & I \\ Q_{12}^T & 0 \end{bmatrix}$$
$$= E(p).$$

If we define the new variables

$$\mathcal{A}(p) = P_{11}(A(p) + B(p)D_c(p)C(p))Q_{11} + P_{12}B_c(p)C(p)Q_{11}$$
$$+ P_{11}B(p)C_c(p)Q_{12}^T + P_{12}A_c(p)Q_{12}^T,$$
$$\mathcal{B}(p) = P_{11}B(p)D_c(p) + P_{12}B_c(p),$$
$$\mathcal{C}(p) = D_c(p)C(p)Q_{11} + C_c(p)Q_{12}^T,$$
$$\mathcal{D}(p) = D_c(p),$$

then the matrix $E(p)$ can be rewritten as

$$E(p) = \begin{bmatrix} A(p)Q_{11} + B(p)\mathcal{C}(p) & A(p) + B(p)\,\mathcal{D}(p)C(p) \\ \mathcal{A}(p) & P_{11}A(p) + \mathcal{B}(p)\,C(p) \end{bmatrix},$$

and the closed-loop stability condition can be expressed as

$$E(p) + E^T(p) < 0$$

or

$$\begin{bmatrix} \begin{pmatrix} \mathcal{L}(A^T(p), Q_{11}) \\ +B(p)\mathcal{C}(p) + \mathcal{C}^T(p)B^T(p) \end{pmatrix} & \begin{pmatrix} A(p) \\ +B(p)\,\mathcal{D}(p)C(p) + \mathcal{A}^T(p) \end{pmatrix} \\ \begin{pmatrix} \mathcal{A}(p) + A^T(p) \\ +C^T(p)\,\mathcal{D}^T(p)B^T(p) \end{pmatrix} & \begin{pmatrix} \mathcal{L}(A(p), P_{11}) \\ +\mathcal{B}(p)C(p) + C^T(p)\,\mathcal{B}^T(p) \end{pmatrix} \end{bmatrix} < 0$$

together with the constraint that

$$P_{cl} > 0.$$

This last condition holds if and only if

$$\Pi_1^T P_{cl} \Pi_1 > 0,$$

or

$$\Pi_2^T \Pi_1 = \begin{bmatrix} I & 0 \\ P_{11} & P_{12} \end{bmatrix} \begin{bmatrix} Q_{11} & I \\ Q_{12}^T & 0 \end{bmatrix} \quad (12.15)$$

$$= \begin{bmatrix} Q_{11} & I \\ I & P_{11} \end{bmatrix} > 0. \quad (12.16)$$

We also have the constraint that

$$P_{11} Q_{11} + P_{12} Q_{12}^T = I. \quad (12.17)$$

We will now assume that the parameter-dependent plant can be described by a fuzzy T-S model using r model rules. In this case, the parameter-dependent plant can be described by the equation

$$\begin{bmatrix} A(p) & B(p) \\ C(p) & 0 \end{bmatrix} = \sum_{i=1}^{r} h_i(p) \begin{bmatrix} A_i & B_i \\ C_i & 0 \end{bmatrix},$$

where $h(p)$ satisfies the normalization condition,

$$h_i(p) \geq 0 \quad \text{and} \quad \sum_{i=1}^{r} h_i(p) = 1.$$

The matrix $E(p)$ can be written as

$$E = \begin{bmatrix} E_{11}(p) & E_{12}(p) \\ E_{21}(p) & E_{22}(p) \end{bmatrix}, \quad (12.18)$$

where

$$E_{11}(p) = \sum_{i=1}^{r} \sum_{j=1}^{r} h_i(p) h_j(p) \left((A_i + B_i D_c(p) C_j) Q_{11} \right.$$

$$\left. + B_i C_c(p) Q_{12}^T \right), \quad (12.19)$$

$$E_{12}(p) = \sum_{i=1}^{r} \sum_{j=1}^{r} h_i(p) h_j(p) (A_i + B_i D_c(p) C_j), \quad (12.20)$$

$$E_{21}(p) = \sum_{i=1}^{r} \sum_{j=1}^{r} h_i(p) h_j(p) \left(P_{11}(A_i + B_i D_c(p) C_j) Q_{11} \right.$$

$$+ P_{12} B_c(p) C_i Q_{11} + P_{11} B_i C_c(p) Q_{12}^T$$

$$\left. + P_{12} A_c(p) Q_{12}^T \right), \quad (12.21)$$

$$E_{22}(p) = \sum_{i=1}^{r} \sum_{j=1}^{r} h_i(p) h_j(p) \left(P_{11}(A_i + B_i D_c(p) C_j) \right.$$

$$\left. + P_{12} B_c(p) C_i \right). \quad (12.22)$$

We are now ready to introduce dynamic parallel distributed compensators for this system. In general, a DPDC can have cubic, quadratic, or linear parameterization. For a given T-S model, the choice of a particular DPDC parameterization will be influenced by the structure of the T-S subsystems. In the following subsections, we discuss each of these three parameterizations.

12.2.1 Cubic Parameterization

Controller Synthesis. In this section, we will assume that the controller has the form

$$\dot{x}_c(t) = \sum_{i=1}^{r} \sum_{j=1}^{r} \sum_{k=1}^{r} h_i(p) h_j(p) h_k(p) A_c^{ijk} x_c(t)$$

$$+ \sum_{i=1}^{r} \sum_{j=1}^{r} h_i(p) h_j(p) B_c^{ij} y(t), \quad (12.23)$$

$$u(t) = \sum_{i=1}^{r} \sum_{j=1}^{r} h_i(p) h_j(p) C_c^{ij} x_c(t) + \sum_{i=1}^{r} h_i(p) D_c^{i} y(t), \quad (12.24)$$

or equivalently that

$$A_c(p) = \sum_{i=1}^{r} \sum_{j=1}^{r} \sum_{i=1}^{r} h_i(p) h_j(p) h_k(p) A_c^{ijk}, \quad (12.25)$$

$$B_c(p) = \sum_{i=1}^{r} \sum_{j=1}^{r} h_i(p) h_j(p) B_c^{ij}, \quad (12.26)$$

$$C_c(p) = \sum_{i=1}^{r} \sum_{j=1}^{r} h_i(p) h_j(p) C_c^{ij}, \quad (12.27)$$

$$D_c(p) = \sum_{i=1}^{r} h_i(p) D_c^{i}. \quad (12.28)$$

Using this controller form, we can rewrite the equations for the matrix

$E(p)$ as

$$E_{11}(p) = \sum_{i=1}^{r} \sum_{j=1}^{r} \sum_{k=1}^{r} h_i(p) h_j(p) h_k(p)$$
$$\times \left((A_i + B_i D_c^j C_k) Q_{11} + B_i C_c^{jk} Q_{12}^T \right), \qquad (12.29)$$

$$E_{12}(p) = \sum_{i=1}^{r} \sum_{j=1}^{r} \sum_{k=1}^{r} h_i(p) h_j(p) h_k(p) (A_i + B_i D_c^j C_k), \qquad (12.30)$$

$$E_{21}(p) = \sum_{i=1}^{r} \sum_{j=1}^{r} \sum_{k=1}^{r} h_i(p) h_j(p) h_k(p) \big(P_{11}(A_i + B_i D_c^j C_k) Q_{11}$$
$$+ P_{12} B_c^{ij} C_k Q_{11} + P_{11} B_i C_c^{jk} Q_{12}^T + P_{12} A_c^{ijk} Q_{12}^T \big), \qquad (12.31)$$

$$E_{22}(p) = \sum_{i=1}^{r} \sum_{j=1}^{r} \sum_{k=1}^{r} h_i(p) h_j(p) h_k(p)$$
$$\times \big(P_{11}(A_i + B_i D_c^j C_k) + P_{12} B_c^{ij} C_k \big), \qquad (12.32)$$

and we have that

$$\mathcal{A}(p) = \sum_{i=1}^{r} \sum_{j=1}^{r} \sum_{k=1}^{r} h_i(p) h_j(p) h_k(p) \mathcal{A}_{ijk}$$
$$\triangleq \sum_{i=1}^{r} \sum_{j=1}^{r} \sum_{k=1}^{r} h_i(p) h_j(p) h_k(p) \big(P_{11}(A_i + B_i D_c^j C_k) Q_{11}$$
$$+ P_{12} B_c^{ij} C_k Q_{11} + P_{11} B_i C_c^{jk} Q_{12}^T + P_{12} A_c^{ijk} Q_{12}^T \big),$$

$$\mathcal{B}(p) = \sum_{i=1}^{r} \sum_{j=1}^{r} h_i(p) h_j(p) \mathcal{B}_{ij}$$
$$\triangleq \sum_{i=1}^{r} \sum_{j=1}^{r} h_i(p) h_j(p) \big(P_{11} B_i D_c^j + P_{12} B_c^{ij} \big),$$

$$\mathcal{C}(p) = \sum_{i=1}^{r} \sum_{j=1}^{r} h_i(p) h_j(p) \mathcal{C}_{ij}$$
$$\triangleq \sum_{i=1}^{r} \sum_{j=1}^{r} h_i(p) h_j(p) \big(D_c^i C_j Q_{11} + C_c^{ij} Q_{12}^T \big),$$

$$\mathcal{D}(p) = \sum_{i=1}^{r} h_i(p) \mathcal{D}_i$$
$$\triangleq \sum_{i=1}^{r} h_i(p) D_c^i.$$

The matrix $E(p)$ then becomes

$$E(p) = \sum_{i=1}^{r}\sum_{j=1}^{r}\sum_{k=1}^{r} h_i(p)h_j(p)h_k(p) E_{ijk}$$

$$= \sum_{i=1}^{r}\sum_{j=1}^{r}\sum_{k=1}^{r} h_i(p)h_j(p)h_k(p)$$

$$\times \begin{bmatrix} A_i Q_{11} + B_i \mathcal{C}_{jk} & A_i + B_i \mathcal{D}_j C_k \\ \mathcal{A}_{ijk} & P_{11} A_i + \mathcal{B}_{ij} C_k \end{bmatrix}. \quad (12.33)$$

The closed-loop stability condition then becomes

$$\sum_{i=1}^{r}\sum_{j=1}^{r}\sum_{k=1}^{r} h_i(p)h_j(p)h_k(p)$$

$$\times \begin{bmatrix} \mathcal{L}(A_i^T, Q_{11}) + B_i \mathcal{C}_{jk} + \mathcal{C}_{jk}^T B_i^T & A_i + B_i \mathcal{D}_j C_k + \mathcal{A}_{ijk}^T \\ \mathcal{A}_{ijk} + (A_i + B_i \mathcal{D}_j C_k)^T & \mathcal{L}(A_i, P_{11}) + \mathcal{B}_{ij} C_k + C_k^T \mathcal{B}_{ij}^T \end{bmatrix} < 0. \quad (12.34)$$

So the system will be stable if the following LMI holds.

$$\begin{bmatrix} \mathcal{L}(A_i^T, Q_{11}) + B_i \mathcal{C}_{jk} + \mathcal{C}_{jk}^T B_i^T & A_i + B_i \mathcal{D}_j C_k + \mathcal{A}_{ijk}^T \\ \mathcal{A}_{ijk} + (A_i + B_i \mathcal{D}_j C_k)^T & \mathcal{L}(A_i, P_{11}) + \mathcal{B}_{ij} C_k + C_k^T \mathcal{B}_{ij}^T \end{bmatrix} < 0,$$

$$\forall i, j, k. \quad (12.35)$$

THEOREM 46 *The T-S model (2.3) and (2.4) is globally quadratically stabilizable via a DPDC controller (12.25)–(12.28) if the LMI conditions (12.16) and (12.35) are feasible with LMI variables Q_{11}, P_{11}, $\mathcal{A}_{ijk}$, $\mathcal{B}_{ij}$, $\mathcal{C}_{jk}$, and $\mathcal{D}_j$. The controller is given by*

$$A_c^{ijk} = P_{12}^{-1}\Big(\mathcal{A}_{ijk} - P_{12} B_{ij}^c C_k Q_{11} - P_{11} B_i C_{jk}^c Q_{12}^T$$

$$- P_{11}(A_i + B_i D_j^c C_k) Q_{11}\Big) Q_{12}^{-1}, \quad (12.36)$$

$$B_c^{ij} = P_{12}^{-1}\Big(\mathcal{B}_{ij} - P_{11} B_i D_j^c\Big), \quad (12.37)$$

$$C_c^{ij} = \Big(\mathcal{C}_{ij} - D_i^c C_j Q_{11}\Big) Q_{12}^{-T}, \quad (12.38)$$

$$D_c^i = \mathcal{D}_i, \quad (12.39)$$

where P_{11}, P_{12}, Q_{11}, and Q_{12} satisfy the constraint $P_{11} Q_{11} + P_{12} Q_{12}^T = I$.

Reduction of LMI Equations (12.34) can be simplified by permuting indices. To this end, we first note that the controller equations can be rewritten as

$$A_c(p) = \sum_{i=1}^{r} h_i^3(p) A_c^{iii}$$

$$+ \sum_{i=1}^{r} \sum_{j<i} 3h_i^2(p) h_j(p) \frac{1}{3} \left(A_c^{iij} + A_c^{iji} + A_c^{jii} \right)$$

$$+ \sum_{i=1}^{r} \sum_{j<i} 3h_j^2(p) h_i(p) \frac{1}{3} \left(A_c^{ijj} + A_c^{jij} + A_c^{jji} \right)$$

$$+ \sum_{i=1}^{r} \sum_{j<i} \sum_{k<j} 6 h_i(p) h_j(p) h_k(p)$$

$$\times \frac{1}{6} \left(A_c^{ijk} + A_c^{ikj} + A_c^{jik} + A_c^{jki} + A_c^{kij} + A_c^{kji} \right),$$

$$B_c(p) = \sum_{i=1}^{r} h_i^2(p) B_c^{ii}$$

$$+ \sum_{i=1}^{r} \sum_{j<i} 2 h_i(p) h_j(p) \frac{1}{2} \left(B_c^{ij} + B_c^{ji} \right),$$

$$C_c(p) = \sum_{i=1}^{r} h_i^2(p) C_c^{ii}$$

$$+ \sum_{i=1}^{r} \sum_{j<i} 2 h_i(p) h_j(p) \frac{1}{2} \left(C_c^{ij} + C_c^{ji} \right),$$

$$D_c(p) = \sum_{i=1}^{r} h_i(p) D_c^{i}.$$

From these equations, we can define an "average" set of vertex variables

$$\overline{A}_c^{ijk} = \frac{1}{6} \left(A_c^{ijk} + A_c^{ikj} + A_c^{jik} + A_c^{jki} + A_c^{kij} + A_c^{kji} \right), \quad (12.40)$$

$$\overline{B}_c^{ij} = \frac{1}{2} \left(B_c^{ij} + B_c^{ji} \right), \quad (12.41)$$

$$\overline{C}_c^{ij} = \frac{1}{2} \left(C_c^{ij} + C_c^{ji} \right), \quad (12.42)$$

$$\overline{D}_c^{i} = D_c^{i}. \quad (12.43)$$

Our aim is to show that the stability conditions (12.34) can also be written in terms of this set of variables. In fact, the controller (12.25)–(12.28) is the same as the following controller:

$$A_c(p) = \sum_{i=1}^{r} h_i^3(p) \overline{A}_c^{iii} + \sum_{i=1}^{r}\sum_{j<i} 3h_i^2(p) h_j(p) \overline{A}_c^{iij}$$

$$+ \sum_{i=1}^{r}\sum_{j<i} 3h_j^2(p) h_i(p) \overline{A}_c^{ijj}$$

$$+ \sum_{i=1}^{r}\sum_{j<i}\sum_{k<j} 6h_i(p)h_j(p)h_k(p) \overline{A}_c^{ijk}, \qquad (12.44)$$

$$B_c(p) = \sum_{i=1}^{r} h_i^2(p) \overline{B}_c^{ii} + \sum_{i=1}^{r}\sum_{j<i} 2h_i(p)h_j(p) \overline{B}_c^{ij}, \qquad (12.45)$$

$$C_c(p) = \sum_{i=1}^{r} h_i^2(p) \overline{C}_c^{ii} + \sum_{i=1}^{r}\sum_{j<i} 2h_i(p)h_j(p) \overline{C}_c^{ij}, \qquad (12.46)$$

$$D_c(p) = \sum_{i=1}^{r} h_i(p) \overline{D}_c^{i}. \qquad (12.47)$$

Consequently, the number of unknowns which must be determined can be reduced considerably. In terms of $E(p)$ our stability condition can be written as

$$E(p) + E^T(p) < 0$$

or

$$\sum_{i=1}^{r}\sum_{j=1}^{r}\sum_{k=1}^{r} h_i(p)h_j(p)h_k(p)\left(E_{ijk} + E_{ijk}^T\right) < 0.$$

This can be rewritten as

$$\sum_{i=1}^{r} h_i^3(p)\left(E_{iii} + E_{iii}^T\right)$$

$$+ \sum_{i=1}^{r}\sum_{j<i} 3h_i^2(p)h_j(p)\left[\frac{1}{3}(E_{iij} + E_{iji} + E_{jii})\right.$$

$$\left. + \frac{1}{3}(E_{iij} + E_{iji} + E_{jii})^T\right]$$

$$+ \sum_{i=1}^{r}\sum_{j<i} 3h_j^2(p)h_i(p)\left[\frac{1}{3}(E_{jji} + E_{jij} + E_{ijj})\right.$$

$$\left. + \frac{1}{3}(E_{jji} + E_{jij} + E_{ijj})^T\right]$$

$$+ \sum_{i=1}^{r} \sum_{j<i} \sum_{k<j} 6h_i(p)h_j(p)h_k(p)$$

$$\times \left[\frac{1}{6}(E_{ijk} + E_{ikj} + E_{jik} + E_{jki} + E_{kij} + E_{kji}) \right.$$

$$\left. + \frac{1}{6}(E_{ijk} + E_{ikj} + E_{jik} + E_{jki} + E_{kij} + E_{kji})^T \right] < 0.$$

We will define the matrices

$$\overline{W}_i = (E_{iii} + E_{iii}^T), \tag{12.48}$$

$$\overline{W}_{ij} = \left(\frac{1}{3}(E_{iij} + E_{iji} + E_{jii}) + \frac{1}{3}(E_{iij} + E_{iji} + E_{jii})^T \right), \tag{12.49}$$

$$\overline{W}_{ijk} = \left(\frac{1}{6}(E_{ijk} + E_{ikj} + E_{jik} + E_{jki} + E_{kij} + E_{kji}) \right.$$

$$\left. + \frac{1}{6}(E_{ijk} + E_{ikj} + E_{jik} + E_{jki} + E_{kij} + E_{kji})^T \right). \tag{12.50}$$

In terms of the decision variables, W_{ijk} can be written as

$$\overline{W}_{ijk} = \frac{1}{6} \begin{bmatrix} \begin{pmatrix} (A_i Q_{11} + B_i \mathcal{C}_{jk}) \\ + Q_{11} A_i^T + \mathcal{C}_{jk}^T B_i^T \end{pmatrix} & \left(A_i + B_i \mathcal{D}_j C_k + \mathcal{A}_{ijk}^T \right) \\ \mathcal{A}_{ijk} + \left(A_i + B_i \mathcal{D}_j C_k \right)^T & \begin{pmatrix} (P_{11} A_i + \mathcal{B}_{ij} C_k) \\ + A_i^T P_{11} + C_k^T \mathcal{B}_{ij}^T \end{pmatrix} \end{bmatrix}$$

$$+ \begin{bmatrix} \begin{pmatrix} (A_i Q_{11} + B_i \mathcal{C}_{kj}) \\ + Q_{11} A_i^T + \mathcal{C}_{kj}^T B_i^T \end{pmatrix} & \left(A_i + B_i \mathcal{D}_k C_j + \mathcal{A}_{ikj}^T \right) \\ \mathcal{A}_{ikj} + \left(A_i + B_i \mathcal{D}_k C_j \right)^T & \begin{pmatrix} (P_{11} A_i + \mathcal{B}_{ik} C_j) \\ + A_i^T P_{11} + C_j^T \mathcal{B}_{ik}^T \end{pmatrix} \end{bmatrix}$$

$$+ \begin{bmatrix} \begin{pmatrix} (A_j Q_{11} + B_j \mathcal{C}_{ik}) \\ + Q_{11} A_j^T + \mathcal{C}_{ik}^T B_j^T \end{pmatrix} & \left(A_j + B_j \mathcal{D}_i C_k + \mathcal{A}_{jik}^T \right) \\ \mathcal{A}_{jik} + \left(A_j + B_j \mathcal{D}_i C_k \right)^T & \begin{pmatrix} (P_{11} A_j + \mathcal{B}_{ji} C_k) \\ + A_j^T P_{11} + C_k^T \mathcal{B}_{ji}^T \end{pmatrix} \end{bmatrix}$$

$$+ \begin{bmatrix} \begin{pmatrix} (A_j Q_{11} + B_j C_{ki} \\ + Q_{11} A_j^T + C_{ki}^T B_j^T) \end{pmatrix} & \left(A_j + B_j \mathcal{D}_k C_i + \mathcal{A}_{jki}^T \right) \\ \mathcal{A}_{jki} + \left(A_j + B_j \mathcal{D}_k C_i \right)^T & \begin{pmatrix} (P_{11} A_j + \mathcal{B}_{jk} C_i \\ + A_j^T P_{11} + C_i^T \mathcal{B}_{jk}^T) \end{pmatrix} \end{bmatrix}$$

$$+ \begin{bmatrix} \begin{pmatrix} (A_k Q_{11} + B_k C_{ji} \\ + Q_{11} A_k^T + C_{ji}^T B_k^T) \end{pmatrix} & \left(A_k + B_k \mathcal{D}_j C_i + \mathcal{A}_{kji}^T \right) \\ \mathcal{A}_{kji} + \left(A_k + B_k \mathcal{D}_j C_i \right)^T & \begin{pmatrix} (P_{11} A_k + \mathcal{B}_{kj} C_i \\ + A_k^T P_{11} + C_i^T \mathcal{B}_{kj}^T) \end{pmatrix} \end{bmatrix}$$

$$+ \begin{bmatrix} \begin{pmatrix} (A_k Q_{11} + B_k C_{ij} \\ + Q_{11} A_k^T + C_{ij}^T B_k^T) \end{pmatrix} & \left(A_k + B_k \mathcal{D}_i C_j + \mathcal{A}_{kij}^T \right) \\ \mathcal{A}_{kij} + \left(A_k + B_k \mathcal{D}_i C_j \right)^T & \begin{pmatrix} (P_{11} A_k + \mathcal{B}_{ki} C_j \\ + A_k^T P_{11} + C_j^T \mathcal{B}_{ki}^T) \end{pmatrix} \end{bmatrix}.$$

We also define variables as

$$\overline{\mathcal{A}}_{ijk} = \frac{1}{6} \left(\mathcal{A}_{ijk} + \mathcal{A}_{ikj} + \mathcal{A}_{jik} + \mathcal{A}_{jki} + \mathcal{A}_{kij} + \mathcal{A}_{kji} \right), \quad (12.51)$$

$$\overline{\mathcal{B}}_{ij} = \frac{1}{2} \left(\mathcal{B}_{ij} + \mathcal{B}_{ji} \right), \quad (12.52)$$

$$\overline{C}_{ij} = \frac{1}{2} \left(C_{ij} + C_{ji} \right), \quad (12.53)$$

$$\overline{\mathcal{D}}_i = \mathcal{D}_i. \quad (12.54)$$

It is noted that W_i, W_{ij}, and W_{ijk} can be represented by these variables, so we have the following corollary:

COROLLARY 6 *The fuzzy control system of the T-S model (2.3) and (2.4) is globally quadratically stabilizable via a DPDC controller (12.44)–(12.47) if the following LMIs are feasible with LMI variables Q_{11}, P_{11}, $\overline{\mathcal{A}}_i$, $\overline{\mathcal{A}}_{ij}$, $\overline{\mathcal{A}}_{ijk}$, $\overline{\mathcal{B}}_i$, $\overline{\mathcal{B}}_{ij}$, $\overline{C}_i$, C_{ij}, and $\overline{\mathcal{D}}_i$:*

$$\begin{bmatrix} Q_{11} & I \\ I & P_{11} \end{bmatrix} > 0, \qquad (12.55)$$

$$\overline{W}_i < 0, \qquad (12.56)$$

$$\overline{W}_{ij} < 0, \qquad (12.57)$$

$$\overline{W}_{ijk} < 0. \qquad (12.58)$$

The controller is given in a similar way as (12.36)–(12.39).

12.2.2 Quadratic Parameterization

Choose the form of the controller as

$$\dot{x}_c = \sum_{i=1}^{r} \sum_{j=1}^{r} h_i(p) h_j(p) A_c^{ij} x_c + \sum_{i=1}^{r} h_i B_c^i y, \qquad (12.59)$$

$$u = \sum_{i=1}^{r} h_i(p) C_c^i x_c + D_c y, \qquad (12.60)$$

or equivalently that

$$A_c(p) = \sum_{i=1}^{r} \sum_{j=1}^{r} h_i(p) h_j(p) A_c^{ij}, \qquad (12.61)$$

$$B_c(p) = \sum_{i=1}^{r} h_i(p) B_c^i, \qquad (12.62)$$

$$C_c(p) = \sum_{i=1}^{r} h_i(p) C_c^i, \qquad (12.63)$$

$$D_c(p) = D_c. \qquad (12.64)$$

So the closed-loop system for the T-S model (2.3) and (2.4) with this controller can be written as

$$\dot{x}_{cl} = \sum_{i=1}^{r} \sum_{j=1}^{r} h_i(p) h_j(p) A_{cl}^{ij} x_{cl}, \qquad (12.65)$$

where

$$A_{cl}^{ij} = \begin{pmatrix} A_i + B_i D_c C_j & B_i C_c^j \\ B_c^i C_j & A_c^{ij} \end{pmatrix}.$$

We can rewrite the equations for the matrix $E(p)$ as

$$E_{11}(p) = \sum_{i=1}^{r}\sum_{j=1}^{r} h_i(p)h_j(p)\bigl((A_i + B_i D_c C_j)Q_{11} + B_i C_c^j Q_{12}^T\bigr), \quad (12.66)$$

$$E_{12}(p) = \sum_{i=1}^{r}\sum_{j=1}^{r} h_i(p)h_j(p)(A_i + B_i D_c C_j), \quad (12.67)$$

$$E_{21}(p) = \sum_{i=1}^{r}\sum_{j=1}^{r} h_i(p)h_j(p)\bigl(P_{11}(A_i + B_i D_c C_j)Q_{11} + P_{12} B_c^i C_j Q_{11}$$
$$+ P_{11} B_i C_c^j Q_{12}^T + P_{12} A_c^{ij} Q_{12}^T\bigr), \quad (12.68)$$

$$E_{22}(p) = \sum_{i=1}^{r}\sum_{j=1}^{r} h_i(p)h_j(p)\bigl(P_{11}(A_i + B_i D_c C_j) + P_{12} B_c^i C_j\bigr), \quad (12.69)$$

and we have that

$$\mathcal{A}(p) = \sum_{i=1}^{r}\sum_{j=1}^{r} h_i(p)h_j(p)\,\mathcal{A}_{ij}$$

$$\triangleq \sum_{i=1}^{r}\sum_{j=1}^{r} h_i(p)h_j(p)\bigl(P_{11}(A_i + B_i D_c C_j)Q_{11}$$

$$+ P_{12} B_c^i C_j Q_{11} + P_{11} B_i C_c^j Q_{12}^T + P_{12} A_c^{ij} Q_{12}^T\bigr),$$

$$\mathcal{B}(p) = \sum_{i=1}^{r} h_i(p)\,\mathcal{B}_i$$

$$\triangleq \sum_{i=1}^{r} h_i(p)(P_{11} B_i D_c + P_{12} B_c^i),$$

$$\mathcal{C}(p) = \sum_{i=1}^{r} h_i(p)\mathcal{C}_i$$

$$\triangleq \sum_{i=1}^{r} h_i(p)(D_c C_i Q_{11} + C_c^i Q_{12}^T),$$

$$\mathcal{D}(p) = \mathcal{D}$$
$$\triangleq D_c.$$

The matrix $E(p)$ then becomes

$$E(p) = \sum_{i=1}^{r} \sum_{j=1}^{r} h_i(p) h_j(p) E_{ij}, \qquad (12.70)$$

where

$$E_{ij} \triangleq \begin{bmatrix} E_{11}^{ij} & E_{12}^{ij} \\ E_{21}^{ij} & E_{22}^{ij} \end{bmatrix} = \begin{bmatrix} A_i Q_{11} + B_i \mathcal{C}_j & A_i + B_i \mathcal{D} C_j \\ \mathcal{A}_{ij} & P_{11} A_i + \mathcal{B}_i C_j \end{bmatrix}. \qquad (12.71)$$

The closed-loop stability condition in terms of $E(p)$ is

$$E(p) + E(p)^T < 0$$

or

$$\sum_{i=1}^{r} \sum_{j=1}^{r} h_i(p) h_j(p) \left(E_{ij} + E_{ij}^T \right) < 0,$$

which is

$$\sum_{i=1}^{r} \sum_{j=1}^{r} h_i(p) h_j(p)$$

$$\times \begin{bmatrix} \begin{pmatrix} \mathcal{L}(A_i^T, Q_{11}) + B_i \mathcal{C}_j \\ + \mathcal{C}_j^T B_i^T \end{pmatrix} & A_i + B_i \mathcal{D} C_j + \mathcal{A}_{ij}^T \\ \mathcal{A}_{ij} + (A_i + B_i \mathcal{D} C_j)^T & \begin{pmatrix} \mathcal{L}(A_i, P_{11}) + \mathcal{B}_i C_j \\ + C_j^T \mathcal{B}_i^T \end{pmatrix} \end{bmatrix} < 0. \qquad (12.72)$$

In this case, we can have a similar theorem as Theorem 46 for the quadratic parameterization case.

In the following we simplify the stability condition by means of permutation. As discussed above, the controller (12.61)–(12.64) is equivalent to

$$A_c(p) = \sum_{i=1}^{r} \sum_{j=1}^{r} h_i(p) h_j(p) \bar{A}_c^{ij}, \qquad (12.73)$$

$$B_c(p) = \sum_{i=1}^{r} h_i(p) \bar{B}_c^i, \qquad (12.74)$$

$$C_c(p) = \sum_{i=1}^{r} h_i(p) \bar{C}_c^i, \qquad (12.75)$$

$$D_c(p) = D_c, \qquad (12.76)$$

where

$$\overline{A}_c^{ij} = \tfrac{1}{2}(A_c^{ij} + A_c^{ji}), \qquad (12.77)$$

$$\overline{B}_c^i = B_c^i, \qquad (12.78)$$

$$\overline{C}_c^i = C_c^i, \qquad (12.79)$$

$$\overline{D}_c = D_c. \qquad (12.80)$$

Also define variables:

$$\overline{\mathcal{A}}_{ij} \triangleq \tfrac{1}{2}(\mathcal{A}_{ij} + \mathcal{A}_{ji}), \qquad (12.81)$$

$$\overline{\mathcal{B}}_i \triangleq \mathcal{B}_i, \qquad (12.82)$$

$$\overline{\mathcal{C}}_i \triangleq \mathcal{C}_i, \qquad (12.83)$$

$$\overline{\mathcal{D}} \triangleq \mathcal{D}. \qquad (12.84)$$

Now we are ready to permute the stability condition

$$E(p) + E^T(p) < 0$$

or

$$\sum_{i=1}^{r} \sum_{j=1}^{r} h_i(p) h_j(p) \left(E_{ij} + E_{ij}^T \right) < 0,$$

which can be permuted to be

$$\frac{1}{2} \sum_{i=1}^{r} \sum_{j=1}^{r} h_i(p) h_j(p) \left(\left(E_{ij} + E_{ij}^T \right) + \left(E_{ji} + E_{ji}^T \right) \right) < 0.$$

Expressing the permuted stability condition in terms of the permuted system variables, we can arrive at the following conditions:

$$\mathbf{T} = \begin{bmatrix} T_{11} & \cdots & T_{1r} \\ \vdots & \ddots & \vdots \\ T_{1r} & \cdots & T_{rr} \end{bmatrix} < 0 \qquad (12.85)$$

and

$$\begin{pmatrix} E_{11}^{ij} + (E_{11}^{ij})^T + E_{11}^{ji} + (E_{11}^{ji})^T & E_{12}^{ij} + (E_{21}^{ij})^T + E_{12}^{ji} + (E_{21}^{ji})^T \\ (E_{12}^{ij})^T + E_{21}^{ij} + (E_{12}^{ji})^T + E_{21}^{ji} & E_{22}^{ij} + (E_{22}^{ij})^T + E_{22}^{ji} + (E_{22}^{ji})^T \end{pmatrix} < T_{ij},$$

$$\forall i \leq j, \quad (12.86)$$

where

$$E_{11}^{ij} + (E_{11}^{ij})^T + E_{11}^{ji} + (E_{11}^{ji})^T = \mathcal{L}(A_i, Q_{11}) + \mathcal{L}(A_j, Q_{11}) + B_i \overline{C}_j$$
$$+ (B_i \overline{C}_j)^T + B_j \overline{C}_i + (B_j \overline{C}_i)^T,$$

$$E_{12}^{ij} + (E_{21}^{ij})^T + E_{12}^{ji} + (E_{21}^{ji})^T = A_i + A_j + B_i \overline{\mathcal{D}} C_j + B_j \overline{\mathcal{D}} C_i + 2\overline{\mathcal{A}}_{ij}^T,$$

$$E_{22}^{ij} + (E_{22}^{ij})^T + E_{22}^{ji} + (E_{22}^{ji})^T = \mathcal{L}(A_i^T, P_{11}) + \mathcal{L}(A_j^T, P_{11}) + \overline{\mathcal{B}}_i C_j$$
$$+ \overline{\mathcal{B}}_j C_i + (\overline{\mathcal{B}}_i C_j)^T + (\overline{\mathcal{B}}_j C_i)^T.$$

The result is summarized in the following theorem:

THEOREM 47 *The fuzzy control system of the T-S model (2.3) and (2.4) is globally quadratically stabilizable via a DPDC controller (12.73)–(12.76) if the LMI conditions (12.16), (12.85), and (12.86) are feasible with LMI variables Q_{11}, P_{11}, T_{ij}, $\overline{\mathcal{A}}_{ij}$, $\overline{\mathcal{B}}_i$, $\overline{\mathcal{C}}_i$, and $\overline{\mathcal{D}}$. The controller is given by*

$$\overline{A}_c^{ij} = \tfrac{1}{2} P_{12}^{-1} \Big(2\overline{\mathcal{A}}_{ij} - P_{12} \overline{\mathcal{B}}_c^i C_j Q_{11} - P_{12} \overline{\mathcal{B}}_c^j C_i Q_{11} - P_{11} B_i \overline{C}_j^c Q_{12}^T$$
$$- P_{11} B_j \overline{C}_i^c Q_{12}^T - P_{11} (A_i + B_i \overline{D}_c C_j) Q_{11}$$
$$- P_{11} (A_j + B_j \overline{D}_c C_i) Q_{11} \Big) Q_{12}^{-1}, \quad (12.87)$$

$$\overline{B}_c^i = P_{12}^{-1} \Big(\overline{\mathcal{B}}_i - P_{11} B_i \overline{D}_c \Big), \quad (12.88)$$

$$\overline{C}_c^i = \Big(\overline{\mathcal{C}}_i - \overline{D}_c C_i Q_{11} \Big) Q_{12}^{-T}, \quad (12.89)$$

$$D_c = \overline{\mathcal{D}}, \quad (12.90)$$

where P_{11}, P_{12}, Q_{11}, and Q_{12} satisfy the constraint $P_{11} Q_{11} + P_{12} Q_{12}^T = I$.

12.2.3 Linear Parameterization

In this section, we consider linear parameterization dynamic feedback designs for system (2.3) and (2.4).

Linear Parameterization: Common B Assume that $B_1 = B_2 = \cdots = B_r = B$ in system (2.3) and (2.4). We have

$$\dot{x} = \sum_{i=1}^{r} h_i(p) A_i x + Bu, \tag{12.91}$$

$$y = \sum_{i=1}^{r} h_i(p) C_i x. \tag{12.92}$$

In general, system (2.3) and (2.4) can be transformed into the common B form (12.91)–(12.92) by the following system augmentation: Introduce $v = \dot{u}$, and augment the system (2.3) and (2.4) as

$$\begin{bmatrix} \dot{x} \\ \dot{u} \end{bmatrix} = \sum_{i=1}^{r} h_i(p) \begin{bmatrix} A_i & B_i \\ 0 & 0 \end{bmatrix} \begin{bmatrix} x \\ u \end{bmatrix} + \begin{bmatrix} 0 \\ I \end{bmatrix} v, \tag{12.93}$$

$$y = \sum_{i=1}^{r} h_i(p) \begin{bmatrix} C_i & 0 \end{bmatrix} \begin{bmatrix} x \\ u \end{bmatrix}. \tag{12.94}$$

Therefore, without loss of generality, let us consider system (12.91). To design a dynamic compensator for system (12.91), instead of using the general cubic or quadratic parameterization, we can employ the following linear parameterization:

Dynamic Part: Rule i

IF $p_1(t)$ is M_{i1} and $\cdots$ and $p_l(t)$ is M_{il},

THEN $\dot{x}_c(t) = A_c^i x_c(t) + B_c y(t)$.

Output Part: Rule i

IF $p_1(t)$ is M_{i1} and $\cdots$ and $p_l(t)$ is M_{il},

THEN $u(t) = C_c^i x_c(t) + D_c y(t)$.

The controller can be written as

$$\dot{x}_c = \sum_{i=1}^{r} h_i(p) A_c^i x_c + B_c y, \tag{12.95}$$

$$u = \sum_{i=1}^{r} h_i(p) C_c^i x_c + D_c y. \tag{12.96}$$

The controller parameters are

$$A_c(p) = \sum_{i=1}^{r} h_i(p) A_c^i, \quad (12.97)$$

$$B_c(p) = B_c, \quad (12.98)$$

$$C_c(p) = \sum_{i=1}^{r} h_i(p) C_c^i, \quad (12.99)$$

$$D_c(p) = D_c. \quad (12.100)$$

The closed-loop system will be

$$\dot{x}_{cl} = \sum_{i=1}^{r} h_i(p) A_{cl}^i x_{cl}, \quad (12.101)$$

where

$$A_{cl}^i = \begin{bmatrix} A_i + BD_c C_i & BC_c^i \\ B_c C_i & A_c^i \end{bmatrix}. \quad (12.102)$$

The closed-loop system (12.101) will be stable with quadratic Lyapunov function if there exists a symmetric positive matrix P such that

$$\mathcal{L}(A_{cl}^i, P) < 0, \quad \forall i. \quad (12.103)$$

Define

$$\mathcal{A}_i = P_{12} A_c^i Q_{12}^T + P_{12} B_c C_i Q_{11} + P_{11} BC_i Q_{12}^T$$
$$+ P_{11}(A_i + BD_c C_i) Q_{11},$$
$$\mathcal{B} = P_{12} B_c + P_{11} BD_c,$$
$$\mathcal{C}_i = C_c Q_{12}^T + D_c C_i X,$$
$$\mathcal{D} = D_c.$$

We have the following theorem:

THEOREM 48 *The fuzzy T-S model (12.93)–(12.94) is globally quadratically stabilizable via a DPDC controller (12.95) and (12.96) if the following LMI conditions are feasible with LMI variables Q_{11}, P_{11}, $\mathcal{A}_i$, $\mathcal{B}$, $\mathcal{C}_i$, and $\mathcal{D}$:*

$$\begin{pmatrix} Q_{11} & I \\ I & P_{11} \end{pmatrix} > 0, \quad (12.104)$$

$$\begin{pmatrix} \mathcal{L}(A_i^T, Q_{11}) + B\mathcal{C}_i + \mathcal{C}_i^T B^T & A_i + B\mathcal{D}C_i + \mathcal{A}_i^T \\ (A_i + B\mathcal{D}C_i)^T + \mathcal{A}_i & \mathcal{L}(A_i, P_{11}) + \mathcal{B}C_i + C_i^T \mathcal{B}^T \end{pmatrix} < 0, \quad \forall i.$$

(12.105)

The controller is given by

$$A_c^i = P_{12}^{-1}\left(\mathcal{A}_i - P_{12}B_c C_i Q_{11} - P_{11}BC_c^i Q_{12}^T\right.$$
$$\left. - P_{11}(A_i + BD_c^i C_i)Q_{11}\right)Q_{12}^{-1}, \quad (12.106)$$

$$B_c = P_{12}^{-1}(\mathcal{B} - P_{11}BD_c), \quad (12.107)$$

$$C_c^i = (\mathcal{C}_i - D_c C_i Q_{11})Q_{12}^{-T}, \quad (12.108)$$

$$D_c = \mathcal{D}, \quad (12.109)$$

where P_{11}, P_{12}, Q_{11}, and Q_{12} satisfy the constraint $P_{11}Q_{11} + P_{12}Q_{12}^T = I$.

Linear Parameterization: Common C The case corresponding to common C, that is, $C_1 = C_2 = \cdots = C_r = C$ in system (2.3) and (2.4) can be handled analogous to the Common B case. Consider

$$\dot{x} = \sum_{i=1}^{r} h_i(p)(A_i x + B_i u), \quad (12.110)$$

$$y = Cx. \quad (12.111)$$

As in the previous case, a common C matrix case can always be obtained by augmenting the outputs of the system with integrators and using the augmented states as a new set of outputs.

In this case, a linear parameterization dynamic controller takes the following form:

Dynamic Part: Rule i

    IF $p_1(t)$ is M_{i1} $\cdots$ and $p_l(t)$ is M_{il},

        THEN $\dot{x}_c(t) = A_c^i x_c(t) + B_c^i y(t)$.

Output Part: Rule i

    IF $p_1(t)$ is M_{i1} $\cdots$ and $p_l(t)$ is M_{il},

        THEN $u(t) = C_c x_c(t) + D_c y(t)$.

The controller can be written as

$$\dot{x}_c = \sum_{i=1}^{r} h_i(p)(A_c^i x_c + B_c^i y), \qquad (12.112)$$

$$u = C_c x_c + D_c y. \qquad (12.113)$$

The stabilizing LMIs in this case are given in the following theorem.

THEOREM 49 *The fuzzy T-S model* (12.110) *and* (12.111) *is globally quadratically stabilizable via the DPDC controller* (12.112) *and* (12.113) *if the following LMI conditions are feasible in* Q_{11}, P_{11}, $\mathcal{A}_i$, $\mathcal{B}_i$, $\mathcal{C}$, *and* $\mathcal{D}$:

$$\begin{pmatrix} Q_{11} & I \\ I & P_{11} \end{pmatrix} > 0, \qquad (12.114)$$

$$\begin{pmatrix} \mathcal{L}(A_i^T, Q_{11}) + B_i \mathcal{C} + \mathcal{C}^T B_i^T & A_i + B_i \mathcal{D} C + \mathcal{A}_i^T \\ (A_i + B_i \mathcal{D} C)^T + \mathcal{A}_i & \mathcal{L}(A_i, P_{11}) + \mathcal{B}_i C + C^T \mathcal{B}_i^T \end{pmatrix} < 0, \quad \forall i. \qquad (12.115)$$

The controller is given by:

$$A_c^i = P_{12}^{-1}\big(\mathcal{A}_i - P_{12} B_c^i C Q_{11} - P_{11} B_i C_c Q_{12}^T$$
$$\qquad - P_{11}(A_i + B_i D_c C) Q_{11}\big) Q_{12}^{-1}, \qquad (12.116)$$

$$B_c^i = P_{12}^{-1}(\mathcal{B}_i - P_{11} B_i D_c), \qquad (12.117)$$

$$C_c = (\mathcal{C} - D_c C Q_{11}) Q_{12}^{-T}, \qquad (12.118)$$

$$D_c = \mathcal{D}, \qquad (12.119)$$

where P_{11}, P_{12}, Q_{11}, and Q_{12} satisfy the constraint $P_{11} Q_{11} + P_{12} Q_{12}^T = I$.

Linear Parameterization: Common B and Common C Consider the case of $B_i = B$ and $C_i = C$:

$$\dot{x} = \sum_{i=1}^{r} h_i(p) A_i x + Bu, \qquad (12.120)$$

$$y = Cx. \qquad (12.121)$$

In this case, a linear parameterization dynamic controller takes the following form:

Dynamic Part: Rule i

IF $p_1(t)$ is M_{i1} $\cdots$ and $p_l(t)$ is M_{il},

THEN $\dot{x}_c(t) = A_c^i x_c(t) + B_c^i y(t)$.

Output Part: Rule i

IF $p_1(t)$ is M_{i1} $\cdots$ and $p_l(t)$ is M_{il},

THEN $u(t) = C_c^i x_c(t) + D_c^i y(t)$.

The controller can be written as

$$\dot{x}_c = \sum_{i=1}^{r} h_i(p)(A_c^i x_c + B_c^i y), \qquad (12.122)$$

$$u = \sum_{i=1}^{r} h_i(p)(C_c^i x_c + D_c^i y). \qquad (12.123)$$

The design conditions are given in the following theorem:

THEOREM 50 *The fuzzy T-S model* (12.120) *and* (12.121) *is globally quadratically stabilizable via the DPDC controller* (12.122) *and* (12.123) *if the following LMI conditions are feasible in the LMI variables* Q_{11}, P_{11}, $\mathcal{A}_i$, $\mathcal{B}_i$, $\mathcal{C}_i$, *and* $\mathcal{D}_i$:

$$\begin{pmatrix} Q_{11} & I \\ I & P_{11} \end{pmatrix} > 0, \qquad (12.124)$$

$$\begin{pmatrix} \mathcal{L}(A_i^T, Q_{11}) + B\mathcal{C}_i + \mathcal{C}_i^T B^T & A_i + B\mathcal{D}_i C + \mathcal{A}_i^T \\ (A_i + B\mathcal{D}_i C)^T + \mathcal{A}_i & \mathcal{L}(A_i, P_{11}) + \mathcal{B}_i C + C^T \mathcal{B}_i^T \end{pmatrix} < 0, \quad \forall i. \qquad (12.125)$$

The controller is given by

$$A_c^i = P_{12}^{-1}(\mathcal{A}_i - P_{12} B_c^i C Q_{11} - P_{11} B C_c^i Q_{12}^T$$
$$- P_{11}(A_i + B D_c^i C) Q_{11}) Q_{12}^{-1}, \qquad (12.126)$$

$$B_c^i = P_{12}^{-1}(\mathcal{B}_i - P_{11} B D_c^i), \qquad (12.127)$$

$$C_c^i = (\mathcal{C}_i - D_c^i C Q_{11}) Q_{12}^{-T}, \qquad (12.128)$$

$$D_c^i = \mathcal{D}_i, \qquad (12.129)$$

where P_{11}, P_{12}, Q_{11}, *and* Q_{12} *satisfy the constraint* $P_{11} Q_{11} + P_{12} Q_{12}^T = I$.

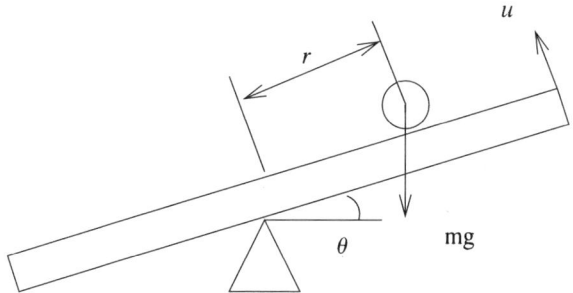

Fig. 12.1 The ball and beam system.

12.3 EXAMPLE

In this section, we consider a ball-and-beam system which is commonly used as an illustrative application of various control schemes. The system is shown in Figure 12.1. To begin with, we represent the original model exactly using a T-S model via sector nonlinearity.

The beam is made to rotate in a vertical plane by applying a torque at the center of rotation and the ball is free to roll along the beam. Assume no slipping between the ball and the beam. Let $x = (r, \dot{r}, \theta, \dot{\theta})$ be the state of the system and $y = r$ is the system output. The system can be expressed by the state-space model:

$$\dot{x} = f(x) + g(x)u, \qquad (12.130)$$

where

$$f(x) = \begin{bmatrix} x_2 \\ B(x_1 x_4^2 - G \sin x_3) \\ x_4 \\ 0 \end{bmatrix}$$

and

$$g(x) = \begin{bmatrix} 0 \\ 0 \\ 0 \\ 1 \end{bmatrix}.$$

There are two nonlinearities in (12.130), the $x_1 x_4^2$ term and the $\sin x_3$ term. As we know, most nonlinearity can be bounded by sector. In this example, assume $x_3 \in [-\pi/2 \quad \pi/2]$ and $x_1 x_4 \in [-d \quad d]$. This is the region that we assume the system will operate within. It follows that

$$\left| \frac{2}{\pi} x \right| \leq |\sin(x)| \leq |x|, \qquad (12.131)$$

$$-dx_4 \leq x_1 x_4^2 \leq dx_4. \qquad (12.132)$$

Define

$$M_{12}(x_3) = \frac{1 - \sin(x_3)/x_3}{1 - 2/\pi}$$

and

$$M_{11}(x_3) = 1 - M_{12}(x_3),$$

$$M_{22}(x_1 x_4) = \begin{cases} 1, & x_1 x_4 \geq d, \\ \dfrac{x_1 x_4}{d}, & 0 < x_1 x_4 < d, \\ 0, & x_1 x_4 \leq 0, \end{cases}$$

$$M_{23}(x_1 x_4) = \begin{cases} 0, & x_1 x_4 \geq 0, \\ \dfrac{x_1 x_4}{-d}, & -d < x_1 x_4 < 0, \\ 1, & x_1 x_4 \leq -d, \end{cases}$$

and

$$M_{21}(x_1 x_4) = 1 - M_{22}(x_1 x_4) - M_{23}(x_1 x_4).$$

Therefore within the region $|x_3| \leq \pi/2$, $|x_1 x_4| \leq d$, we can write $f(x)$ as

$$f(x) = M_{11} M_{21} \begin{bmatrix} x_2 \\ -BGx_3 \\ x_4 \\ 0 \end{bmatrix} + M_{11} M_{22} \begin{bmatrix} x_2 \\ -BGx_3 + Bdx_4 \\ x_4 \\ 0 \end{bmatrix}$$

$$+ M_{11} M_{23} \begin{bmatrix} x_2 \\ -BGx_3 - Bdx_4 \\ x_4 \\ 0 \end{bmatrix} + M_{12} M_{21} \begin{bmatrix} x_2 \\ -\dfrac{2BG}{\pi} x_3 \\ x_4 \\ 0 \end{bmatrix}$$

$$+ M_{12} M_{22} \begin{bmatrix} x_2 \\ -\dfrac{2BG}{\pi} x_3 - Bdx_4 \\ x_4 \\ 0 \end{bmatrix} + M_{12} M_{22} \begin{bmatrix} x_2 \\ -\dfrac{2BG}{\pi} x_3 - Bdx_4 \\ x_4 \\ 0 \end{bmatrix}.$$

The T-S model follows directly as follows:

Rule ij

IF $|x_3|$ is M_{1i} and $x_1 x_4$ is M_{2j},

THEN $\dot{x}(t) = A_{ij} x(t) + B_{ij} u(t)$, $i = 1, 2$, $j = 1, 2, 3$.

For example,

$$A_{11} = \begin{bmatrix} 0 & 1 & 0 & 0 \\ 0 & 0 & -BG & 0 \\ 0 & 0 & 0 & 1 \\ 0 & 0 & 0 & 0 \end{bmatrix}, \quad B_{11} = \begin{bmatrix} 0 \\ 0 \\ 0 \\ 1 \end{bmatrix}.$$

Since the ball-and-beam system is a common B and common C case as discussed in Section 12.2.3.3, we will apply DPDC with linear parameterization for the system. The simulation result is shown in Figure 12.2. The system parameters for simulation are chosen as $B = 0.7143$, $G = 9.81$, $d = 5$, and the initial condition is $[1, 0, 0.0564, 0]$.

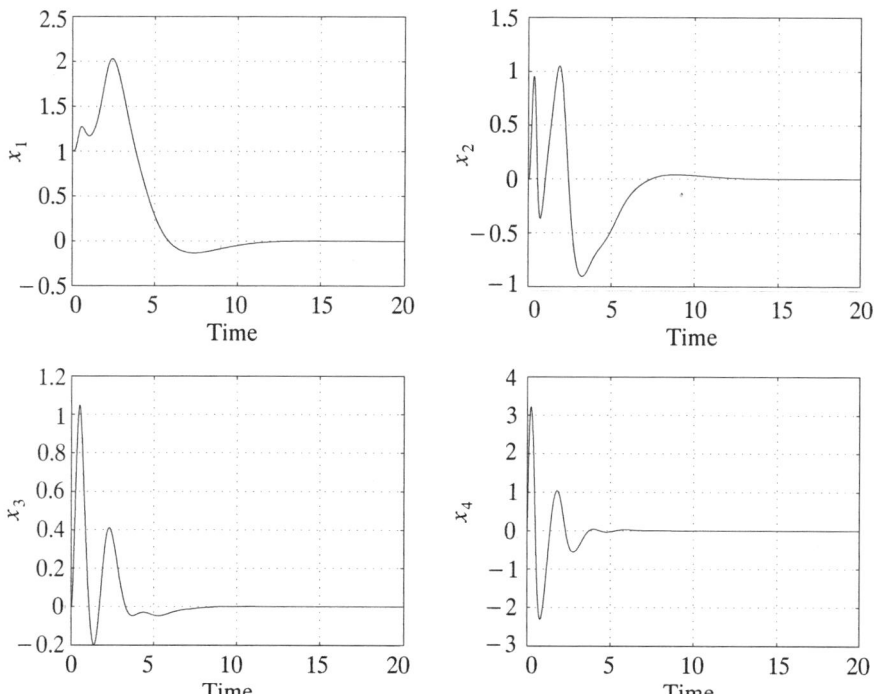

Fig. 12.2 Response of Ball and Beam using DPDC with linear parameterization.

Remark 40 From the simulation results, we know that x_3 and $x_1 x_4$ do not exceed the bound limit assumed in the modeling. A more systematic approach is to incorporate the constraints as performance specifications in the controller design. This issue is addressed in the next chapter.

BIBLIOGRAPHY

1. J. Li, D. Niemann, H. O. Wang, and K. Tanaka, "Multiobjective Dynamic Feedback Control of Takagi-Sugeno Model via LMIs" *Proc. 4th Joint Conference of Information Sciences, Durham*, Vol. 1, Oct. 1998, pp. 159–162.
2. J. Li, D. Niemann, H. O. Wang, and K. Tanaka, "Parallel Distributed Compensation for Takagi-Sugeno Fuzzy Models: Multiobjective Controller Design," *Proc. 1999 American Control Conference*, San Diego, June 1999, pp.1832–1836.
3. D. Niemann, J. Li, H. O. Wang, and K. Tanaka, "Parallel Distributed Compensation for Takagi-Sugeno Fuzzy Models: New Stability Conditions and Dynamic Feedback Designs," *Proc. 1999 International Federation of Automatic Control (IFAC) World Congress*, Beijing, July 1999, pp. 207–212.
4. J. Li, H. O. Wang, D. Niemann, and K. Tanaka, "Synthesis of Gain-Scheduled Controller for a Class of LPV Systems," *Proc. 38th IEEE Conference on Decision and Control*, Phoenix, Dec. 1999, pp. 2314–2319.
5. J. Li, H. O. Wang, D. Niemann, and K. Tanaka, "Dynamic Parallel Distributed Compensation for Takagi-Sugeno Fuzzy Systems: An LMI Approach," *Inform. Sci.*, Vol. 123, pp. 201–221 (2000).
6. P. Apkarian, P. Gahinet, and G. Becker, " Self-Scheduled H_∞ Control of Linear Parameter Varying Systems: A Design Example," *Automatica*, Vol. 31, No. 9, pp. 1251–1261 (1995).
7. S. G. Cao, N. W. Rees, and G. Feng, "Fuzzy Control of Nonlinear Continuous-Time Systems," in *Proc. 35th IEEE Conf. Decision and Control*, Kobe, Japan, 1996, pp. 592–597.
8. G. Chen and H. Ying, "On the Stability of Fuzzy Control Systems," in *Proc. 3rd IFIS*, Houston, 1993.
9. S. S. Farinwata and G. Vachtsevanos, "Stability Analysis of the Fuzzy Logic Controller," *Proc. IEEE CDC*, San Antonio, 1993.
10. P. Gahinet and P. Apkarian, "A Linear Matrix Inequality Approach to H_∞ Control," *Int. J. Robust Nonlinear Control*, Vol. 4, No. 4, pp. 421–428 (1994).
11. M. Johansson and A. Rantzer, "On the Computation of Piecewise Quadratic Lyapunov Function," in *Proc. 36th CDC*, 1997, pp. 3515–3520.
12. R. Langari and M. Tomizuka, "Analysis and Synthesis of Fuzzy Linguistic Control Systems," in *Proc. 1990 ASME Winter Annual Meet.*, 1990, pp. 35–42.
13. J. Li, H. O. Wang, and K. Tanaka, " Stable Fuzzy Control of the Benchmark Nonlinear Control Problem: A System-Theoretic Approach," in *Joint Conf. of Information Science*, 1997, pp. 263–266.
14. C. Scherer, P. Gahinet, and M. Chilali, "Multiobjective Output-Feedback Control via LMI Optimization," *IEEE Trans. Automatic Control*, Vol. 42, No. 7, pp. 896–911 (1997).

15. J. Shamma and M. Athans, "Analysis of Nonlinear Gain Scheduled Control Systems," *IEEE Trans. Automatic Control*, Vol. 35, pp. 898–907 (1990).
16. T. Takagi and M. Sugeno, "Fuzzy Identification of Systems and Its Applications to Modeling and Control," *IEEE Trans. Syst. Man and Cybernet.*, Vol. 15, pp. 116–132 (1985).
17. K. Tanaka, T. Ikeda, and H. O. Wang, "Fuzzy Regulators and Fuzzy Observers: Relaxed Stability Conditions and LMI-Based Designs," *IEEE Trans. Fuzzy Syst.*, Vol. 6, No. 2, pp. 250–265 (1998).
18. K. Tanaka and M. Sugeno, "Stability Analysis and Design of Fuzzy Control Systems," *Fuzzy Sets Syst.*, Vol. 45, No. 2, pp. 135–156 (1992).
19. H. O. Wang, K. Tanaka, and M. F. Griffin, "Parallel Distributed Compensation of Nonlinear Systems by Takagi-Sugeno Fuzzy Model," in *Proc. FUZZ-IEEE/IFES '95*, 1995, pp. 531–538.
20. H. O. Wang, K. Tanaka, and M. F. Griffin, " An Approach to Fuzzy Control of Nonlinear Systems: Stability and Design Issues," *IEEE Trans. Fuzzy Syst.*, Vol. 4, No. 1, pp. 14–23 (1996).
21. K. Tanaka, T. Ikeda, and H. O. Wang, " Robust Stabilization of a Class of Uncertain Nonlinear Systems via Fuzzy Control: Quadratic Stabilizability, H_∞ Control Theory and Linear Matrix Inequalities," *IEEE Trans. Fuzzy Syst.*, Vol. 4, No. 1, pp. 1–13 (1996).
22. J. Zhao, V. Wertz, and R. Gorez, " Fuzzy Gain Scheduling Controllers Based on Fuzzy Models," in *Proc. Fuzzy-IEEE'96*, 1996, pp. 1670–1676.
23. K. Zhou, P. P. Khargonekar, J. Stoustrup, and H. H. Niemann, "Robust Stability and Performance of Uncertain Systems in State Space," in *Proc. 31st IEEE Conf. Decision and Control*, 1992, pp. 662–667.
24. S. Boyd, L. E. Ghaoui, E. Feron, and V. Balakrishnan, *Linear Matrix Inequalities in Systems and Control Theory*, SIAM, Philadelphia, PA, 1994.
25. P. Gahinet, A. Nemirovski, A. J. Laub, and M. Chilali, *LMI Control Toolbox*, Math Works, 1995.
26. L. X. Wang, *Adaptive Fuzzy Systems and Control: Design and Stability Analysis*, Prentice-Hall, Englewood Cliffs, NJ, 1993.

CHAPTER 13

MULTIOBJECTIVE CONTROL VIA DYNAMIC PARALLEL DISTRIBUTED COMPENSATION

This chapter treats the multiobjective control synthesis problems [1–5] via the dynamic parallel distributed compensation (DPDC). It is often the case in the practice of control engineering that a number of design objectives have to be achieved concurrently. The associated synthesis problems are formulated as linear matrix inequality (LMI) problems, that is, the parameters of the DPDC controllers are obtained from a set of LMI conditions. The approach in this chapter can also be applied to hybrid or switching systems.

We present the performance-oriented controller synthesis of DPDCs to incorporate a number of practical design objectives such as disturbance attenuation, passivity, and output constraint. Performance specifications presented in this chapter include L_2 gain, general quadratic constraints, generalized H_2 performance, and output and input constraints. The controller synthesis procedures are formulated as LMI problems. In the case of meeting multiple design objectives, we only need to group these LMI conditions together and find a feasible solution to the augmented LMI problem [15].

First we introduce some notation: $\Re^+ = [0, \infty)$; $L_2^p(\Re^+)$ is defined as the set of all p-dimensional vector valued functions $u(t)$, $t \in \Re^+$, such that $\|u\|_2 = (\int_0^\infty \|u(t)\|^2 \, dt)^{1/2} < \infty$ and $L_2^e(\Re^+)$ is its extended space, which is defined as the set of the vector-valued functions $u(t)$, $t \in \Re^+$, such that $\|u\|_2^e = (\int_0^T \|u(t)\|^2 \, dt)^{1/2} < \infty$ for all $T \in \Re^+$.

As discussed in Chapter 12, in general, the choice of a particular DPDC parameterization will be influenced by the structure of the T-S subsystems. In this chapter, we will only discuss DPDC in the quadratic parameterization form. It is easy to extend the results in this chapter to the cubic parameterization case. Recall the quadratic parameterization is represented as

$$\dot{x}_c = \sum_{i=1}^{r}\sum_{j=1}^{r} h_i(p)h_j(p) A_c^{ij} x_c + \sum_{i=1}^{r} h_i B_c^i y, \qquad (13.1)$$

$$u = \sum_{i=1}^{r} h_i(p) C_c^i x_c + D_c y, \qquad (13.2)$$

or equivalently that

$$A_c(p) = \sum_{i=1}^{r}\sum_{j=1}^{r} h_i(p)h_j(p) A_c^{ij}, \quad B_c(p) = \sum_{i=1}^{r} h_i(p) B_c^i,$$

$$C_c(p) = \sum_{i=1}^{r} h_i(p) C_c^i, \quad D_c(p) = D_c. \qquad (13.3)$$

As in Chapter 12, we use p as premise variables and z as performance variables.

13.1 PERFORMANCE-ORIENTED CONTROLLER SYNTHESIS

This section presents LMI conditions which can be used to design DPDC controllers which satisfy a variety of useful performance criteria. The presentation is divided into two subsections. In the first subsection, we assume only a linear parameter-dependent controller structure and derive a collection of parameter-dependent conditions expressed in inequalities. Each condition corresponds to a different performance criterion. In the second subsection, we restrict our consideration to a DPDC controller structure. This restriction allows us to convert the parameter-dependent inequalities to parameter-free LMIs which can be solved numerically

13.1.1 Starting from Design Specifications

We will consider the class of systems G which can be described by the equations

$$\dot{x}_{cl}(t) = A_{cl}(p) x_{cl}(t) + B_{cl}(p) w(t),$$
$$z(t) = C_{cl}(p) x_{cl}(t) + D_{cl}(p) w(t), \qquad (13.4)$$

where $x(t)$, $w(t)$, and $z(t)$ stand for state, input, and performance variables correspondingly; $p(t)$ is the system parameter which may be affected by both the system states or some exogenous input variables.

L_2 Gain Performance

Definition 2 [14]: For a casual NLTI (nonlinear time-invariant operator) G: $w \in L_2^e(\Re^+) \to z \in L_2^e(\Re^+)$ with $G(0) = 0$, G is L_2 stable if $w \in L_2(\Re)$ implies $z \in L_2(\Re)$. Here, G is said to have L_2 gain less than or equal to $\gamma \geq 0$ if and only if

$$\int_0^T \|z(t)\|^2 \, dt \leq \gamma^2 \int_0^T \|w(t)\|^2 \, dt \tag{13.5}$$

for all $T \in \Re^+$.

The well-known Bounded Real Lemma is given below [25].

LEMMA 1 For system G: $(A_{cl}(p), B_{cl}(p), C_{cl}(p), D_{cl}(p))$, the L_2 gain will be less than $\gamma > 0$ if there exists a matrix $P = P^T > 0$ such that

$$\begin{bmatrix} \mathcal{L}(A_{cl}(p), P) & PB_{cl}(p) & C_{cl}(p)^T \\ B_{cl}(p)^T P & -\gamma I & D_{cl}(p)^T \\ C_{cl}(p) & D_{cl}(p) & -\gamma I \end{bmatrix} < 0. \tag{13.6}$$

General Quadratic Constraint

Definition 3 [15]: For a casual NLTI, G: $w \to z$ with $G(0) = 0$. Given fixed matrices $U = S\Sigma^{-1}S^T$, $V = V^T$, and W, where $\Sigma > 0$. The variables $z(t)$ and $w(t)$ need to satisfy the following constraint:

$$\int_0^T \begin{pmatrix} z(t) \\ w(t) \end{pmatrix}' \begin{pmatrix} U & W \\ W^T & V \end{pmatrix} \begin{pmatrix} z(t) \\ w(t) \end{pmatrix} dt < 0, \quad \forall T \geq 0, \tag{13.7}$$

for $x_{cl}(0) = 0$ and $w(t) \in L_2(\Re^+)$.

Remark 41 [15]: Many performance specifications (such as L_2 gain, passivity, and sector constraint) can be incorporated into this general quadratic constraint framework by choosing different U, V, and W.

Define the function $V(x_{cl}) = x'_{cl} P x_{cl}$, where $P = P^T > 0$. Suppose

$$\begin{pmatrix} \mathcal{L}(A_{cl}(p), P) & PB_{cl}(p) + C_{cl}^T W \\ B_{cl}^T(p) P + W^T C_{cl} & D_{cl}^T W + W^T D_{cl} + V \end{pmatrix}$$

$$+ \begin{pmatrix} C_{cl}^T(p) \\ D_{cl}^T(p) \end{pmatrix}^T U (C_{cl}(p) \quad D_{cl}(p)) < 0. \tag{13.8}$$

Then

$$\frac{d}{dt}V(x_{cl}(t)) = \begin{pmatrix} x_{cl}(t) \\ w(t) \end{pmatrix}' \begin{pmatrix} \mathcal{L}(A_{cl}(p), P) & PB_{cl}(p) \\ B_{cl}^T(p)P & 0 \end{pmatrix} \begin{pmatrix} x_{cl}(t) \\ w(t) \end{pmatrix}$$

$$< - \begin{pmatrix} z(t) \\ w(t) \end{pmatrix}' \begin{pmatrix} U & W \\ W^T & V \end{pmatrix} \begin{pmatrix} z(t) \\ w(t) \end{pmatrix}. \qquad (13.9)$$

Inequality (13.7) will result by integrating both sides of (13.9).

Applying the Schur complement to (13.8), we get the following lemma:

LEMMA 2 *For system G: $(A_{cl}(p), B_{cl}(p), C_{cl}(p), D_{cl}(p))$, the general quadratic constraint (13.7) will be satisfied if there exists a matrix $P = P^T > 0$ such that*

$$\begin{bmatrix} \mathcal{L}(A_{cl}(p), P) & PB_{cl}(p) + C_{cl}^T(p)W & C_{cl}^T(p)S \\ B_{cl}^T(p)P + W^T C_{cl} & W^T D_{cl} + D_{cl}^T W + V & D_{cl}^T(p)S \\ S^T C_{cl}(p) & S^T D_{cl}(p) & -\Sigma \end{bmatrix} < 0. \qquad (13.10)$$

Generalized H_2 Performance

Definition 4 [15]: A causal NLTI G: $w \to z$ with $G(0) = 0$ is said to have generalized H_2 performance less than or equal to ζ if and only if

$$\|z(T)\| \le \zeta, \qquad \forall T \ge 0, \qquad (13.11)$$

where $x_{cl}(0) = 0$ and $\int_0^T \|w(t)\|^2 \, dt \le 1$.

Define the function $V(x_{cl}(t)) = x_{cl}' P x_{cl}$, where $P > 0$. Suppose

$$\begin{pmatrix} \mathcal{L}(A_{cl}(p), P) & PB_{cl}(p) \\ B_{cl}^T(p)P & -\zeta I \end{pmatrix} < 0. \qquad (13.12)$$

Then $(d/dt)V(x_{cl}(t)) < \zeta w'(t)w(t)$. We will suppose $D_{cl}(p) = 0$. In this case, if the equation

$$\begin{pmatrix} P & C_{cl}^T(p) \\ C_{cl}(p) & \zeta I \end{pmatrix} > 0 \qquad (13.13)$$

is satisfied, then $z'(t)z(t) < \zeta V(x_{cl}(t))$. This leads to the following lemma:

LEMMA 3 *For system $G: (A_{cl}(p), B_{cl}(p), C_{cl}(p), 0)$, the generalized H_2 performance will be less than ζ if there exists a matrix $P = P^T > 0$ such that (13.12) and (13.13) are feasible.*

Constraint on System Output

Definition 5 [24]: A casual NLTI $G: \dot{x}_{cl} = A_{cl}(p)x_{cl}$ and $z = C_{cl}(p)x_{cl}$ satisfies an exponential constraint on the output if

$$\|z(T)\| \leq \zeta e^{-\alpha T}, \quad \forall T \geq 0, \tag{13.14}$$

where $x_{cl}(0) = x_0$.

Define the function $V(x_{cl}) = x'_{cl} P x_{cl}$, where $P = P^T > 0$. Suppose that the equation

$$\mathcal{L}(A_{cl}, P) + 2\alpha P < 0 \tag{13.15}$$

holds. In this case, the inequality $V(x_{cl}(t)) < e^{-2\alpha t} V(x_{cl}(0))$ will be satisfied. Furthermore, if the equations

$$\begin{pmatrix} P & Px_{cl}(0) \\ x'_{cl}(0)P & \zeta I \end{pmatrix} > 0 \tag{13.16}$$

and

$$\begin{pmatrix} P & C_{cl}^T(p) \\ C_{cl}(p) & \zeta I \end{pmatrix} > 0 \tag{13.17}$$

hold, then the inequality

$$z'(t)z(t) < \zeta(x'_{cl}(t)Px_{cl}(t)) < \zeta e^{-2\alpha t}(x'_{cl}(0)Px_{cl}(0)) < \zeta^2 e^{-2\alpha t}$$

will also be satisfied. Combining these results, we have the following lemma:

LEMMA 4 *For the system $G: \dot{x}_{cl} = A_{cl}(p)x_{cl}$ and $z = C_{cl}(p)x_{cl}$, the exponential constraint $\|z(T)\| \leq \zeta e^{-\alpha T}, \forall T \geq 0$, will be satisfied if there exists a matrix $P = P^T > 0$ such that (13.15), (13.16), and (13.17) are feasible.*

Constraints on Control Input

Definition 6 [24]: A casual NLTI $G: \dot{x}_{cl} = A_{cl}(p)x_{cl}$ and $u = K(p)x_{cl}$ with a specified initial condition $x_{cl}(0)$ satisfies an exponential constraint on the input if

$$\|u(T)\| \leq \zeta e^{-\alpha T}, \quad \forall T \geq 0. \tag{13.18}$$

Similar to the discussion for exponential constraint on the system output, we have the following lemma:

LEMMA 5 *For system* $G: \dot{x}_{cl} = A_{cl}(p)x_{cl}$ *and* $u = K(p)x_{cl}$, *the exponential constraint* $\|u(T)\| \leq \zeta e^{-\alpha T}, \forall T \geq 0$, *will be satisfied if there exists a matrix* $P = P^T > 0$ *such that* (13.15), (13.16), *and*

$$\begin{pmatrix} P & K(p)^T \\ K(p) & \zeta I \end{pmatrix} < 0. \tag{13.19}$$

13.1.2 Performance-Oriented Controller Synthesis

In this subsection, we consider T-S models which are represented by a set of fuzzy rules in the following form:

Dynamic Part

Rule i

IF $p_1(t)$ is M_{i1} $\cdots$ and $p_l(t)$ is M_{il},

THEN $\dot{x}(t) = A_i x(t) + B_i u(t) + B_w^i w(t)$.

Output Part

Rule i

IF $p_1(t)$ is M_{i1} $\cdots$ and $p_l(t)$ is M_{il},

THEN

$$y(t) = C_i x(t) + D_w^i w(t),$$
$$z(t) = C_z^i x(t) + D_z^i u(t) + D_{zw}^i w(t).$$

Here, $p_i(t)$ are some fuzzy variables, $x(t)$ are the system states, $u(t)$ are the control inputs, $w(t)$ are exogenous inputs such as disturbance signals, noises, or reference signals, $y(t)$ represent the measurements, and $z(t)$ stand for performance variables of the control systems.

We can simplify the expressions of the T-S model as

$$\dot{x} = \sum_{i=1}^{r} h_i(p)(A_i x + B_i u + B_w^i w), \tag{13.20}$$

$$z = \sum_{i=1}^{r} h_i(p)(C_z^i x + D_z^i u + D_{zw}^i w), \tag{13.21}$$

$$y = \sum_{i=1}^{r} h_i(p)(C_i x + D_w^i w). \tag{13.22}$$

The closed-loop system equations for a T-S model (13.20)–(13.22) with DPDC controller (13.1) and (13.2) have the form

$$\dot{x}_{cl} = \sum_{i=1}^{r}\sum_{j=1}^{r} h_i(p)h_j(p)\left(A_{cl}^{ij} x_{cl} + B_{cl}^{ij} w\right), \qquad (13.23)$$

$$z_{cl} = \sum_{i=1}^{r}\sum_{j=1}^{r} h_i(p)h_j(p)\left(C_{cl}^{ij} x_{cl} + D_{cl}^{ij} w\right), \qquad (13.24)$$

where

$$A_{cl}^{ij} = \begin{pmatrix} A_i + B_i D_c C_j & B_i C_c^j \\ B_c^i C_j & A_c^{ij} \end{pmatrix}, \quad B_{cl}^{ij} = \begin{bmatrix} B_w^i + B_i D_c D_w^j \\ B_c^i D_w^j \end{bmatrix},$$

$$C_{cl}^{ij} = \begin{bmatrix} C_z^i + D_z^i D_c C_j & D_z^i C_c^j \end{bmatrix}, \quad D_{cl}^{ij} = \begin{bmatrix} D_{zw}^i + D_z^i D_c D_w^j \end{bmatrix}.$$

Now, we are ready to apply the results in Section 13.1.1 to (13.23) and (13.24).

L_2 Gain Performance We begin by applying a congruence transformation on (13.6) using the matrix

$$\begin{pmatrix} \Pi_1 & 0 & 0 \\ 0 & I & 0 \\ 0 & 0 & I \end{pmatrix},$$

where the closed-loop system is defined as in (13.23) and (13.24). By utilizing the notation in the quadratic parametrization discussed in Chapter 12, (13.6) becomes

$$\sum_{i=1}^{r}\sum_{j=1}^{r} h_i(p)\, h_j(p)\, \mathbf{E}_{ij} < 0, \qquad (13.25)$$

where

$$\mathbf{E}_{ij} = \begin{pmatrix} E_{11}^{ij} & E_{12}^{ij} & E_{13}^{ij} & E_{14}^{ij} \\ \left(E_{12}^{ij}\right)^T & E_{22}^{ij} & E_{23}^{ij} & \left(E_{42}^{ij}\right)^T \\ \left(E_{13}^{ij}\right)^T & \left(E_{23}^{ij}\right)^T & -\gamma I & \left(E_{43}^{ij}\right)^T \\ \left(E_{14}^{ij}\right)^T & E_{42}^{ij} & E_{43}^{ij} & -\gamma I \end{pmatrix}$$

and

$$E_{11}^{ij} = \mathcal{L}(A_i, Q_{11}) + B_i \mathcal{C}_j + \left(B_i \mathcal{C}_j\right)^T, \qquad E_{12}^{ij} = A_i + B_i \mathcal{D} C_j + \mathcal{A}_{ij}^T,$$

$$E_{13}^{ij} = B_w^i + B_i \mathcal{D} D_w^j, \qquad E_{14}^{ij} = \left(C_z^i Q_{11} + D_z^i \mathcal{C}_j\right)^T,$$

$$E_{22}^{ij} = \mathcal{L}(A_i^T, P_{11}) + \mathcal{B}_i C_j + \left(\mathcal{B}_i C_j\right)^T, \qquad E_{23}^{ij} = P_{11} B_w^i + \mathcal{B}_i D_w^j,$$

$$E_{42}^{ij} = C_z^i + D_z^i \mathcal{D} C_j, \qquad E_{43}^{ij} = D_z^i \mathcal{D} D_w^j + D_{zw}^i.$$

Condition (13.25) is equivalent to

$$\sum_{i=1}^{r} \sum_{j=1}^{r} h_i(p) h_j(p) (\mathbf{E}_{ij} + \mathbf{E}_{ji}) < 0. \tag{13.26}$$

The inequality (13.26) will hold true according to Theorem 45 if there exist symmetric matrices T_{ij} satisfying (12.85) and $(\mathbf{E}_{ij} + \mathbf{E}_{ji}) < T_{ij}$.

We will express the resulting theorem using the notation in the previous section:

THEOREM 51 *Given a T-S model of the form* (13.20)–(13.22) *with DPDC controller* (13.1) *and* (13.2), *the L_2 gain performance will be less than γ if the LMI conditions* (12.16), (13.27), *and* (12.85) *are feasible with LMI variables* Q_{11}, P_{11}, T_{ij}, $\mathcal{A}_{ij}$, $\mathcal{B}_i$, $\mathcal{C}_i$, *and* $\mathcal{D}$:

$$\begin{pmatrix} \bar{E}_{11}^{ij} & \bar{E}_{12}^{ij} & \bar{E}_{13}^{ij} & \bar{E}_{14}^{ij} \\ \left(\bar{E}_{12}^{ij}\right)^T & \bar{E}_{22}^{ij} & \bar{E}_{23}^{ij} & \left(\bar{E}_{42}^{ij}\right)^T \\ \left(\bar{E}_{13}^{ij}\right)^T & \left(\bar{E}_{23}^{ij}\right)^T & -2\gamma I & \left(\bar{E}_{43}^{ij}\right)^T \\ \left(\bar{E}_{14}^{ij}\right)^T & \bar{E}_{42}^{ij} & \bar{E}_{43}^{ij} & -2\gamma I \end{pmatrix} < T_{ij}, \quad \forall i \leq j, \tag{13.27}$$

where

$$\bar{E}_{11}^{ij} = \mathcal{L}(A_i, Q_{11}) + \mathcal{L}(A_j, Q_{11}) + B_i \bar{\mathcal{C}}_j + \left(B_i \bar{\mathcal{C}}_j\right)^T + B_j \bar{\mathcal{C}}_i + \left(B_j \bar{\mathcal{C}}_i\right)^T,$$

$$\bar{E}_{12}^{ij} = A_i + A_j + B_i \bar{\mathcal{D}} C_j + B_j \bar{\mathcal{D}} C_i + 2\mathcal{A}_{ij}^T,$$

$$\bar{E}_{13}^{ij} = B_w^i + B_w^j + B_i \bar{\mathcal{D}} D_w^j + B_j \bar{\mathcal{D}} D_w^i,$$

$$\bar{E}_{14}^{ij} = \left(C_z^i Q_{11} + C_z^j Q_{11} + D_z^i \bar{\mathcal{C}}_j + D_z^j \bar{\mathcal{C}}_i\right)^T,$$

$$\bar{E}_{22}^{ij} = \mathcal{L}(A_i^T, P_{11}) + \mathcal{L}(A_j^T, P_{11}) + \bar{\mathcal{B}}_i C_j + \bar{\mathcal{B}}_j C_i + \left(\bar{\mathcal{B}}_i C_j\right)^T + \left(\bar{\mathcal{B}}_j C_i\right)^T,$$

$$\overline{E}_{23}^{ij} = P_{11}B_w^i + P_{11}B_w^j + \overline{\mathcal{B}}_i D_w^j + \overline{\mathcal{B}}_j D_w^i,$$

$$\overline{E}_{42}^{ij} = C_z^i + C_z^j + D_z^i \overline{\mathcal{D}} C_j + D_z^j \overline{\mathcal{D}} C_i,$$

$$\overline{E}_{43}^{ij} = D_z^i \overline{\mathcal{D}} D_w^j + D_z^j \overline{\mathcal{D}} D_w^i + D_{zw}^i + D_{zw}^j.$$

The resulting dynamic controller is given by (12.87)–(12.90) where P_{11}, P_{12}, Q_{11}, and Q_{12} satisfy the constraint $P_{11}Q_{11} + P_{12}Q_{12}^T = I$.

General Quadratic Performance Similarly, we get the following theorem by applying a congruence transform on (13.10) using the matrix

$$\begin{pmatrix} \Pi_1 & 0 & 0 \\ 0 & I & 0 \\ 0 & 0 & I \end{pmatrix}.$$

THEOREM 52 *For a T-S model* (13.20)–(13.22) *with a DPDC controller* (13.1) *and* (13.2), *the generalized quadratic constraint* (13.7) *will be satisfied if the LMI conditions* (12.16), (12.85), *and* (13.28) *are feasible with LMI variables* Q_{11}, P_{11}, T_{ij}, $\overline{\mathcal{A}}_{ij}$, $\overline{\mathcal{B}}_i$, $\overline{\mathcal{C}}_i$ *and* $\overline{\mathcal{D}}$.

$$\begin{pmatrix} \overline{E}_{11}^{ij} & \overline{E}_{12}^{ij} & \overline{E}_{13}^{ij} & \overline{E}_{14}^{ij} \\ (\overline{E}_{12}^{ij})^T & \overline{E}_{22}^{ij} & \overline{E}_{23}^{ij} & \overline{E}_{24}^{ij} \\ (\overline{E}_{13}^{ij})^T & (\overline{E}_{23}^{ij})^T & \overline{E}_{33}^{ij} & \overline{E}_{34}^{ij} \\ (\overline{E}_{14}^{ij})^T & (\overline{E}_{24}^{ij})^T & (\overline{E}_{34}^{ij})^T & \overline{E}_{44}^{ij} \end{pmatrix} < T_{ij}, \quad \forall i \leq j, \quad (13.28)$$

where

$$\overline{E}_{11}^{ij} = \mathcal{L}(A_i, Q_{11}) + \mathcal{L}(A_j, Q_{11}) + B_i \overline{\mathcal{C}}_j + (B_i \overline{\mathcal{C}}_j)^T + B_j \overline{\mathcal{C}}_i + (B_j \overline{\mathcal{C}}_i)^T,$$

$$\overline{E}_{12}^{ij} = A_i + A_j + B_i \overline{\mathcal{D}} C_j + B_j \overline{\mathcal{D}} C_i + 2\overline{\mathcal{A}}_{ij}^T,$$

$$\overline{E}_{13}^{ij} = B_w^i + B_w^j + B_i \overline{\mathcal{D}} D_w^j + B_j \overline{\mathcal{D}} D_w^i$$

$$+ \left(C_z^i Q_{11} + D_z^i \overline{\mathcal{C}}_j + C_z^j Q_{11} + D_z^j \overline{\mathcal{C}}_i\right)^T W,$$

$$\overline{E}_{14}^{ij} = \left(C_z^i Q_{11} + C_z^j Q_{11} + D_z^i \overline{\mathcal{C}}_j + D_z^j \overline{\mathcal{C}}_i\right)^T S,$$

$$\bar{E}_{22}^{ij} = \mathcal{L}(A_i^T, P_{11}) + \mathcal{L}(A_j^T, P_{11}) + \overline{\mathcal{B}}_i C_j + \overline{\mathcal{B}}_j C_i + \left(\overline{\mathcal{B}}_j C_i\right)^T + \left(\overline{\mathcal{B}}_j C_i\right)^T,$$

$$\bar{E}_{23}^{ij} = P_{11} B_w^i + P_{11} B_w^j + \overline{\mathcal{B}}_i D_w^j + \overline{\mathcal{B}}_j D_w^i$$

$$+ \left(C_z^i + C_z^j + D_z^i \overline{\mathcal{D}} C_j + D_z^j \overline{\mathcal{D}} C_i\right)^T W,$$

$$\bar{E}_{24}^{ij} = \left(C_z^i + C_z^j + D_z^i \overline{\mathcal{D}} C_j + D_z^j \overline{\mathcal{D}} C_i\right)^T S,$$

$$\bar{E}_{33}^{ij} = 2V + W^T \left(D_{zw}^i + D_{zw}^j + D_z^i \overline{\mathcal{D}} D_w^j + D_z^j \overline{\mathcal{D}} D_w^i\right)$$

$$+ \left(D_{zw}^i + D_{zw}^j + D_z^i \overline{\mathcal{D}} D_w^j + D_z^j \overline{\mathcal{D}} D_w^i\right)^T W,$$

$$\bar{E}_{34}^{ij} = \left(D_{zw}^i + D_{zw}^j + D_z^i \overline{\mathcal{D}} D_w^j + D_z^j \overline{\mathcal{D}} D_w^i\right)^T S,$$

$$\bar{E}_{44}^{ij} = -2\Sigma.$$

The controller is given by (12.87)–(12.90).

Generalized H_2 Performance If we apply a congruence transform on both (13.12) and (13.13) using the matrix

$$\begin{pmatrix} \Pi_1 & 0 \\ 0 & I \end{pmatrix},$$

we get the following theorem:

THEOREM 53 *For a T-S model (13.20)–(13.22) with PDC controller (13.1) and (13.2), the generalized H_2 performance will be less than ζ if the LMI conditions (12.16), (12.85), (13.20), (13.30), and (13.31) are feasible with LMI variables Q_{11}, P_{11}, T_{ij}, S_{ij}, $\overline{\mathcal{A}}_{ij}$, $\overline{\mathcal{B}}_i$, $\overline{\mathcal{C}}_i$, and $\overline{\mathcal{D}}$ for all $i \leq j$:*

$$\begin{pmatrix} 2Q_{11} & 2I & \left(\begin{array}{c}((C_z^i + C_z^j)Q_{11} \\ +D_z^i \overline{\mathcal{C}}_j + D_z^j \overline{\mathcal{C}}_i)\end{array}\right)^T \\ 2I & 2P_{11} & \left(\begin{array}{c}(C_z^i + C_z^j + D_z^i \overline{\mathcal{D}} C_j \\ +D_z^j \overline{\mathcal{D}} C_i)\end{array}\right)^T \\ \left(\begin{array}{c}(C_z^i + C_z^j)Q_{11} \\ +D_z^i \overline{\mathcal{C}}_j + D_z^j \overline{\mathcal{C}}_i\end{array}\right) & \left(\begin{array}{c}(C_z^i + C_z^j) \\ +D_z^i \overline{\mathcal{D}} C_j + D_z^j \overline{\mathcal{D}} C_i\end{array}\right) & 2\zeta I \end{pmatrix} > S_{ij},$$

(13.29)

$$\mathbf{S} = \begin{bmatrix} S_{11} & \cdots & S_{1r} \\ \vdots & \ddots & \vdots \\ S_{1r} & \cdots & S_{rr} \end{bmatrix} > 0, \tag{13.30}$$

$$\begin{pmatrix} \overline{E}_{11}^{ij} & \overline{E}_{12}^{ij} & \overline{E}_{13}^{ij} \\ \left(\overline{E}_{12}^{ij}\right)^T & \overline{E}_{22}^{ij} & \overline{E}_{23}^{ij} \\ \left(\overline{E}_{13}^{ij}\right)^T & \left(\overline{E}_{23}^{ij}\right)^T & \overline{E}_{33}^{ij} \end{pmatrix} < T_{ij}, \tag{13.31}$$

where

$$\overline{E}_{11}^{ij} = \mathcal{L}(A_i, Q_{11}) + \mathcal{L}(A_j, Q_{11}) + B_i \overline{\mathcal{C}}_j + \left(B_i \overline{\mathcal{C}}_j\right)^T + B_j \overline{\mathcal{C}}_i + \left(B_j \overline{\mathcal{C}}_i\right)^T,$$

$$\overline{E}_{12}^{ij} = A_i + A_j + B_i \overline{\mathcal{D}} C_j + B_j \overline{\mathcal{D}} C_i + 2 \overline{\mathcal{A}}_{ij}^T,$$

$$\overline{E}_{13}^{ij} = B_w^i + B_w^j + B_i \overline{\mathcal{D}} D_w^j + B_j \overline{\mathcal{D}} D_w^i,$$

$$E_{22}^{ij} = \mathcal{L}(A_i^T, P_{11}) + \mathcal{L}(A_j^T, P_{11}) + \overline{\mathcal{B}}_i C_j + \overline{\mathcal{B}}_j C_i + \left(\overline{\mathcal{B}}_i C_j\right)^T + \left(\overline{\mathcal{B}}_j C_i\right)^T,$$

$$\overline{E}_{23}^{ij} = P_{11} B_w^i + P_{11} B_w^j + \overline{\mathcal{B}}_i D_w^j + \overline{\mathcal{B}}_j D_w^i,$$

$$\overline{E}_{33}^{ij} = -2\zeta I,$$

and

$$D_{zw}^i + D_{zw}^j + D_z^i \overline{\mathcal{D}} D_w^j + D_z^j \overline{\mathcal{D}} D_w^i = 0, \quad \forall i \le j. \tag{13.32}$$

The controller is given by (12.87)–(12.90).

Constraints on the Outputs Applying a congruence transform on (13.15) using the matrix Π and on (13.16) and (13.17) using the matrix

$$\begin{pmatrix} \Pi_1 & 0 \\ 0 & I \end{pmatrix},$$

we get the following theorem.

THEOREM 54 *Consider a T-S model* (13.20)–(13.22) *(suppose $D_{zw}^i = 0$, $D_w^i = 0$ and $B_w^i = 0$) with DPDC controller* (13.1) *and* (13.2). *Suppose the initial state is given by* $[x(0)\ x_c(0)]'$; *then* $\|z(t)\| < \zeta e^{-\alpha t}$ *for all $t \ge 0$ if the LMI conditions* (13.33)–(13.35) *and* (12.85) *and* (13.31) *are feasible with LMI variables* $Q_{11}, P_{11}, P_{12}, T_{ij}, S_{ij}, \overline{\mathcal{A}}_{ij}, \overline{\mathcal{B}}_i, \overline{\mathcal{C}}_i$ *and* $\overline{\mathcal{D}}$:

$$\begin{pmatrix} \overline{E}_{11}^{ij} & \overline{E}_{12}^{ij} \\ \left(\overline{E}_{12}^{ij}\right)^T & \overline{E}_{22}^{ij} \end{pmatrix} < T_{ij}, \quad \forall i \le j, \tag{13.33}$$

where

$$\overline{E}_{11}^{ij} = \mathcal{L}(A_i, Q_{11}) + \mathcal{L}(A_j, Q_{11}) + B_i \overline{C}_j + \left(B_i \overline{C}_j\right)^T$$
$$+ B_j \overline{C}_i + \left(B_j \overline{C}_i\right)^T + 2\alpha Q_{11},$$

$$\overline{E}_{12}^{ij} = A_i + A_j + B_i \overline{\mathcal{D}} C_j + B_j \overline{\mathcal{D}} C_i + \overline{\mathcal{A}}_{ij}^T + 2\alpha I,$$

$$\overline{E}_{22}^{ij} = \mathcal{L}(A_i^T, P_{11}) + \mathcal{L}(A_j^T, P_{11}) + \overline{\mathcal{B}}_i C_j + \overline{\mathcal{B}}_j C_i$$
$$+ \left(\overline{\mathcal{B}}_i C_j\right)^T + \left(\overline{\mathcal{B}}_j C_i\right)^T + 2\alpha P_{11},$$

$$\begin{pmatrix} Q_{11} & I & x(0) \\ I & P_{11} & \begin{pmatrix} P_{11}x(0) \\ +P_{12}x_c(0) \end{pmatrix} \\ x'(0) & \begin{pmatrix} x'(0)P_{11} \\ +x'_c(0)P_{12}^T \end{pmatrix} & \zeta I \end{pmatrix} > 0,$$

(13.34)

$$\begin{pmatrix} 2Q_{11} & 2I & \begin{pmatrix} (C_z^i + C_z^j)Q_{11} \\ +D_z^i \overline{C}_j + D_z^j \overline{C}_i \end{pmatrix}^T \\ 2I & 2P_{11} & \begin{pmatrix} C_z^i + C_z^j + D_z^i \overline{\mathcal{D}} C_j \\ +D_z^j \overline{\mathcal{D}} C_i \end{pmatrix}^T \\ \begin{pmatrix} (C_z^i + C_z^j)Q_{11} \\ +D_z^i \overline{C}_j + D_z^j \overline{C}_i \end{pmatrix} & \begin{pmatrix} C_z^i + C_z^j + D_z^i \overline{\mathcal{D}} C_j \\ +D_z^j \overline{\mathcal{D}} C_i \end{pmatrix} & 2\zeta I \end{pmatrix} > S_{ij}.$$

(13.35)

The controller is given by (12.87)–(12.90).

Constraints on the Inputs Applying a congruence transform on (13.15) using the matrix Π_1 and on (13.16) and (13.19) using the matrix

$$\begin{pmatrix} \Pi_1 & 0 \\ 0 & I \end{pmatrix},$$

we get the following theorem:

THEOREM 55 *Consider a T-S model* (13.20)–(13.22) *(suppose $D_{zw}^i = 0$, $D_w^i = 0$, and $B_w^i = 0$) with PDC controller* (13.1) *and* (13.2). *Suppose the initial state is given by* $[x(0) \ x_c(0)]$; *then* $\|u(t)\| < \zeta e^{-\alpha t}$ *for all* $t \geq 0$ *if the LMI conditions* (13.33), (13.34), (13.36), *and* (12.85) *are feasible with LMI variables* Q_{11}, P_{11}, T_{ij}, $\overline{\mathcal{A}}_{ij}$, $\overline{\mathcal{B}}_i$, $\overline{\mathcal{C}}_i$, *and* $\mathcal{D}$.

$$\begin{pmatrix} Q_{11} & I & \overline{C}_i^T \\ I & P_{11} & \left(\overline{\mathcal{D}C_i}\right)^T \\ \overline{C}_i & \overline{\mathcal{D}C_i} & \zeta I \end{pmatrix} > 0. \tag{13.36}$$

The controller is given by (12.87)–(12.90).

13.2 EXAMPLE

To illustrate the DPDC approach, consider the problem of balancing an inverted pendulum on a cart. Recall the equations of motion for the pendulum [26]:

$$\dot{x}_1 = x_2,$$
$$\dot{x}_2 = \frac{g \sin(x_1) - amlx_2^2 \sin(2x_1)/2 - a \cos(x_1)u}{4l/3 - aml \cos^2(x_1)}, \tag{13.37}$$

where x_1 denotes the angle (in radians) of the pendulum from the vertical and x_2 is the angular velocity; $g = 9.8$ m/s^2 is the gravity constant, m is the mass of the pendulum, M is the mass of the cart, $2l$ is the length of the pendulum, and u is the force applied to the cart (in newtons); $a = 1/(m + M)$. We choose $m = 2.0$ kg, $M = 8.0$ kg, $2l = 1.0$ m in this study.

The control objective is to balance the inverted pendulum for the approximate range $x_1 \in (-\pi/2, \pi/2)$. In order to use the DPDC approach, we first represent the system (13.37) by a Takagi-Sugeno fuzzy model. Notice that when $x_1 = \pm \pi/2$, the system is uncontrollable. Hence we use the following two-rule fuzzy model as shown in Chapter 2.

Model Rule 1

  IF x_1 is about 0,

   THEN $\dot{x} = A_1 x + B_1 u$.

Model Rule 2

IF x_1 is about $\pm \pi/2$ ($|x_1| < \pi/2$),

THEN $\dot{x} = A_2 x + B_2 u$.

Here,

$$A_1 = \begin{bmatrix} 0 & 1 \\ \dfrac{g}{4l/3 - aml} & 0 \end{bmatrix}, \quad B_1 = \begin{bmatrix} 0 \\ -\dfrac{a}{4l/3 - aml} \end{bmatrix},$$

$$A_2 = \begin{bmatrix} 0 & 1 \\ \dfrac{2g}{\pi(4l/3 - aml\beta^2)} & 0 \end{bmatrix}, \quad B_2 = \begin{bmatrix} 0 \\ -\dfrac{a\beta}{4l/3 - aml\beta^2} \end{bmatrix},$$

and $\beta = \cos(88°)$.

Membership functions for Rules 1 and 2 are shown in Figure 13.1.

Now we apply the DPDC design to the pendulum system. Assume that only x_1 is measurable, that is, $y = Cx = [1 \ 0]x$. We employ the following DPDC controller:

$$\dot{x}_c = \sum_{i=1}^{2} \sum_{j=1}^{2} h_i(y) h_j(y) \overline{A}_c^{ij} x_c + \sum_{i=1}^{2} h_i \overline{B}_c^i y,$$

$$u = \sum_{i=1}^{2} h_i(y) \overline{C}_c^i x_c + \overline{D}_c y.$$

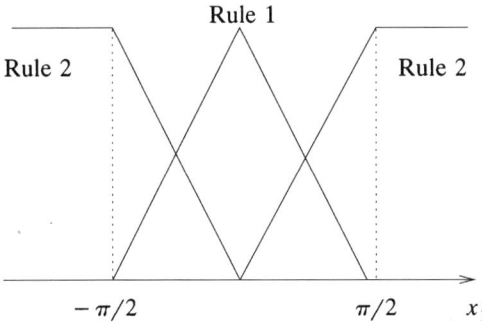

Fig. 13.1 Membership functions of the fuzzy model.

Employing Theorem 47, we obtain the following control parameters for the DPDC controller:

$$\bar{A}_c^{11} = \begin{bmatrix} -13.2565 & -1.4197 \\ -79.0890 & -21.9652 \end{bmatrix},$$

$$\bar{A}_c^{12} = \bar{A}_c^{21} = \begin{bmatrix} -30.7121 & -4.8468 \\ -173.9217 & -51.7244 \end{bmatrix},$$

$$\bar{A}_c^{22} = \begin{bmatrix} -4.0809 & 0.9859 \\ -24.5808 & -6.5791 \end{bmatrix},$$

$$\bar{B}_c^1 = \begin{bmatrix} 5.0666 \\ 20.8530 \end{bmatrix},$$

$$\bar{B}_c^2 = \begin{bmatrix} 3.4824 \\ 12.5320 \end{bmatrix},$$

$$\bar{C}_c^1 = [388.9291 \quad 113.6926],$$

$$\bar{C}_c^2 = [794.6242 \quad 247.5543],$$

$$\bar{D}_c = 4.4624.$$

Figure 13.2 illustrates the closed-loop system response with the DPDC controller for initial conditions $x_1 = \pi/4$ and $x_2 = 0.1$. A number of performance-oriented DPDC designs have also been carried out according to the principles of Section 13.1.

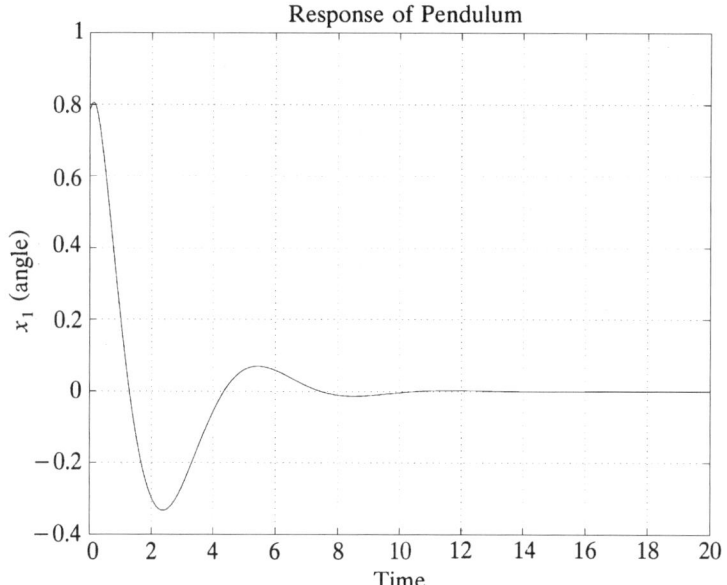

Fig. 13.2 Angle response using the DPDC controller.

If variable p comes from the output of the system, the dynamic feedback controller will become a dynamic output feedback controller which is essential for practical applications when only the system output is available.

The framework used in this chapter can also be applied to generate nonlinear controllers for uncertain systems. One of the basic tools for robustness analysis of such uncertain systems is the small-gain theorem which can be related to the L_2 gain. Thus by making the gain of the nominal plant sufficiently small, we can guarantee the robust stability. The results in this chapter are also applicable to hybrid and switching systems.

BIBLIOGRAPHY

1. J. Li, D. Niemann, H. O. Wang, and K. Tanaka, "Multiobjective Dynamic Feedback Control of Takagi-Sugeno Model via LMIs," *Proc. 4th Joint Conference of Information Sciences*, Vol. 1, Durham, Oct. 1998, pp. 159–162.
2. J. Li, D. Niemann, H. O. Wang, and K. Tanaka, "Parallel Distributed Compensation for Takagi-Sugeno Fuzzy Models: Multiobjective Controller Design," *Proc. 1999 American Control Conference*, San Diego, June 1999, pp. 1832–1836.
3. D. Niemann, J. Li, H. O. Wang, and K. Tanaka, "Parallel Distributed Compensation for Takagi-Sugeno Fuzzy Models: New Stability Conditions and Dynamic Feedback Designs," *Proc. 1999 International Federation of Automatic Control (IFAC) World Congress*, Beijing, July 1999, pp. 207–212.
4. J. Li, H. O. Wang, D. Niemann, and K. Tanaka, "Synthesis of Gain-Scheduled Controller for a Class of LPV Systems," *Proc. 38th IEEE Conference on Decision and Control*, Phoenix, Dec. 1999, pp. 2314–2319.
5. J. Li, H. O. Wang, D. Niemann, and K. Tanaka, "Dynamic Parallel Distributed Compensation for Takagi-Sugeno Fuzzy Systems: An LMI Approach," *Inform. Sci.*, Vol. 123, pp. 201–221 (2000).
6. P. Apkarian, P. Gahinet, and G. Becker, "Self-Scheduled H_∞ Control of Linear Parameter Varying Systems: A Design Example," *Automatica*, Vol. 31, No. 9, pp. 1251–1261 (1995).
7. S. G. Cao, N. W. Rees, and G. Feng, "Fuzzy Control of Nonlinear Continuous-Time Systems," in *Proc. 35th IEEE Conf. Decision and Control*, New York, 1996, pp. 592–597.
8. G. Chen and H. Ying, "Stability Analysis of Nonlinear Fuzzy PI Control Systems," in *Proc. of the 3rd Int. Conf. on Industrial Fuzzy Control and Intelligent Systems*, Kobe, Japan, 1993, pp. 128–133.
9. P. Gahinet and P. Apkarian, "A Linear Matrix Inequality Approach to H_∞ Control," *Int. J. Robust Nonlinear Control*, Vol. 4, No. 4, pp. 421–428 (1994).
10. M. Johansson and A. Rantzer, "On the Computation of Piecewise Quadratic Lyapunov Function," in *Proc. of the 36th IEEE Conf. Decision and Control*, San Diego, CA, 1997, pp. 3515–3520.
11. R. Langari and M. Tomizuka, "Stability of Fuzzy Linguistic Control Systems," in *Proceedings of the 29th IEEE Conf. Decision and Control*, Honolulu, HI, 1990, pp. 35–42.

12. J. Li, H. O. Wang, and K. Tanaka, "Stable Fuzzy Control of the Benchmark Nonlinear Control Problem: A System-Theoretic Approach," Joint Conf. of Information Science, Triangle Park, NC, 1997, pp. 263–266.
13. D. Niemann, J. Li, and H. O. Wang, "Parallel Distributed Compensation for Takagi-Sugeno Fuzzy Models: New Stability Conditions and Dynamic Feedback Designs, *Proc. IFAC 1999*, Beijing, 1999, to appear.
14. A. J. van der Schaft, "L_2-Gain Analysis of Nonlinear Systems and Nonlinear State Feedback H_∞ Control," *IEEE Trans. Automatic Control*, Vol. 37, No. 6, pp. 770–784 (1992).
15. C. Scherer, P. Gahinet, and M. Chilali, "Multiobjective Output-Feedback Control Via LMI Optimization," *IEEE Trans. Automatic Control*, Vol. 42, No. 7, pp. 896–911 (1997).
16. T. Takagi and M. Sugeno, "Fuzzy Identification of Systems and Its Applications to Modeling and Control," *IEEE Trans. Syst. Man. Cybernet.*, Vol. 15, No. 1, pp. 116–132 (1985).
17. K. Tanaka, T. Ikeda, and H. O. Wang, "Robust Stabilization of a Class of Uncertain Nonlinear Systems via Fuzzy Control: Quadratic Stabilizability, H_∞ Control Theory and Linear Matrix Inequalities," *IEEE Trans. Fuzzy Syst.* Vol. 4, No. 1, pp. 1–13 (1996).
18. K. Tanaka, T. Ikeda, and H. O. Wang, "Fuzzy Regulators and Fuzzy Observers: Relaxed Stability Conditions and LMI-Based Designs," *IEEE Trans. Fuzzy Syst.*, Vol. 6, No. 2, pp. 250–265 (1998).
19. K. Tanaka and M. Sugeno, "Stability Analysis and Design of Fuzzy Control Systems," *Fuzzy Sets Syst.*, Vol. 45, No. 2, pp. 135–156 (1992).
20. H. O. Wang, K. Tanaka, and M. F. Griffin, "Parallel Distributed Compensation of Nonlinear Systems by Takagi-Sugeno Fuzzy Model," *Proc. of the FUZZ-IEEE/IFES'95*, 1995, pp. 531–538.
21. H. O. Wang, K. Tanaka, and M. F. Griffin, "An Approach to Fuzzy Control of Nonlinear Systems: Stability and Design Issues," *IEEE Trans. Fuzzy Syst.* Vol. 4, No. 1, pp. 14–23 (1996).
22. J. Zhao, V. Wertz, and R. Gorez, "Fuzzy Gain Scheduling Controllers Based on Fuzzy Models," in *Proc. of the FUZZ-IEEE'96*, New Orleans, LA, 1996, pp. 1670–1676.
23. S. Boyd and C. H. Barratt, *Linear Controller Design: Limits of Performance*, Prentice-Hall, Englewood Cliffs, NJ, 1991.
24. S. Boyd, L. E. Ghaoui, E. Feron, and V. Balakrishnan, *Linear Matrix Inequalities in Systems and Control Theory*, SIAM, Philadelphia, PA, 1994.
25. P. Gahinet, A. Nemirovski, A. J. Laub, and M. Chilali, *LMI Control Toolbox*, The Math Works, 1995.
26. R. H. Cannon, *Dynamics of Physical Systems*, McGraw-Hill, New York, 1967.

CHAPTER 14

T-S FUZZY MODEL AS UNIVERSAL APPROXIMATOR

In this chapter, we present two results concerning the fuzzy modeling and control of nonlinear systems [1]. First, we prove that any smooth nonlinear control systems can be approximated by Takagi-Sugeno fuzzy models with linear rule consequence. Then, we prove that any smooth nonlinear state feedback controller can be approximated by the parallel distributed compensation (PDC) controller.

Among various fuzzy modeling themes, the Takagi-Sugeno (T-S) model [2] has been one of the most popular modeling frameworks. A general T-S model employs an affine model with a constant term in the consequent part for each rule. This is often referred as an *affine* T-S model. In this book, we focus on the special type of T-S fuzzy model in which the consequent part for each rule is represented by a linear model (without a constant term). We refer to this type of T-S fuzzy model as a T-S model with linear rule consequence, or simply a linear T-S model. As evident throughout this book, the appeal of a T-S model with linear rule consequence is that it renders itself naturally to Lyapunov based system analysis and design techniques [12, 15]. A commonly held view is that a T-S model with linear rule consequence has limited capability in representing a nonlinear system in comparison with an affine T-S model [9].

In Chapter 2, the PDC controller structure was introduced [11, 12]. This structure utilizes a fuzzy state feedback controller which mirrors the structure of the associated T-S model with linear rule consequence. As shown throughout this book, T-S models together with PDC controllers form a powerful framework for fuzzy control systems resulting in many successful applications [10, 13, 14].

278 T-S FUZZY MODEL AS UNIVERSAL APPROXIMATOR

In this chapter, we attempt to address the fundamental capabilities of T-S models with linear rule consequence and PDC controllers. To this end, two results are presented. The first result is that a linear Takagi-Sugeno fuzzy model can be a universal approximator of any smooth nonlinear control system. It has been known that smooth nonlinear dynamic systems can be approximated by T-S models with affine models as fuzzy rule consequences [4, 7]. However, most results on stability analysis and controller design of T-S models are based on T-S models with linear rule consequence. The question needed to be addressed is: "Is it possible to approximate any smooth nonlinear systems with Takagi-Sugeno models having linear models as rule consequences?" Reference [6] gave an answer to this question for the simple one-dimensional case. This chapter tries to answer this question for the n-dimensional nonlinear dynamic system by constructing T-S model to approximate the original nonlinear system. The answer is yes. That is, the original vector field plus its velocity can be accurately approximated if enough fuzzy rules are used.

The second result is that the PDC controller can be a universal approximator of any nonlinear state feedback controller. Therefore linear T-S models and PDC controllers together provide a universal framework for the modeling and control of nonlinear control systems.

In this chapter, $\mathbb{R}^n$ is used to denote the n-dimensional vector spaces of real vectors; C_n^m is used to represent the set of n-dimension functions whose mth derivative is continuous on the defined region; x_i stands for the ith component of vector x and $\|\ \|$ stands for the standard vector norm or matrix norm; $O(x)$ is the set of numbers y such that $|y/x| < M$, where M is a constant.; and $\sum_{j_1 j_2 \cdots j_n}$ is used to represent the summation with all the possible combinations of $j_1, j_2, \ldots, j_n$. We will often drop the x and just write h_i, but it should be kept in mind that h_i's are functions of the variable x.

14.1 APPROXIMATION OF NONLINEAR FUNCTIONS USING LINEAR T-S SYSTEMS

14.1.1 Linear T-S Fuzzy Systems

The main feature of linear Takagi-Sugeno fuzzy systems is to express the local properties of each fuzzy implication (rule) by a linear function. The overall fuzzy system is achieved by fuzzy "blending" of these linear functions. Specifically, the linear Takagi-Sugeno fuzzy system is of the following form:

Rule i

  IF x_1 is M_{i1} $\cdots$ and x_n is M_{in},

   THEN $y = a_i x$,

APPROXIMATION OF NONLINEAR FUNCTIONS USING LINEAR T-S SYSTEMS   279

where $x^T = [x_1, x_2, \ldots, x_n]$ are the function variables; $i = 1, 2, \ldots, r$ and r is the number of IF-THEN rules; and M_{ij} are fuzzy sets. The linear function $y = a_i x$ is the consequence of the ith IF-THEN rule, where $a_i \in \mathbb{R}^{1 \times n}$.

The possibility that the ith rule will fire is given by the product of all the membership functions associated with the ith rule:

$$h_i(x) = \prod_{j=1}^{n} M_{ij}(x_j).$$

We will assume that h_i's have already been normalized, that is, $h_i(x) \geq 0$ and $\sum_{i=1}^{r} h_i(x) = 1$. Then by using the center-of-gravity method for defuzzification, we can represent the T-S system as

$$y = \hat{f}(x) = \sum_{i=1}^{r} h_i(x) a_i x. \qquad (14.1)$$

The summation process associated with the center of gravity defuzzification in system (14.1) can also be viewed as an interpolation between the functions $a_i x$ based on the value of the parameter x.

14.1.2 Construction Procedure of T-S Fuzzy Systems

Suppose that the nonlinear function $f(x): \mathbb{R}^n \to \mathbb{R}$ is defined over the compact region $D \subset \mathbb{R}^n$ with the following assumptions:

1. $f(0) = 0$.
2. $f \in C_1^2$. Therefore, f, $\partial f/\partial x$, and $\partial^2 f/\partial x^2$ are continuous and therefore bounded over D.

Next, we will construct the T-S system $\hat{f}(x) = \sum_{i=1}^{r} h_i(x) a_i x$ to approximate $f(x)$. The objective is to make the approximation error $e(x) = f(x) - \hat{f}(x)$ and its derivative $\partial e/\partial x$ small for all $x \in D$.

Construction Procedures:

1. In region $D_0 = \{x \mid |x_i| < \epsilon_0\}$ where ϵ_0 is a chosen positive number, choose $a_0 = \partial f/\partial x|_{x=0}$.
2. Define the projection operator $P|_x$ mapping $\mathbb{R}^n$ to $n-1$ dimensional subspace $\mathbb{R}^n/x$ as

$$P|_x y = y - \frac{\langle y, x \rangle}{\|x\|^2} x.$$

In region $D \setminus D_0$, choose $x_{j_1 j_2 \ldots j_n}$ as $[j_1 \epsilon \; j_2 \epsilon \ldots j_n \epsilon]^T$, where ϵ is a positive number and j_i are integers. Build the linear model $a_{j_1 j_2 \ldots j_n}$ as the solution of the following linear equations:

$$a_{j_1 j_2 \ldots j_n} x_{j_1 j_2 \ldots j_n} = f(x_{j_1 j_2 \ldots j_n}), \tag{14.2}$$

$$a_{j_1 j_2 \ldots j_n} P|_{x_{j_1 j_2 \ldots j_n}} = \frac{\partial f}{\partial x}\bigg|_{x_{j_1 j_2 \ldots j_n}} P|_{x_{j_1 j_2 \ldots j_n}}. \tag{14.3}$$

For fixed $x_{j_1 j_2 \ldots j_n}$, (14.2)–(14.3) are n linear equations with the component of $a_{j_1 j_2 \ldots j_n}$ as the variables. Equation (14.2) implies that f and $\hat{f}$ have the same value at point $x_{j_1 j_2 \ldots j_n}$. Equation (14.3) implies that $a_{j_1 j_2 \ldots j_n}$ agree with $\partial f / \partial x$ in the $n-1$ dimensional space $\mathbb{R}^n / x_{j_1 j_2 \ldots j_n}$. They are always solvable since x and P are independent of each other, that is, the matrices $[x_{j_1 j_2 \ldots j_n} P|x_{j_1 j_2 \ldots j_n}]$ are always invertible.

3. Choose the fuzzy rules as following:

Rule 0

IF x_1 is about $0 \cdots$ and x_n is about 0,

THEN $\hat{f}(x) = a_0 x$.

Rule $j_1 j_2 \ldots j_n$

IF x_1 is about $j_1 \epsilon \cdots$ and x_n is about $j_n \epsilon$,

THEN $\hat{f}(x) = a_{j_1 j_2 \ldots j_n} x$.

For Rule 0, choose the possibility of firing $h_0(x)$ as 1 inside D_0 and 0 outside. The possibility of firing for the $(j_1 j_2 \ldots j_n)$th rule is given by the product of all the membership functions associated with the $(j_1 j_2 \ldots j_n)$th rule:

$$h_{j_1 j_2 \ldots j_n}(x) = \prod_{i=1}^{n} M_{j_i}(x_i), \tag{14.4}$$

where the membership function for x_i is given as

$$M_{j_i}(x_i) = \begin{cases} 1 - \dfrac{|x_i - j_i \epsilon|}{\epsilon}, & |x_i - j_i \epsilon| < \epsilon, \\ 0, & \text{elsewhere}. \end{cases} \tag{14.5}$$

It is noted that $h_{j_1 j_2 \ldots j_n}(x)$ have already been normalized, that is, $h_{j_1 j_2 \ldots j_n}(x) \geq 0$ and $\sum_{j_1 j_2 \ldots j_n} h_{j_1 j_2 \ldots j_n}(x) = 1$.

Therefore, we can write $\hat{f}(x)$ as

$$\hat{f}(x) = h_0 a_0 x + \sum_{j_1 j_2 \ldots j_n} h_{j_1 j_2 \ldots j_n} a_{j_1 j_2 \ldots j_n} x. \tag{14.6}$$

APPROXIMATION OF NONLINEAR FUNCTIONS USING LINEAR T-S SYSTEMS

Remark 42 It should be pointed out that the specific membership function constructed above is only needed when we want to approximate both the nonlinear function and its derivative. There will be much more freedom if we only want to approximate the function itself.

14.1.3 Analysis of Approximation

In this subsection, we will prove the fact that any smooth nonlinear function satisfying the assumptions outlined in the previous subsection can be approximated, to any degree of accuracy, using the linear T-S fuzzy systems constructed above. This fact forms the foundation of the two statements in this chapter.

First, we divide region $D \setminus D_0$ into many small regions:

$$D_{j_1 j_2 \ldots j_n} = \{x \mid x \in D, j_i \epsilon \leq x_i \leq (j_i + 1)\epsilon \; \forall i\}.$$

In the following discussions, we concentrate on one such region ($D_{j_1 j_2 \ldots j_n}$), which is shown in Figure 14.1, by assuming that $x \in D_{j_1 j_2, \ldots, j_n}$. From the construction procedure above, we know that only the fuzzy rules centered at the vertices of $D_{j_1 j_2 \ldots j_n}$ can be activated at x. That is, $h_{l_1 l_2 \ldots l_n}(x) \neq 0$ only if $x_{l_1 l_2 \ldots l_n}$ is one of the vertex points of $D_{j_1 j_2 \ldots j_n}$.

Consider $e(x)$, the approximation error between $f(x)$ and $\hat{f}(x)$:

$$\|e(x)\| = \left\| f(x) - \sum_{j_1 j_2 \ldots j_n} h_{j_1 j_2 \ldots j_n}(x) a_{j_1 j_2 \ldots j_n} x \right\|$$

$$= \left\| f(x) - \sum_{j_1 j_2 \ldots j_n} h_{j_1 j_2 \ldots j_n}(x) a_{j_1 j_2 \ldots j_n} x_{j_1 j_2 \ldots j_n} \right.$$

$$\left. - \sum_{j_1 j_2 \ldots j_n} h_{j_1 j_2 \ldots j_n}(x) a_{j_1 j_2 \ldots j_n} (x - x_{j_1 j_2 \ldots j_n}) \right\|$$

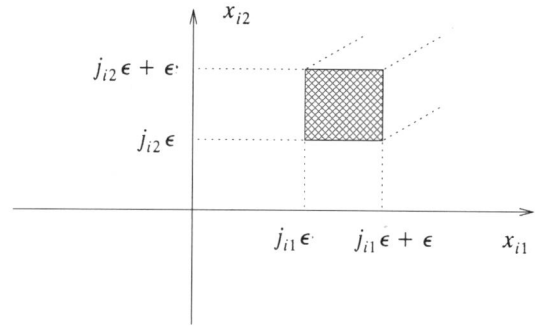

Fig. 14.1 Projection of $D_{j_1 j_2 \ldots j_n}$ on $x_{i_1} x_{i_2}$ plane.

$$= \left\| f(x) - \sum_{j_1 j_2 \cdots j_n} h_{j_1 j_2 \cdots j_n}(x) f(x_{j_1 j_2 \cdots j_n}) \right.$$

$$\left. - \sum_{j_1 j_2 \cdots j_n} h_{j_1 j_2 \cdots j_n}(x) a_{j_1 j_2 \cdots j_n}(x - x_{j_1 j_2 \cdots j_n}) \right\|$$

$$\leq \sum_{j_1 j_2 \cdots j_n} h_{j_1 j_2 \cdots j_n}(x) \| f(x) - f(x_{j_1 j_2 \cdots j_n}) \|$$

$$+ \sum_{j_1 j_2 \cdots j_n} h_{j_1 j_2 \cdots j_n}(x) \| a_{j_1 j_2 \cdots j_n}(x - x_{j_1 j_2 \cdots j_n}) \|$$

$$\leq \max_{l_1 l_2 \cdots l_n} \| f(x) - f(x_{l_1 l_2 \cdots l_n}) \| + \max_{l_1 l_2 \cdots l_n} \| a_{l_1 l_2 \cdots l_n}(x - x_{l_1 l_2 \cdots l_n}) \|.$$

Note that

$$a_{l_1 l_2 \cdots l_n}(x - x_{l_1 l_2 \cdots l_n})$$

$$= \frac{\partial f}{\partial x}\bigg|_{x_{l_1 l_2 \cdots l_n}} \left((x - x_{l_1 l_2 \cdots l_n}) - \frac{\langle (x - x_{l_1 l_2 \cdots l_n}), x_{l_1 l_2 \cdots l_n} \rangle}{\| x_{l_1 l_2 \cdots l_n} \|^2} x_{l_1 l_2 \cdots l_n} \right)$$

$$+ \frac{\langle (x - x_{l_1 l_2 \cdots l_n}), x_{l_1 l_2 \cdots l_n} \rangle}{\| x_{l_1 l_2 \cdots l_n} \|^2} f(x_{l_1 l_2 \cdots l_n}).$$

Since $x \in D_{j_1 j_2 \cdots j_n}$, the distance between x and any vertex point of $D_{j_1 j_2 \cdots j_n}$ is less than $\sqrt{n}\, \epsilon$, that is, $|x - x_{l_1 l_2 \cdots l_n}| \leq \sqrt{n}\, \epsilon$, we can make $e(x)$ arbitrarily small by just reducing ϵ.

Now consider the approximation of $\partial f / \partial x$. Before doing that, three facts for the membership functions are presented.

LEMMA 6 *Define*

$$\frac{\partial h_{j_1 j_2 \cdots j_n}}{\partial x}\bigg|_x = \left[\frac{\partial h_{j_1 j_2 \cdots j_n}}{\partial x_1}\bigg|_x \quad \frac{\partial h_{j_1 j_2 \cdots j_n}}{\partial x_2}\bigg|_x \quad \cdots \quad \frac{\partial h_{j_1 j_2, \cdots, j_n}}{\partial x_n}\bigg|_x \right]$$

where it exists; then

$$\sum_{j_1 j_2 \cdots j_n} \frac{\partial h_{j_1 j_2 \cdots j_n}}{\partial x}\bigg|_x = 0. \tag{14.7}$$

Proof. Take the derivatives of $\sum_{j_1 j_2 \cdots j_n} h_{j_1 j_2 \cdots j_n}$. Since $\sum_{j_1 j_2 \cdots j_n} h_{j_1 j_2 \cdots j_n} = 1$, its derivatives with respect to x_i will be 0. (Q.E.D.)

APPROXIMATION OF NONLINEAR FUNCTIONS USING LINEAR T-S SYSTEMS   283

LEMMA 7

$$\sum_{j_1 j_2 \ldots j_n} (x - x_{j_1 j_2 \ldots j_n}) \left. \frac{\partial h_{j_1 j_2 \ldots j_n}}{\partial x} \right|_x = -I.$$

Proof. For vertex point $x_{l_1 l_2 \ldots l_n} \in D_{j_1 j_2 \ldots j_n}$, define $\bar{l}_i = 2j_i + 1 - l_i$; then it can be proven that

$$(x - x_{l_1 l_2 \ldots \bar{i}_i \ldots l_n})_i \left. \frac{\partial h_{l_1 l_2 \ldots \bar{i}_i \ldots l_n}}{\partial x_i} \right|_x + (x - x_{l_1 l_2 \ldots l_n})_i \left. \frac{\partial h_{l_1 l_2 \ldots l_n}}{\partial x_i} \right|_x$$

$$= -\left(h_{l_1 l_2 \ldots l_n} + h_{l_1 l_2 \ldots \bar{i}_i \ldots l_n} \right),$$

$$(x - x_{l_1 l_2 \ldots \bar{i}_i \ldots l_n})_i \left. \frac{\partial h_{l_1 l_2 \ldots \bar{i}_i \ldots l_n}}{\partial x_j} \right|_x + (x - x_{l_1 l_2 \ldots l_n})_i \left. \frac{\partial h_{l_1 l_2 \ldots l_n}}{\partial x_j} \right|_x = 0,$$

$$i \neq j.$$

Summing up these equations for all the rules $l_1 l_2 \ldots l_n$ that are effective in region $D_{j_1 j_2 \ldots j_n}$, the fact is proved. (Q.E.D.)

LEMMA 8 *Define a_x as the solution of the following linear equations*:

$$a_x x = f(x), \tag{14.8}$$

$$a_x P|_x = \left. \frac{\partial f}{\partial x} \right|_x P. \tag{14.9}$$

Then $\forall \delta, \exists \epsilon$ such that $\|a_x - a_{j_1 j_2 \ldots j_n}\| \leq \delta$ if $\|x - x_{j_1 j_2 \ldots j_n}\| \leq \epsilon \ll 1$.

Proof. Since a_x is the solution of the linear equations (14.8) and (14.9) and all the parameters of the equations ($f(x)$, $\partial f/\partial x$, and $P|_x$) are continuous functions of x, a_x will depend continuously on x. Consequently, $\|a - a_{j_1 j_2 \ldots j_n}\|$ can be made arbitrarily small by choosing a small enough value for ϵ.
(Q.E.D.)

Now consider $\partial e/\partial x$, the difference between $\partial f/\partial x$ and $\partial \bar{f}/\partial x$.

$$\left\| \frac{\partial e}{\partial x} \right\| = \left\| \left. \frac{\partial f}{\partial x} \right|_x - \frac{\partial (\sum_{j_1 j_2 \ldots j_n} h_{j_1 j_2 \ldots j_n} a_{j_1 j_2 \ldots j_n} x)}{\partial x} \right\|$$

$$= \left\| \left. \frac{\partial f}{\partial x} \right|_x - \sum_{j_1 j_2 \ldots j_n} a_{j_1 j_2 \ldots j_n} x \left. \frac{\partial h_{j_1 j_2 \ldots j_n}}{\partial x} \right|_x \right.$$

$$\left. - \sum_{j_1 j_2 \ldots j_n} h_{j_1 j_2 \ldots j_n}(x) a_{j_1 j_2 \ldots j_n} \right\|$$

$$= \left\| \frac{\partial f}{\partial x}\bigg|_x - \sum_{j_1 j_2 \cdots j_n} a_{j_1 j_2 \cdots j_n}(x - x_{j_1 j_2 \cdots j_n}) \frac{\partial h_{j_1 j_2 \cdots j_n}}{\partial x}\bigg|_x \right.$$

$$- \sum_{j_1 j_2 \cdots j_n} a_{j_1 j_2 \cdots j_n} x_{j_1 j_2 \cdots j_n} \frac{\partial h_{j_1 j_2 \cdots j_n}}{\partial x}\bigg|_x$$

$$\left. - \sum_{j_1 j_2 \cdots j_n} h_{j_1 j_2 \cdots j_n}(x) a_{j_1 j_2 \cdots j_n} \right\|$$

$$= \left\| \frac{\partial f}{\partial x}\bigg|_x - \sum_{j_1 j_2 \cdots j_n} a_{j_1 j_2 \cdots j_n}(x - x_{j_1 j_2 \cdots j_n}) \frac{\partial h_{j_1 j_2 \cdots j_n}}{\partial x}\bigg|_x \right.$$

$$\left. - \sum_{j_1 j_2 \cdots j_n} f(x_{j_1 j_2 \cdots j_n}) \frac{\partial h_{j_1 j_2 \cdots j_n}}{\partial x}\bigg|_x - \sum_{j_1 j_2 \cdots j_n} h_{j_1 j_2 \cdots j_n}(x) a_{j_1 j_2 \cdots j_n} \right\|$$

$$= \left\| \frac{\partial f}{\partial x}\bigg|_x - \sum_{j_1 j_2 \cdots j_n} a_{j_1 j_2 \cdots j_n}(x - x_{j_1 j_2 \cdots j_n}) \frac{\partial h_{j_1 j_2 \cdots j_n}}{\partial x}\bigg|_x \right.$$

$$- \sum_{j_1 j_2 \cdots j_n} \left(f(x) + \frac{\partial f}{\partial x}\bigg|_x (x_{j_1 j_2 \cdots j_n} - x) + O(\epsilon^2) \right) \frac{\partial h_{j_1 j_2 \cdots j_n}}{\partial x}\bigg|_x$$

$$\left. - \sum_{j_1 j_2 \cdots j_n} h_{j_1 j_2 \cdots j_n}(x) a_{j_1 j_2 \cdots j_n} \right\|$$

$$= \left\| \frac{\partial f}{\partial x}\bigg|_x - \sum_{j_1 j_2 \cdots j_n} a_{j_1 j_2 \cdots j_n}(x - x_{j_1 j_2 \cdots j_n}) \frac{\partial h_{j_1 j_2 \cdots j_n}}{\partial x}\bigg|_x \right.$$

$$- \sum_{j_1 j_2 \cdots j_n} \frac{\partial f}{\partial x}\bigg|_x (x_{j_1 j_2 \cdots j_n} - x) \frac{\partial h_{j_1 j_2 \cdots j_n}}{\partial x}\bigg|_x$$

$$\left. - \sum_{j_1 j_2 \cdots j_n} h_{j_1 j_2 \cdots j_n}(x) a_{j_1 j_2 \cdots j_n} \right\| + O(\epsilon) \quad \text{(from Fact 6)}$$

$$= \left\| - \sum_{j_1 j_2 \cdots j_n} a_{j_1 j_2 \cdots j_n}(x - x_{j_1 j_2 \cdots j_n}) \frac{\partial h_{j_1 j_2 \cdots j_n}}{\partial x}\bigg|_x \right.$$

$$\left. - \sum_{j_1 j_2 \cdots j_n} h_{j_1 j_2 \cdots j_n}(x) a_{j_1 j_2 \cdots j_n} \right\| + O(\epsilon) \quad \text{(from Fact 7)}$$

$$= \left\| \sum_{j_1 j_2 \cdots j_n} a_{j_1 j_2 \cdots j_n}(x - x_{j_1 j_2 \cdots j_n}) \frac{\partial h_{j_1 j_2 \cdots j_n}}{\partial x}\bigg|_x + a_x \right.$$

$$\left. + \sum_{j_1 j_2 \cdots j_n} h_{j_1 j_2 \cdots j_n} a_{j_1 j_2 \cdots j_n} - \sum_{j_1 j_2 \cdots j_n} h_{j_1 j_2 \cdots j_n}(x) a_x \right\| + O(\epsilon)$$

$$\leq \left\| \sum_{j_1 j_2 \ldots j_n} (a_{j_1 j_2 \ldots j_n} - a_x)(x - x_{j_1 j_2 \ldots j_n}) \frac{\partial h_{j_1 j_2 \ldots j_n}}{\partial x} \bigg|_x \right\|$$

$$+ \left\| \sum_{j_1 j_2 \ldots j_n} h_{j_1 j_2 \ldots j_n}(a_{j_1 j_2 \ldots j_n} - a_x) \right\| + O(\epsilon) \quad \text{(from Fact 7)}.$$

From Fact 8, it is known that $\partial e/\partial x$ can be made arbitrarily small by reducing ϵ.

Next consider region D_0. In region D_0, it is known from Taylor series that $e(x)$ and $\partial e/\partial x$ can also be made arbitrarily small by reducing ϵ_0. Therefore, we have the following theorem by summarizing the results above:

THEOREM 56 *For any smooth nonlinear function $f(x): \mathbb{R}^n \to \mathbb{R}^1$ defined on a compact region, satisfying $f(0) = 0$ and $f \in C_n^2$, both the function and its derivatives can be approximated, to any degree of accuracy, by linear T-S fuzzy systems.*

Remark 43 It may be argued that the condition $f(0) = 0$ is too restrictive. However, in the case of $f(0) \neq 0$, we argue that f can still be approximated by a linear T-S model through a simple coordination transformation, that is, the function f is now represented by a linear T-S model in the new coordinate system. A coordination transformation might puzzle the mind of a purist of function approximation. However, for control system analysis and design, which is the sole focus of this book, this is not a problem at all. It is well known that for the stability analysis and design of nonlinear control systems, it can be assumed without loss of generality that the origin is an equilibrium point of the system.

$$f(x_1, x_2) = 8x_1 + 10x_2 \sin(4x_1) + x_1^3 - 4x_1 x_2$$

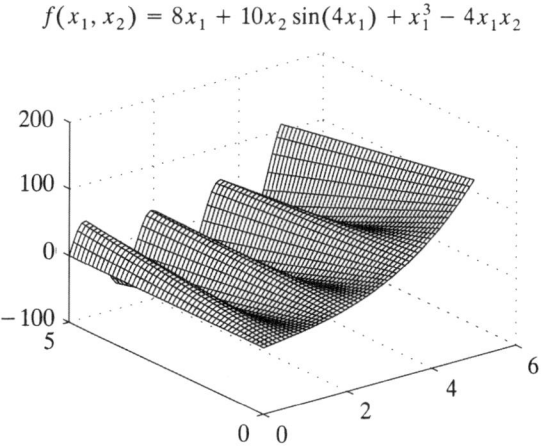

Fig. 14.2 Nonlinear function $f(x_1, x_2) = 8x_1 + 10x_2 \sin(4x_1) + x_1^3 - 4x_1 x_2$.

Remark 44 It may be argued that the membership function is not continuous on the boundary between D_0 and $D_{j_1 j_2 \ldots j_n}$. To overcome the discontinuity, some bumper functions can be included to smooth the membership function without affecting the approximation accuracy [16].

14.1.4 Example

An example is given in this subsection for illustration. Consider the approximation of a two-dimensional nonlinear function $f(x_1, x_2) = 8x_1 + 10x_2 \sin(4x_1) + x_1^3 - 4x_1 x_2$ as shown in Figure 14.2. The constructed T-S fuzzy model is shown in Figure 14.3. A 25×40 grid is used. The maximum approximation error is 1.38. We also plot the approximation error in Figure 14.4. It should be pointed out that the approximation error could be further reduced by using more fuzzy rules.

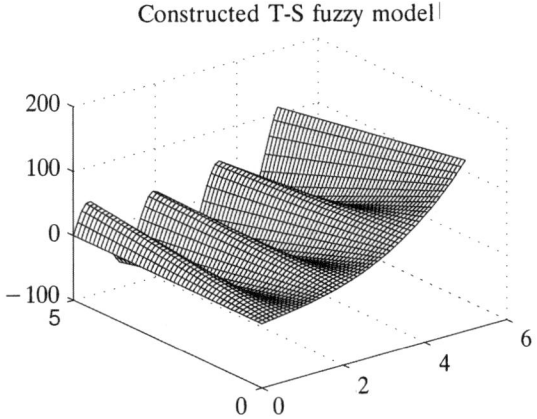

Fig. 14.3 Constructed T-S fuzzy model.

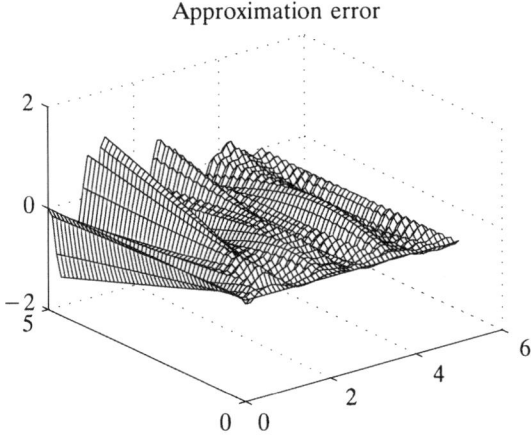

Fig. 14.4 Approximation error of nonlinear function.

14.2 APPLICATIONS TO MODELING AND CONTROL OF NONLINEAR SYSTEMS

14.2.1 Approximation of Nonlinear Dynamic Systems Using Linear Takagi-Sugeno Fuzzy Models

The following dynamic linear Takagi-Sugeno fuzzy model is used to describe dynamic systems:

Rule i

IF $x_1(t)$ is $M_{i1}, \cdots$ and $x_n(t)$ is M_{in},

THEN $\dot{x}(t) = A_i x(t)$,

where $x^T(t) = [x_1(t), x_2(t), \ldots, x_n(t)]$ are the system states; $i = 1, 2, \ldots, r$ and r is the number of IF-THEN rules; M_{ij} are fuzzy sets; and $\dot{x}(t) = A_i x(t)$ are the consequences of the ith IF-THEN rule.

By using the center-of-gravity method for defuzzification, we can represent the T-S model as

$$\dot{x} = \hat{f}(x) = \sum_{i=1}^{r} h_i(x) A_i x, \qquad (14.10)$$

where $h_i(x)$ is the possibility for the ith rule to fire.

Consider the nonlinear system

$$\dot{x} = f(x), \qquad (14.11)$$

where $f(x)$ is a vector field defined over the compact region $D \subset \mathbb{R}^n$ with the following assumptions:

1. $f(0) = 0$, that is, the origin is an equilibrium point.
2. $f \in C_n^2$. Therefore, f, $\partial f/\partial x$, and $\partial^2 f/\partial x^2$ are continuous and bounded over D.

Suppose $f(x)$ can be written as $[f_1(x) \ldots f_n(x)]^T$. What we mean by approximation is finding a T-S fuzzy model $\hat{f}(x) = [\hat{f}_1(x) \ldots \hat{f}_n(x)]^T$ such that $\|f(x) - \hat{f}(x)\|$ is small. Since $\|f(x) - \hat{f}(x)\|$ is small if and only if each of its components (which are nonlinear functions) are small, then by applying Theorem 56, we obtain the following corollary:

COROLLARY 7 *For any smooth nonlinear system* (14.11) *satisfying the assumptions stated above, it can be approximated, to any degree of accuracy, by a T-S model* (14.10).

Similarly, a smooth nonlinear control system $\dot{x} = f(x) + g(x)u$ can also be approximated using a T-S fuzzy model $\dot{x} = \sum_{i=1}^{r} h_i(x)(A_i x + B_i u)$. By treating u as an extraneous system state, we can also approximate the smooth nonlinear control system $\dot{x} = f(x, u)$ by a T-S fuzzy model $\dot{x} = \sum_{i=1}^{r} \hat{h}_i(x, u)(A_i x + B_i u)$. In this case, the fuzzy rule is of the following form:

Rule i

    IF $x_1(t)$ is $M_{i1}, \ldots, x_n(t)$ is M_{in}, $u_1(t)$ is $N_{i1}, \ldots,$ and $u_m(t)$ is N_{im},

    THEN $\dot{x}(t) = A_i x(t) + B_i u(t)$,

where $x^T(t) = [x_1(t), x_2(t), \ldots, x_n(t)]$ are the system states and $u^T(t) = [u_1(t), u_2(t), \ldots, u_m(t)]$ are the system inputs; $i = 1, 2, \ldots, r$ and r is the number of IF-THEN rules; M_{ij}, N_{ij} are fuzzy sets and $\dot{x}(t) = A_i x(t) + B_i u(t)$ is the consequence of the ith IF-THEN rule; and

$$\hat{h}_i(x, u) = \prod_{j=1}^{n} M_{ij}(x_i(t)) \prod_{k=1}^{m} N_{ik}(u_k(t))$$

is the possibility for the ith rule to fire.

14.2.2 Approximation of Nonlinear State Feedback Controller Using PDC Controller

In this chapter, we consider the special form of the fuzzy controller introduced in [12] where it was termed parallel distributed compensation (PDC). The PDC controller structure consists of the following fuzzy rules:

Rule j

    IF $x_1(t)$ is $M_{j1} \cdots$ and $x_n(t)$ is M_{jn}

    THEN $u(t) = K_j x(t)$,

where $j = 1, 2, \ldots, s$. The output of the PDC controller is

$$u = \sum_{j=1}^{s} h_j(x) K_j x. \tag{14.12}$$

Following a similar argument as in the above subsection, we obtain the following theorem:

THEOREM 57 *For a smooth nonlinear state feedback controller, $u = K(x)$ defined over a compact region ($u(0) = 0$) can be approximated, to any degree of accuracy, by a PDC controller* (14.12).

BIBLIOGRAPHY

1. H. O. Wang, J. Li, D. Niemann, and K. Tanaka, "T-S Fuzzy Model with Linear Rule Consequence and PDC Controller: A Universal Framework for Nonlinear Control Systems," *Proc. FUZZ-IEEE'2000*, San Antonio, TX, 2000, pp. 549–554.
2. T. Takagi and M. Sugeno, "Fuzzy Identification of Systems and Its Applications to Modeling and Control," *IEEE Trans. Syst. Man, Cybernet.*, Vol. 15, pp. 116–132 (1985).
3. J. J. Buckley, "Universal Fuzzy Controllers," *Automatica*, Vol. 28, pp. 1245–1248 (1992).
4. S. G. Cao, N. W. Rees, and G. Feng, "Fuzzy Control of Nonlinear Continuous-Time Systems," in *Proc. 35th IEEE Conf. Decision and Control*, Kobe, Japan, 1996, pp. 592–597.
5. J. L. Castro, "Fuzzy Logic Controllers are Universal Approximators," *IEEE Trans. Syst. Man Cybernet.*, Vol. 25, No. 4, pp. 629–635 (1998).
6. C. Fantuzzi and R. Rovatti, "On the Approximation Capabilities of the Homogeneous Takagi-Sugeno Model," *Proc. FUZZ-IEEE'96*, 1996, pp. 1067–1072.
7. H. Ying, "Sufficient Conditions on Uniform Approximation of Multivariate Functions by General Takagi-Sugeno Fuzzy Systems with Linear Rule Consequence," *IEEE Trans. Syst. Man Cybern.* Vol. 28, No. 4, pp. 515–521 (1998).
8. X. J. et and M. G. Singh, "Approximation Theory of Fuzzy Systems—SISO Case," *IEEE Trans. Fuzzy Syst.*, Vol. 2, pp. 162–176 (1994).
9. G. Kang, W. Lee, and M. Sugeno, "Design of TSK Fuzzy Controller Based on TSK Fuzzy Model Using Pole Placement," *Proc. FUZZ-IEEE'98*, 1998, pp. 246–251.
10. S. K. Hong and R. Langari, "Synthesis of an LMI-Based Fuzzy Control System with Guaranteed Optimal H_∞ performance," *Proc. FUZZ-IEEE'98*, 1998, pp. 422–427.
11. H. O. Wang, K. Tanaka, and M. F. Griffin, "Parallel Distributed Compensation of Nonlinear Systems by Takagi-Sugeno Fuzzy Model," in *Proc. FUZZ-IEEE/IFES'95*, 1995, pp. 531–538.
12. H. O. Wang, K. Tanaka, and M. Griffin, "An Approach to Fuzzy Control of Nonlinear Systems: Stability and Design Issues," *IEEE Trans. Fuzzy Syst.*, Vol. 4, No. 1, pp. 14–23 (1996).
13. J. Li, D. Niemann, and H. O. Wang, "Robust Tracking for High-Rise/High-Speed Elevators," *Proc. 1998 American Control Conference*, 1998, pp. 3445–3449.
14. T. Tanaka and M. Sano, "A Robust Stabilization Problem of Fuzzy Control Systems and Its Applications to Backing Up Control of a Truck-Trailer," *IEEE Trans. Fuzzy Syst.*, Vol. 2, No. 3, pp. 119–134 (1994).
15. J. Zhao, V. Wertz, and R. Gorez, " Fuzzy Gain Scheduling Controllers Based on Fuzzy Models," *Proc. Fuzzy-IEEE'96*, 1996.
16. M. Spivak, *Comprehensive Introduction to Differential Geometry*, Vol. 1, Addison-Wesley, Reading, MA, 1979.
17. R. R. Yager and P. F. Dimitar, *Essential on Fuzzy Modeling and Control*, Wiley, New York, 1994.

CHAPTER 15

FUZZY CONTROL OF NONLINEAR TIME-DELAY SYSTEMS

In this chapter, a class of nonlinear time-delay systems based on the Takagi-Sugeno (T-S) fuzzy model is defined [1]. We investigate the delay-independent stability of this model. A model-based fuzzy stabilization design utilizing the concept of parallel distributed compensation (PDC) is employed. The main idea of the controller design is to derive each control rule to compensate each rule of a fuzzy system. Moreover, the problem of H_∞ control of this class of nonlinear time-delay systems is considered. The associated control synthesis problems are formulated as linear matrix inequality (LMI) problems.

In the original T-S fuzzy model formulation, there is no delay in the control and state. However, time delays often occur in many dynamical systems such as biological systems, chemical systems, metallurgical processing systems, and network systems. Their existence is frequently a cause of instability and poor performance. The study of stability and stabilization for linear time-delay systems has received considerable attention [2–6]. But these efforts were mainly restricted to linear time-delay systems. Thus, it is important to extend the stability and stabilization issues to nonlinear time-delay systems. In this chapter, a particular class of nonlinear time-delay systems is introduced based on the Tagaki-Sugeno fuzzy model. This kind of nonlinear system is represented by a set of linear time-delay systems. We will call this a T-S model with time delays (T-SMTD). In the literature, the problem of stability and stabilization of time-delay systems has been dealt with a number of different ways. There are some results that are independent of the size of the time delays in [2–4], and the stability is satisfied for any value of the time delays. There are also some delay-dependent results, in which the stability is guaranteed up to some maximum value for the time delays [5, 6]. This chapter is concerned with the problems of delay-independent stability and stabilization of T-S fuzzy models with time delays. Particularly, we will employ the concept of parallel distributed compensation to study these problems. Several new results concerned with the stability and stabilization of T-SMTD are derived. Also, a sufficient condition for the H_∞ control of this model is given. All the synthesis problems are formulated as LMIs, thus they are numerically efficient.

Throughout the chapter, the notation $M > 0$ will mean that M is a positive definite symmetric matrix. The symbol p will be used for premise variables as in Chapters 12 and 13.

15.1 T-S FUZZY MODEL WITH DELAYS AND STABILITY CONDITIONS

15.1.1 T-S Fuzzy Model with Delays

To begin with, we represent a given nonlinear plant by the Takagi-Sugeno fuzzy model. Then, we will define a new kind of model, the Takagi-Sugeno fuzzy model with time delays. The main feature of the T-S fuzzy model is to express the joint dynamics of each fuzzy implication (rule) by a linear system model. Specifically, the Takagi-Sugeno fuzzy system is described by fuzzy IF-THEN rules, which locally represent linear input-output relations of a system. The fuzzy system is of the following form:

Dynamic Part: Rule i

  IF $p_1(t)$ is $M_{i1}, \ldots,$ and $p_l(t)$ is M_{il},

  THEN

$$\dot{x}(t) = A_i x(t) + B_i u(t), \qquad i = 1, 2, \ldots, r. \qquad (15.1)$$

Output Part: Rule i

  IF $p_1(t)$ is $M_{i1}, \ldots,$ and $p_l(t)$ is M_{il},

  THEN

$$y(t) = C_i x(t).$$

Here, $x(t)$, $u(t)$, $y(t)$, and $p(t)$ respectively denote the state, input, output, and parameter vectors. The jth component of $p(t)$ is denoted by $p_j(t)$, and the fuzzy membership function associated with the ith rule and jth parameter component is denoted by M_{ij}. Each $p_j(t)$ is a measurable time-varying quantity. In general, these parameters may be functions of the state variables, external disturbances, and/or time.

There are two functions of $p(t)$ associated with each rule. The first function is called the truth value. The truth value for the ith rule is defined by the equation

$$\omega_i(p(t)) = \prod_{j=1}^{l} M_{ij}(p_j(t)).$$

Throughout this chapter, we will assume that each ω_i is a nonnegative function and that the truth value of at least one rule is always nonzero. The second function is called the firing probability. The firing probability for the ith rule is defined by the equation

$$h_i(p(t)) = \frac{\omega_i(p(t))}{\sum_{i=1}^{r} \omega_i(p(t))},$$

where r denotes the number of rules in the rule base. Under the previously stated assumptions, this is always a well-defined function taking values between 0 and 1, and the sum of all the firing probabilities is identically equal to 1.

Now, we introduce time delays into the above T-S fuzzy model. Here, we assume there are time delays in both the state and control of the dynamic part. Then, the i rule of the dynamic part of T-S fuzzy model becomes:

Rule i

  IF $p_1(t)$ is $M_{i1}, \ldots,$ and $p_l(t)$ is M_{il}

  THEN

$$\dot{x}(t) = A_{i0}x(t) + A_{id}x(t-\tau_1) + B_{i0}u(t) + B_{id}u(t-\tau_2),$$

$$i = 1, 2, \ldots, r, \quad (15.2)$$

where $0 \le \tau_1 < \infty$ and $0 \le \tau_2 < \infty$ are the size of the time delays. The initial condition is $x(t) = 0$, where $t < 0$.

We call this model the T-S model with time delays (T-SMTD). In the following we will investigate the stability and design issues, such as delay-independent stabilization and H_∞ control, of this system.

The dynamics described by the T-SMTD evolve according to the system of equations

$$\dot{x}(t) = \sum_{i=1}^{r} h_i(p)\{A_{i0}x(t) + A_{id}x(t - \tau_1)$$
$$+ B_{i0}u(t) + B_{id}u(t - \tau_2)\}, \quad (15.3)$$

$$y(t) = \sum_{i=1}^{r} h_i(p)C_i x(t).$$

The open-loop system is of the form

$$\dot{x}(t) = \sum_{i=1}^{r} h_i(p)\{A_{i0}x(t) + A_{id}x(t - \tau_1)\}. \quad (15.4)$$

Remark 45 Our proposed model description can also be viewed as parameter-dependent interpolation between linear models; however, the exact classification of the resultant system depends on the nature of the parameters. For example, if each p_i is a known function of time, then the T-S model describes a linear time-varying system. If, on the other hand, each p_i is a function of the state variables, then the T-S model describes an autonomous nonlinear system.

15.1.2 Stability Analysis via Lyapunov Approach

A sufficient delay-independent stability condition for the open-loop system (15.4) is given as follows:

THEOREM 58 *The open-loop T-S fuzzy system with time delays (15.4) is globally asymptotically stable if there exist two common positive definite matrices P and R such that*

$$PA_{i0} + A_{i0}^T P + PA_{id}R^{-1}A_{id}^T P + R < 0, \quad i = 1, 2, \ldots, r, \quad (15.5)$$

that is, two common matrices P and R have to exist for all subsystems.

Proof. For the open-loop system (15.4), we define a Lyapunov function as the following:

$$V(x) = x(t)^T Px(t) + \int_{t-\tau_1}^{t} x(s)^T Rx(s)\,ds. \quad (15.6)$$

The derivate of $V(x)$ along the open-loop system (15.4) is

$$\dot{V}(x) = \sum_{i=1}^{r} h_i(p) x(t)^T \left[PA_{i0} + A_{i0}^T P \right] x(t)$$
$$+ 2 \sum_{i=1}^{r} h_i(p) x(t)^T PA_{id} x(t - \tau_1)$$
$$+ x(t)^T Rx(t) - x(t - \tau_1)^T Rx(t - \tau_1). \quad (15.7)$$

Using the fact that

$$2x(t)^T PA_{id} x(t - \tau_1) \leq x(t)^T PA_{id} R^{-1} A_{id}^T Px(t)$$
$$+ x(t - \tau_1)^T Rx(t - \tau_1), \quad (15.8)$$

we have

$$\dot{V}(x) \leq \sum_{i=1}^{r} h_i(p) x(t)^T$$
$$\times \left\{ PA_{i0} + A_{i0}^T P + PA_{id} R^{-1} A_{id}^T P + R \right\} x(t) < 0, \quad \forall x \neq 0. \quad (15.9)$$

(Q.E.D.)

Remark 46 The system (15.4) is also said to be quadratically stable and the function $V(x)$ is called a quadratic Lyapunov function. Theorem 58 thus presents a sufficient condition for quadratic stability of the open-loop system (15.4).

15.1.3 Parallel Distributed Compensation Control

In [7], Wang et al. utilized the concept of parallel distributed compensation (PDC) to design fuzzy controllers to stabilize fuzzy system (15.1). The idea is to design a compensator for each rule of the fuzzy model. The resulting overall fuzzy controller, which is nonlinear in general, is a fuzzy blending of each individual linear controller. The fuzzy controller shares the same fuzzy sets with the fuzzy system (15.1). Here, we will apply the same controller structure to the T-SMTD, so the ith control rule is as follows:

Control Rule i

IF $p_1(t)$ is M_{i1} and, ..., and $p_l(t)$ is M_{il},

THEN $u(t) = -F_i x(t), \quad i = 1, \ldots, r.$

296 FUZZY CONTROL OF NONLINEAR TIME-DELAY SYSTEMS

The output of the PDC controller is determined by the summation

$$u(t) = -\sum_{i=1}^{r} h_i(p) F_i x(t). \qquad (15.10)$$

Note that the controller (15.10) is nonlinear in general.

The Closed-Loop System Substituting (15.10) into (15.3), we obtain the corresponding closed-loop system

$$\dot{x}(t) = \sum_{i=1}^{r} h_i^2(p)\{G_{ii}x(t) + A_{id}x(t-\tau_1) - B_{id}F_i x(t-\tau_2)\}$$

$$+ 2\sum_{i=1}^{r}\sum_{i<j} h_i(p)h_j(p)\left\{\frac{G_{ij}+G_{ji}}{2}x(t) + \frac{A_{id}x(t-\tau_1)+A_{jd}x(t-\tau_1)}{2}\right.$$

$$\left. + \frac{-B_{id}F_j x(t-\tau_2) - B_{jd}F_i x(t-\tau_2)}{2}\right\},$$

$$(15.11)$$

where $G_{ij} = A_{i0} - B_{i0}F_j$.

15.2 STABILITY OF THE CLOSED-LOOP SYSTEMS

Now, we present a delay-independent stability condition for the closed-loop system (15.11).

THEOREM 59 *If there exist matrices $P > 0$, $R_1 > 0$, and $R_2 > 0$ such that the following matrix inequalities are satisfied, the closed-loop system (15.11) is quadratically stable:*

$$PG_{ii} + G_{ii}^T P + PA_{id}R_1^{-1}A_{id}^T P + R_1 + PB_{id}F_i P^{-1}R_2^{-1}P^{-1}F_i^T B_{id}^T P + PR_2 P < 0,$$

$$i = 1,\ldots,r, \quad (15.12)$$

$$P\left(\frac{G_{ij}+G_{ji}}{2}\right) + \left(\frac{G_{ij}+G_{ji}}{2}\right)^T P + \frac{1}{2}P\left(A_{id}R_1^{-1}A_{id}^T + A_{jd}R_1^{-1}A_{jd}^T\right)P$$

$$+ R_1 + \frac{1}{2}P\left(B_{id}F_j P^{-1}R_2^{-1}P^{-1}F_j^T B_{id}^T\right.$$

$$\left. + B_{jd}F_i P^{-1}R_2^{-1}P^{-1}F_i^T B_{jd}^T\right) + PR_2 P \leq 0. \qquad (15.13)$$

Proof. Define the following Lyapunov function for the closed-loop system:

$$V(x) = x(t)^T P x(t) + \int_{t-\tau_1}^{t} x(s)^T R_1 x(s)\, ds$$

$$+ \int_{t-\tau_2}^{t} x(s)^T P R_2 P x(s)\, ds. \quad (15.14)$$

Taking the derivative of $V(x)$ along the closed-loop system and using the fact that for any vector x_1 and x_2 and matrix Y

$$x_1^T Y x_2 + x_2^T Y^T x_1 \leq x_1^T Y R^{-1} Y^T x_1 + x_2^T R x_2, \quad (15.15)$$

where R is a positive definite matrix, we have

$$\dot{V}(x) = \sum_{i=1}^{r} h_i^2(p) x(t)^T \{ P G_{ii} + G_{ii}^T P + P A_{id} R_1^{-1} A_{id}^T P + R_1$$

$$+ P B_{id} F_i P^{-1} R_2^{-1} P^{-1} F_i^T B_{id}^T P + P R_2 P \} x(t)$$

$$+ 2 \sum_{i=1}^{r} \sum_{i<j} h_i h_j x(t)^T \left\{ P \left(\frac{G_{ij} + G_{ji}}{2} \right) + \left(\frac{G_{ij} + G_{ji}}{2} \right)^T P \right.$$

$$+ \frac{1}{2} P \left(A_{id} R_1^{-1} A_{id}^T + A_{jd} R_1^{-1} A_{jd}^T \right) P + R_1$$

$$+ \frac{1}{2} \left(B_{id} F_j P^{-1} R_2^{-1} P^{-1} F_j^T B_{id}^T \right.$$

$$\left. \left. + B_{jd} F_i P^{-1} R_2^{-1} P^{-1} F_i^T B_{jd}^T \right) + P R_2 P \right\} x(t).$$

$$(15.16)$$

Since $\sum_{i=1}^{r} h_i > 0$ and $h_i \geq 0$, we have

$$\dot{V}(x) < 0, \quad \forall x \neq 0. \quad (15.17)$$

(Q.E.D.)

15.3 STATE FEEDBACK STABILIZATION DESIGN VIA LMIs

The state feedback stabilization design problem can be stated as follows: Given a plant described by a T-SMTD model, find a PDC control that quadratically stabilizes the closed-loop system. The design variables in this problem are the gain matrices F_i ($1 \leq i \leq r$). The following theorem states conditions that are sufficient for the existence of such a PDC controller. Taken together, these conditions form an LMI feasibility problem. If this problem is analyzed numerically and a feasible solution is found, then a set of stabilizing gain matrices can be computed directly from the solution data.

THEOREM 60 *A sufficient condition for the existence of a PDC controller that quadratically stabilizes the T-SMTD model* (15.3) *is that there exist matrices $X > 0$, $W_1 > 0$, $W_2 > 0$, and M_i, $1 \leq i \leq r$, such that the following two LMI conditions hold*:

(a) *For every $1 \leq i \leq r$, the following equation is satisfied*:

$$\begin{bmatrix} \begin{pmatrix} A_{i0}X + XA_{i0}^T + A_{id}W_1 A_{id}^T \\ - B_{i0}M_i - M_i^T B_{i0}^T + W_2 \end{pmatrix} & X & B_{id}M_i \\ X & -W_1 & 0 \\ M_i^T B_{id}^T & 0 & -W_2 \end{bmatrix} < 0. \quad (15.18)$$

(b) *For every pair of indices satisfying $1 \leq i \leq j \leq r$, the equation*

$$\begin{bmatrix} U_{ij} + V_{ij} + W_{ij} & X & B_{id}M_j & B_{jd}M_i \\ X & -\frac{1}{2}W_1 & 0 & 0 \\ M_j^T B_{id}^T & 0 & -W_2 & 0 \\ M_i^T B_{jd}^T & 0 & 0 & -W_2 \end{bmatrix} \leq 0 \quad (15.19)$$

holds, where

$$U_{ij} = A_{i0}X + XA_{i0}^T + A_{j0}X + XA_{j0}^T,$$

$$V_{ij} = -B_{i0}M_j - M_j^T B_{i0} - B_{j0}M_i - M_i^T B_{j0},$$

$$W_{ij} = A_{id}W_1 A_{id}^T + A_{jd}W_1 A_{jd}^T + 2W_2.$$

Furthermore, if the matrices exist which satisfy these inequalities, then the feedback gains $F_i = M_i X^{-1}$ will provide a quadratically stabilizing PDC controller.

Proof. Let $P = X^{-1}$, $W_1 = R_1^{-1}$, and $W_2 = R_2$. Then we can get the above results following Theorem 59. (Q.E.D.)

15.4 H_∞ CONTROL

In this section, we will investigate the problem of disturbance rejection for the T-S fuzzy model with time delays. We assume the ith rule of the model is

IF $p_1(t)$ is M_{i1} and, ..., and $p_l(t)$ is M_{il},

THEN

$$\dot{x}(t) = A_{i0}x(t) + A_{id}x(t - \tau_1) + B_{i0}u(t) + B_{id}u(t - \tau_2) + D_i w(t),$$

$$i = 1, 2, \ldots, r, \quad (15.20)$$

$$z(t) = E_i x(t),$$

where $w(t)$ is the square integrable disturbance input vector and $z(t)$ is the controlled output.

Our objective here is to construct an H_∞ controller in the form (15.10) such that (a) the controller is a stabilizer for the nonlinear time-delay system and (b) subject to assumption of zero initial condition, the controlled output z satisfies $\int_0^\infty \|z(t)\|^2 \le \gamma^2 [\int_0^\infty \|w(t)\|^2 dt]$ for all $w \in L_2[0\ \infty]$, where γ is a prespecified positive constant. If this kind of controller exists, the nonlinear time-delay system (15.20) is said to be stabilizable with an H_∞-norm bound γ.

THEOREM 61 *For the system (15.20), a sufficient condition for the existence of a PDC controller that stabilizes the T-SMTD model with an H_∞-norm bound γ is that there exist matrices $X > 0$, $W_1 > 0$, $W_2 > 0$, and M_i, $1 \le i \le r$, such that the following two LMI conditions hold*:

(1) *For every $1 \le i \le r$, the equation*

$$\begin{bmatrix} H_{ii} & X & B_{id}M_i & D_i & XE_i^T \\ X & -W_1 & 0 & 0 & 0 \\ M_i^T B_{id}^T & 0 & -W_2 & 0 & 0 \\ D_i^T & 0 & 0 & -\gamma I & 0 \\ E_i X & 0 & 0 & 0 & -\gamma I \end{bmatrix} < 0, \quad (15.21)$$

where

$$H_{ii} = A_{i0} X + X A_{i0}^T + A_{id} W_1 A_{id}^T - B_{i0} M_i - M_i^T B_{i0}^T + W_2,$$

is satisfied.

(2) *For every pair of indices satisfying $1 \leq i \leq j \leq r$, the equation*

$$\begin{bmatrix} U_{ij} + V_{ij} + W_{ij} & X & B_{id}M_j & B_{jd}M_i & D_{ij} & XE_{ij}^T \\ X & -\frac{1}{2}W_1 & 0 & 0 & 0 & 0 \\ M_j^T B_{id}^T & 0 & -W_2 & 0 & 0 & 0 \\ M_i^T B_{jd}^T & 0 & 0 & -W_2 & 0 & 0 \\ D_{ij}^T X & 0 & 0 & 0 & -\gamma I & 0 \\ E_{ij} X & 0 & 0 & 0 & 0 & -\gamma I \end{bmatrix} \leq 0 \quad (15.22)$$

holds, where

$$U_{ij} = A_{i0}X + XA_{i0}^T + A_{j0}X + XA_{j0}^T,$$

$$V_{ij} = -B_{i0}M_j - M_j^T B_{i0}^T - B_{j0}M_i - M_i^T B_{j0}^T,$$

$$W_{ij} = A_{id}W_1 A_{id}^T + A_{jd}W_1 A_{jd}^T + 2W_2,$$

$$D_{ij} = [D_i D_j^T + D_j D_i^T]^{\frac{1}{2}},$$

$$E_{ij} = [E_i E_j^T + E_j E_i^T]^{\frac{1}{2}}.$$

Furthermore, if matrices exist which satisfy these inequalities, then the feedback gains are given by $F_i = M_i X^{-1}$.

15.5 DESIGN EXAMPLE

Consider the following simple T-S fuzzy model with time delays where the fuzzy rules are given by

Rule 1

IF $x_2(t)$ is M_1 (e.g., Small)

THEN

$$\dot{x}(t) = A_{10}x(t) + A_{1d}x(t - \tau_1) + B_{10}u(t) + B_{1d}u(t - \tau_2).$$

Rule 2

IF $x_2(t)$ is M_2 (e.g., Big)

THEN

$$\dot{x}(t) = A_{20}x(t) + A_{2d}x(t - \tau_1) + B_{20}u(t) + B_{2d}u(t - \tau_2).$$

Here $x(t) = [x_1(t) \quad x_2(t)]^T$ and

$$A_{10} = \begin{bmatrix} 1 & -0.5 \\ 1 & 0 \end{bmatrix}, \qquad A_{20} = \begin{bmatrix} -1 & -0.5 \\ 1 & 0 \end{bmatrix},$$

$$A_{1d} = A_{2d} = \begin{bmatrix} 0 & -0.2 \\ 0.2 & 0 \end{bmatrix}, \qquad B_{10} = B_{20} = \begin{bmatrix} 1 \\ 0 \end{bmatrix},$$

$$B_{1d} = B_{2d} = \begin{bmatrix} 0.2 \\ 0 \end{bmatrix}.$$

This system is unstable for some initial conditions as shown in Figure 15.1 for the initial condition $x(t) = [2 \quad 2]^T$. Now we want to design a PDC controller to stabilize this system. Using Theorem 60, we obtain the feedback gains of the PDC controller:

$$F_1 = [11.22 \quad 12.87],$$
$$F_2 = [8.87 \quad 12.33].$$

The closed-loop response for the initial condition $x(t) = [2 \quad 2]^T$ is shown in Figure 15.2. In the simulations, τ_1 and τ_2 are chosen to 1 though they can be of different values.

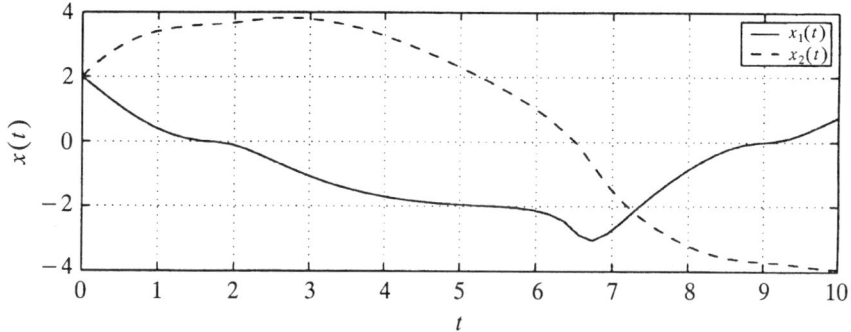

Fig. 15.1 Response of the open-loop system.

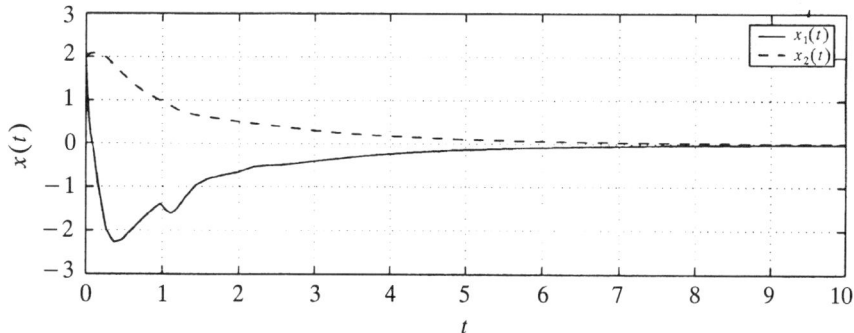

Fig. 15.2 Response of the closed-loop system.

REFERENCES

1. Y. Gu, H. O. Wang, and K. Tanaka. "Fuzzy Control of Nonlinear Time-Delay Systems: Stability and Design Issues," Proceedings of the 2001 American Control Conference, Arlington, VA, June 25–27, 2001.
2. S. Phoojaruenchanachai and K. Furuta, "Memoryless Stabilization of Uncertain Linear Systems Including Time-Varying State Delays," *IEEE Trans. Automat. Control*, Vol. 37, No. 7, pp. 1022–1026 (1992).
3. Y. Gu, C. Geng, J. Qian, and L. Wang, "Robust H_∞ Control for Linear Time-Delay System Subject to Norm-Bounded Nonlinear Uncertainty," *Proceedings of the 1998 American Control Conference*, Philadelphia, pp. 2417–2420, 1998.
4. J. C. Shen, "Designing Stabilizing Controllers and Observers for Uncertain Linear Systems with Time-Varying Delay," *Automatica*, Vol. 33, No. 4, pp. 331–333 (1997).
5. J. H. Su, "Further Result on the Robust Stability of Linear Systems with a Single Time Delay," *Syst. Control Lett.*, Vol. 23, pp. 375–379, 1994.
6. Y. Gu, S. Wang, Q. Li, Z. Cheng, and J. Qian, "On Delay-Dependent Stability and Decay Estimate for Uncertain Systems with Time-Varying Delay," *Automatica*, Vol. 34, No. 8, pp. 1035–1039 (1998).
7. H. O. Wang, K. Tanaka, and M. F. Griffin, "An Approach to Fuzzy Control of Nonlinear Systems: Stability and Design Issues," *IEEE Trans. Fuzzy Syst.*, Vol. 4. No. 1, pp. 14–23 (1996).

INDEX

Affine model, 277
Algebraic Riccati equations, 109
Asymptotically stable in the large, 27
Augmented system, 86, 91

Ball and beam system, 253
Bounded Real Lemma, 261

Cancellation technique, 153–154, 159, 165, 222
Case A, 84
Case B, 84, 90
CFS, 6
Chaos, 153
Chaotic model following control, 153, 182
Chaotic systems, 153
Closed-loop system, 31
CMFC, 153, 182
Common **B** matrix, 32, 51, 78, 141, 154, 157, 160, 204, 210, 213, 223, 225
Common **C** matrix, 92
Common Lyapunov function, 195
Constraint on the control input, 79–80, 163, 269
Constraint on the input, 270
Constraint on the output, 81, 142, 163
Construction of a fuzzy model, 8
Continuous fuzzy system, 6, 50
Controlling chaos, 153

Convex optimization techniques, 27
CT, 153, 165

DC motor, 76
Decay rate 62, 78, 88
Decay rate controller design: CFS, 62
Decay rate controller design using relaxed stability conditions: CFS, 64
Decay rate controller design using relaxed stability conditions: DFS, 64
Decay rate fuzzy controller, 63
Decay rate fuzzy controller design, 160, 163
Decay rate fuzzy controller design: DFS, 63, 160
Decay rate fuzzy controller design using the CT: CFS, 167, 185
Decay rate fuzzy controller design using the CT: DFS, 168, 186
Delay-independent stability, 291
Delay-independent stability condition for the open-loop system, 294
Delay-independent stability condition for the closed-loop system, 296
Descriptor system, 195
DFS, 6
Discrete fuzzy system, 6, 50
Disturbance, 69, 133
Disturbance attenuation, 259
Disturbance rejection, 69, 73, 134, 299
DPDC, 229, 232

303

INDEX

Duffing forced-oscillation, 162
Dynamic compensator, 229
Dynamic parallel distributed compensation, 229, 232, 259
Dynamic output feedback, 168
Dynamic output feedback controller, 26

Equilibrium, 27
EVP, 220, 223
Exact fuzzy model construction, 10

Feedback linearization, 165
Fuzzy controller design using relaxed stability conditions: CFS, 60
Fuzzy controller design using relaxed stability conditions: DFS, 60
Fuzzy control rules, 26
Fuzzy descriptor system, 195, 217
Fuzzy IF-THEN rule, 6
Fuzzy implication, 6
Fuzzy modeling, 9
Fuzzy observer, 83
Fuzzy observer-based control, 83
Fuzzy regulator, 83
Fuzzy set, 6

Generalized eigenvalue minimization problem, 62
Generalized H_2 performance, 259, 262, 268
General quadratic constraint, 259, 261
General Quadratic Performance, 267
GEVP, 62–65, 89
Global design conditions, 26
Global linearization, 165
Global sector, 10

Henon mapping model, 163
H_∞ control, 29, 291

Input constraint, 68
Input vector, 6
Interior-point methods, 27
Inverted pendulum, 14, 23
Inverted pendulum on a cart, 5, 38, 271

Jack-knife, 133

Kang-Sugeno fuzzy modeling method, 8

L_2 gain, 71, 74, 259, 261
L_2 gain performance, 265
Largest Lyapunov exponent, 62
LDI, 38, 230
Linear differential inclusion, 230

Linear matrix inequality, 5, 27, 34
Linear observer, 83
Linear T-S model, 277
LMI, 5, 27, 34
LMI problems, 36
Local approximation, 10, 23
Local design structure, 26
Local sector nonlinearity, 10
Local sector(s), 10
Lorenz's equation, 160, 163, 165, 170, 173–174, 186
Lyapunov stability theorem, 27

Model-based fuzzy controller design, 5
Model Rule, 6
Model rules, 8
Modified PDC, 199
Multi-objective control, 168, 259

New sufficient condition, 229
NLTI, 261
Nondynamic constraints, 195
Nonlinear control benchmark problem, 121
Nonlinear model following control, 217
Nonlinear reference model, 217–218
Nonlinear term, 11
Nonlinear time-invariant operator, 261
Number of model rules, 6

OGY method, 153
Open-loop, 27
Optimal control, 109
Optimality, 97
Optimal fuzzy control, 109, 121
Output and input constraints, 259
Output constraint, 68, 259
Output feedback controller, 26
Output vector, 6

Parallel distributed compensation, 5, 25
Parameter dependent linear model, 232
Parameter dependent plant, 235
Parameter dependent state feedback, 230
Parameter identification, 9
Passivity, 259
PDC, 25, 84
PDC controller, 159
PDC fuzzy controller, 30
Pendulum, 23
Performance-oriented controller synthesis, 259
Premise part uncertainty, 56
Premise uncertainty, 98
Premise variables, 6

INDEX

Quadratic function, 27
Quadratically stable, 31
Quadratically stabilizable, 31, 37
Quadratic Lyapunov function, 27, 89
Quadratic performance function, 109–110
Quadratic stability, 27

Reference chaotic system, 170, 182
Reference fuzzy model, 170, 182, 187, 190
Reference nonlinear model, 225
Regulation problem, 153
Relaxed stability conditions, 52
Robust controller design, 23
Robust fuzzy control, 97, 121
Robustness, 97
Robust-optimal fuzzy control problem, 121
Robust stabilization problem, 105
Rossler's equation, 155, 161, 170, 176–77, 182, 188
Rule reduction, 142

Saturation of the actuator, 133
Schur complement, 37, 59, 166
Sector nonlinearity, 10
Separation principle, 83, 90
Stable fuzzy controller design, 58
Stable fuzzy controller design: CFS, 58, 160
Stable fuzzy controller design: DFS, 59, 160
Stable fuzzy controller design using the CT: CFS, 167, 184
Stable fuzzy controller design using the CT: DFS, 167, 185
Stabilization 153, 159
Stability, 27, 49
Stability analysis, 5, 49
Stability conditions for the open-loop systems, 49
Stability of the closed-loop system, 50
State feedback, 26

State-space representation, 195
State vector, 6
Structure identification, 9
Subsystem, 7
Sufficient stability condition, 27
Switching system, 259
Synchronization, 153

Takagi-Sugeno fuzzy model, 5
Takagi-Sugeno fuzzy systems with uncertainty, 97
Ten-trailer case, 151
Time-varying linear system, 230
TORA, 125
TORA system, 121
TPDC, 217, 219
Trailer-truck, 133
Translational oscillator with an eccentric rotational proof mass actuator, 125
Trial-and-error, 27
Triple trailer, 134
T-S fuzzy model, 5
T-S model with linear rule consequence, 277
T-SMTD, 291, 293
T-S model with time delays, 291, 293
Twin parallel distributed compensation, 217, 219

Uncertainty, 98
Universal, 6
Universal approximator, 278
Upper bound, 110

Van del pol, 226
Vehicle with a trailer, 133
Vehicle with multiple trailers, 133